T0132889

UITGAVEN VAN HET
NEDERLANDS HISTORISCH-ARCHAEOLOGISCH INSTITUUT TE INSTANBUL

Publications de l'Institut historique-archéologique néerlandais de Stamboul
sous la direction de
Machteld J. MELLINK, J. de ROOS,
J.J. ROODENBERG et K.R. VEENHOF

LXXXIII

AN ARCHAEOLOGICAL ETHNOGRAPHY OF A NEOLITHIC COMMUNITY

Space, Place and Social Relations in the Burnt Village at Tell Sabi Abyad, Syria

Virtual reconstruction of the Burnt Village

Cover illustration: people in their surroundings: the circular building VII of the Burnt Village (detail of fig. 7.4 A).

AN ARCHAEOLOGICAL ETHNOGRAPHY OF A NEOLITHIC COMMUNITY

Space, Place and Social Relations in the Burnt Village
at Tell Sabi Abyad, Syria

by

MARC VERHOEVEN

NEDERLANDS HISTORISCH-ARCHAEOLOGISCH INSTITUUT
TE INSTANBUL
1999

Marc Verhoeven

An Archaeological Ethnography of a Neolithic Community. Space, Place and Social Relations in the Burnt Village at Tell Sabi Abyad, Syria. Istanbul: Nederlands Historisch-Archaeologisch Instituut; Leiden: Nederlands Instituut voor het Nabije Oosten [distr.]. (Uitgaven van het Nederlands Historisch-Archaeologisch Instituut te Istanbul, ISSN 0926-9568 ; 83)
ISBN 90-6258-084 X

Printed in Belgium

PREFACE AND ACKNOWLEDGEMENTS

In 1988 I joined the excavations at Tell Sabi Abyad in the Balikh valley in northern Syria. I was then a student of archaeology at the University of Amsterdam. Since that successful campaign, in which we dug up parts of a late sixth millennium B.C. village of the later Neolithic Halaf culture, I have been fascinated by Near Eastern archaeology in general and the prehistory of Sabi Abyad in particular. The director of the excavations, Peter Akkermans, has since become a good friend, and from 1988 onwards we have worked on Sabi Abyad in close cooperation. In 1991 the so-called Burnt Village was discovered. This pre-Halaf settlement was destroyed by a violent fire, resulting in an exceptional preservation of architecture and related artefact assemblages. This discovery was the major impetus for designing a research programme which aimed at a reconstruction of activities and social relationships at prehistoric Tell Sabi Abyad. In 1994 I received a four-year grant to produce the proposed study, i.e. the present dissertation.

This study presents a detailed spatial (and contextual) analysis of the results of the 1986 to 1993 campaigns at Tell Sabi Abyad. Since 1993 excavation has continued and further fieldwork is also planned for the coming years. Moreover, analysis of artefacts and ecological material continues. This book, therefore, should not be regarded as the *final* interpretation. In fact, a major objective of the present study is to present information which will guide further fieldwork and interpretation. This is, however, not to say that the present analysis is wholly preliminary; it is based on a large body of analysed and published data, i.e. stratigraphy, architecture (exposures of up to 875 m^2), pottery, 'small finds', flint and obsidian artefacts and animal bones. Analysis of these assemblages by various specialists has resulted in important new information about the Late Neolithic. It is felt that by combining and contextualizing this information, the present study offers a significant contribution to our knowledge of later Neolithic society in general and a prehistoric community at Sabi Abyad in particular.

Sytze Bottema acted as my Ph.D. supervisor, and I am greatly indebted for his support and advice, without which the present study would not have been possible. Evidently, Peter Akkermans has been deeply involved too, commenting upon my work and continuously stimulating, inspiring and encouraging me, making me aware what archaeology is all about. Edgar Peltenburg is warmly thanked for his critical advice and support. Corrie Bakels, Leendert Louwe Kooijmans, Diederik Meijer and Pieter van de Velde carefully read the manuscript as well and I am most grateful for their comments.

Lorraine Copeland generously allowed me to use her primary data on the Sabi Abyad flint and obsidian artefacts. Likewise, Chiara Cavallo, Olivier Nieuwenhuyse and Marie Le Mière provided their databases (and valuable advice), for which I am most grateful. Under the direction of Annelou van Gijn, Marie-Claire Schallig analysed microwear traces on a selection of flint artefacts. Peter Heavens skillfully produced the virtual reconstruction of the Burnt Village. The photographs were taken by Jan Pauptit, and the

VI

drawings were made by Pieter Collet (axonometric reconstruction), Peter Deunhouwer (cover illustration, virtual reconstructions, distribution maps) and Paul van der Kroft (flint). Hans Kamermans, Jan-Albert Schenk, Rauno de Smit, Paul Verhoeven and Milco Wansleeben helped to solve the numerous computer problems. The layout of this study was prepared by Dick Noordhuizen. Ans Bulles corrected the English text.

Furthermore, I am much obliged to the following people for their advice and their help (in alphabetical order): Bram van As, Jos Bazelmans, Jean-Marie Buijs, Peter van Dommelen, Erick van Driel, Kim Duistermaat, David Fontijn, Nies Huijsmans, Lou Jacobs, Jan Kolen, Gerrit van der Kooij, Peter Kranendonk, Ted LaGro, Henk de Lorm, Pascal van Meurs, Sander Spaans, Richard Spoor, Marcel Vellinga, Lies Verhoeven-De Neve, Alexander Verpoorte, Miguel John Versluys, Marianne Wanders and Loes van Wijngaarden-Bakker.

The research was supported by the Foundation for History, Archaeology and Art History, which is subsidized by the Netherlands Organisation for Scientific Research (NWO). The excavations at Tell Sabi Abyad are conducted under the auspices of the National Museum of Antiquities in Leiden, which is thanked for its permission to study the Sabi Abyad databases. The work has been carried out at, and facilities were provided by, the Faculty of Archaeology of the University of Leiden and the National Museum of Antiquities in Leiden; the staff of both organizations is warmly thanked for their support and help. The Directorate General of Antiquities and Museums of Syria, Damascus, is thanked for its continuous assistance and encouragement concerning the excavations at Tell Sabi Abyad. Prof. Dr. J. de Roos and Dr. J. Roodenberg of the Netherlands Institute of the Near East in Leiden, are thanked for their willingness to publish the present work. At last, I would like to thank the people of Hammam et-Turkman; without their help and often skilled labour the excavations at Sabi Abyad, and subsequently the present analysis, would not have been possible.

Leiden, September 1998

TABLE OF CONTENTS

Chapter 7
ON A DUAL BASIS: NOMADS AND RESIDENTS AT TELL SABI ABYAD

Appendix 1
FUNCTIONS OF ARTEFACTS OF THE BURNT VILLAGE

Appendix 2
EXCAVATION AND POST-EXCAVATION
RESEARCH METHODOLOGY

Appendix 3
NUMBERS AND PERCENTAGES RELATED
TO THE VISUAL INSPECTION IN CHAPTER 6 (SECTION 2)

Appendix 4
EXAMPLE OF FUNCTIONAL ASSESSMENT OF SPATIAL UNITS
(ROOM 1, BUILDING I)

"I'm not sure I can tell the truth ... I can only tell what I know"

The words of a Cree hunter when administered the oath in court.
He came to Montreal to testify concerning the fate of his hunting lands
in the new James Bay hydroelectric scheme (quoted in Clifford 1986:8).

CHAPTER 1

RESEARCH OBJECTIVES

1 Introduction

Until recent times, archaeology in the Near East was mainly concerned with two fundamental developments: the introduction of agriculture (ca. 8000 B.C., 9110 cal BC) and the
process of urbanization (ca. 3500 B.C., 4331 cal BC).[1] The later Neolithic (ca. 6000-4500
B.C., 6860-5360 cal BC), the period between these 'revolutions', actually belongs among
the poorest known periods in the Near East so far. Recent archaeological investigations,
however, have clearly indicated the significance of this period (see e.g. Akkermans 1993,
Akkermans, ed., 1996; Bernbeck 1994, 1995; Campbell 1992; Davidson 1977; Matsutani,
ed., 1991; Yoffee and Clark, eds., 1993).

Intensive research in the Balikh valley in northern Syria, especially the excavations at
Tell Sabi Abyad (e.g. Akkermans 1993; Akkermans and Verhoeven 1995; Akkermans,
ed., 1996), point out that in the later Neolithic period many important socio-economic
and, perhaps, ecological changes took place. The population density and the settlement
organization in the early sixth millennium B.C., for instance, differed markedly from that
of the preceding Pre-Pottery Neolithic B period (ca. 7600-6000 B.C., 8610-6860 cal BC);
many settlements in the Balikh valley were deserted. Occupation contracted to a few sites
only, all located in the northern part of the valley (Akkermans 1993:170). This trend
towards site desertion has also been attested in many other regions in the Levant and
Syria, and is generally referred to as the *Hiatus Palestinien* (Mellaart 1975:67-69; Moore
1983:99).

Desiccation of the environment has often been put forward as a possible cause, but the
mechanisms behind the phenomenon are still a matter of debate (see e.g. Bar-Yosef and
Meadow 1995; Bottema et al., eds., 1990; Köhler Rollefson 1988). Towards the end of
the sixth millennium B.C. the number of settlements and the population density in the
Balikh valley, and in other parts of Syria and the Levant, increased again.

---

[1] In previous publications about Tell Sabi Abyad dates were used in a 'traditional' manner, i.e.
uncalibrated B.C.. In this study both the uncalibrated and the calibrated (cal BC) dates are indicated (e.g.
5200 B.C., 5970 cal BC, the latter based upon the orginal uncalibrated BP dates), according to Stuiver et
al., eds, 1993, and the *Groningen Radiocarbon Calibration Program* (Van Der Plicht 1993). The dates are
according to the 1 sigma (68.3%) confidence level. The calibrated dates have been rounded off, using
intervals of 10 years.

Chronological periods (e.g. 'the early sixth millennium B.C.'), i.e. rough estimations, could of course
not be calibrated.

Historical dates, i.e. those from the Early Bronze Age onwards, have not been calibrated since these
have mainly been based upon exact textual data.

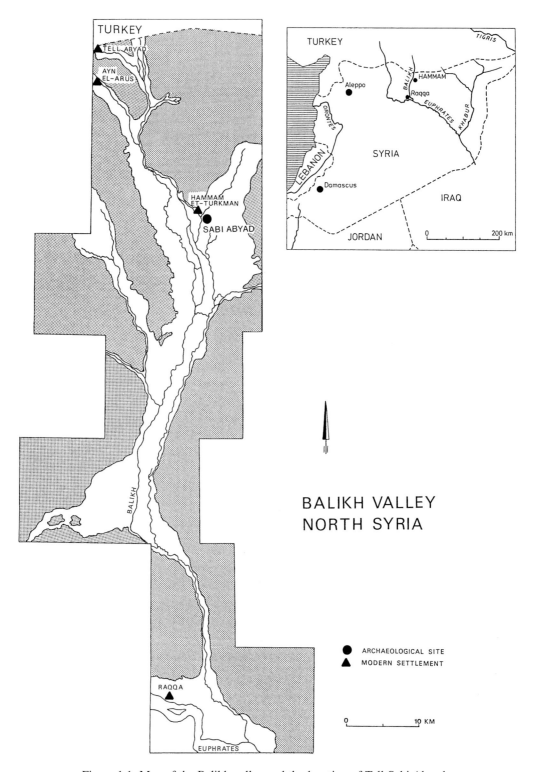

Figure 1.1. Map of the Balikh valley and the location of Tell Sabi Abyad.

At Tell Sabi Abyad, a sequence of prehistoric occupational layers (dated at ca. 5700-5000 B.C., 6460-5760 cal BC) indicates that, apart from the changes in population density and settlement organization, other alterations took place as well in the sixth millennium B.C., especially in the second half of this period. Investigation of the architecture of the various phases of occupation has shown, for instance, that there is a marked difference in the use of space between the early and the later settlements: tightly packed clusters of regular buildings make way for more spaciously structured settlements. An important innovation, furthermore, is the introduction of painted ceramics. The sequence at Sabi Abyad shows a gradual development of busily painted 'Fine Ware' pottery out of thick-walled and unpainted 'Coarse Ware'. Many other alterations in the material culture have been observed. For instance: arrowheads were gradually replaced by so-called sling missiles, stampseals and 'tokens' (counting devices) now appear regularly, and figurines differ in style and number in the various phases of occupation.

Obviously, these alterations in settlement organization and layout and material culture reflect changes in the socio-economic sphere. The cultural developments at Sabi Abyad suggest that the more or less autonomous social groups of the Pre-Pottery Neolithic B period in the early sixth millennium B.C. gradually evolved towards more open societies which maintained interregional contacts, resulting in an exchange of goods and ideas.

Eventually, at the end of the sixth millennium (about 5100 B.C., 5910 cal BC), the Halaf culture emerged. This cultural complex is characterized by its attractive painted pottery, circular buildings (so-called tholoi) and a variety of typical beads, amulets and figurines (see e.g. Akkermans 1993; Perkins 1949; LeBlanc and Watson 1973; Mellaart 1975; Watson 1983). The pre-Halaf Neolithic occupation in Syria and surrounding countries is, in view of the material culture, marked by small groups operating in restricted areas. The Halaf cultural features, however, spread within a short time over the northern part of the 'Fertile Crescent', representing an unpreceded cultural uniformity (LeBlanc and Watson 1973:117). The investigations at Tell Sabi Abyad have resulted in important new insights in the origin, nature and development of this culture (e.g. Akkermans, ed., 1996).

2 Tell Sabi Abyad: The Site and the Excavations

Since 1986, excavations have been carried out at Tell Sabi Abyad ('mound of the white boy'). The site is located in the upper part of the Balikh valley in northern Syria, about 30 km south of the Syro-Turkish border (fig. 1.1). The tell is the largest of a cluster of four prehistoric mounds (Tells Sabi Abyad I to IV), dating from the seventh and sixth millennium B.C.. Locally the cluster is known as Khirbet Sabi Abyad. Tell Sabi Abyad I, the focal point of this study, was primarily occupied in the sixth millennium B.C.. Tell Sabi Abyad II is a small Pre-Pottery Neolithic B site dating from the second half of the seventh millennium B.C. (Verhoeven 1994, 1997a, Verhoeven and Akkermans, eds., in prep.). Tell Sabi Abyad III is another small Pre-Pottery Neolithic mound, more or less contemporaneous with Tell Sabi Abyad II. Tell Sabi Abyad IV, finally, is a Halafian mound which is nowadays used as a graveyard by the inhabitants of the nearby village of Hammam et-Turkman.

The Sabi Abyad tells are located at a short distance from each other in a linear pattern. This situation of the mounds suggests that they lay alongside a prehistoric wadi, possibly the Nahr et-Turkman, a branch of the Balikh. Nowadays the Balikh itself flows ca. 5 km to the west of the tells.

Tell Sabi Abyad I, henceforth simply called Tell Sabi Abyad, is the largest mound of the cluster. The tell measures about 4.5. ha at its base, and its height varies between 5 and 10 m above modern field level. At present, the tell has a rather flat and coherent appearance, but in fact the tell consists of four small, mainly prehistoric, mounds which have merged in the course of time.

The excavations at Tell Sabi Abyad are carried out on the top of the tell, at the north-eastern mound and at the southeastern mound (fig. 1.2). The trenches at the top of the tell have revealed impressive remains of an Assyrian border fortress and governor seat surrounded by domestic structures, all dating from the late second millennium B.C. (the Late Bronze Age) and covering the lower prehistoric strata (Akkermans et al. 1993).

At the northeastern mound, late sixth millennium B.C. remains have been reached in narrow trenches. The stratigraphy and architecture of the northeastern mound awaits further analysis, but some preliminary remarks can be made. Pre-Halaf Neolithic strata, which gave evidence of wall fragments, pits, etc., were excavated on a limited scale only, due to the restricted size of the various trenches. In respect of material culture and radio-carbon dates, the earliest strata thus far reached closely resemble the lower strata at the southeastern mound and date from around 5300 B.C. (6070 cal BC). The excavations have been carried out to a depth of ca. 2 m below the level of the surrounding fields, but virgin soil has not yet been reached (Akkermans 1993:48). It seems that Halaf occupation, represented by at least one well-preserved building, was concentrated on the upper eastern slope of the mound (Nieuwenhuyse 1997:229). The extent of the Halaf occupation must have been rather limited; perhaps it consisted of only one isolated structure for living. The duration of settlement must have been limited as well; only one main Halaf level has been recognized so far. The larger part of the mound seems to have been used for domestic open-air activities (Akkermans 1989a:19-22; 1993:48).

The main area chosen for the prehistoric investigations, and the focal point of the present study, is the relatively low southeastern mound of Sabi Abyad. Here well-preserved prehistoric strata of occupation have been excavated over an area of up to 1300 m² (including the most recent - 1997 - excavations). Eleven 'levels', or main phases of occupation, have been recognized at the mound (Verhoeven and Kranendonk 1996). Virgin soil was reached in a narrow trench (P15) on the southern slope of the mound.

The earliest levels (11-7) at the mound represent the initial stage of the Pottery Neolithic, dated around 6000/5900-5200 B.C. (6860/6610-5970 cal BC). In the local chronological framework (Akkermans 1991, 1993:111-113) these levels fit in the Balikh II period. This phase has been divided into three subphases, viz. Balikh IIA to C. Founded on ceramic evidence, the lowest level 11 is provisionally ascribed to the final Balikh IIA phase, dated around 5700 B.C. (6460 cal BC). However, on the basis of the analysis of the flint and obsidian industries a slightly later date (ca. 5600 B.C., 6380 cal BC), representing the Balikh IIB period, can also be suggested (Copeland 1996:303). Between levels 11 and 10 a break in occupation of 200 to 300 years has possibly been attested in the deep sounding at trench P15. This break seems to indicate that Sabi Abyad, like many other

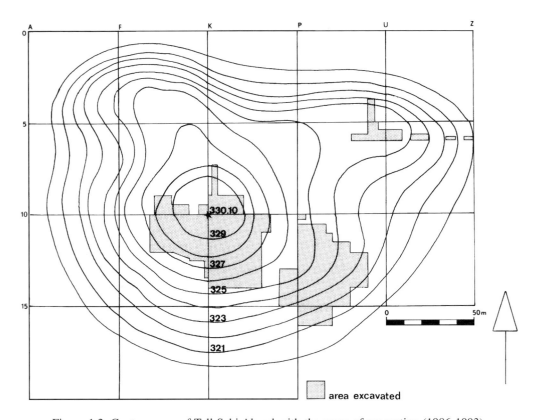

Figure 1.2. Contour map of Tell Sabi Abyad with the areas of excavation (1986-1993).

sites, was deserted at the beginning of the sixth millennium B.C., i.e. during the Hiatus Palestinien. The subsequent levels 10-7, antedating the appearance of fine painted pottery at the site, have been ascribed to the Balikh IIC phase. Levels 10-9 then seem to date from around 5500/5400-5300 B.C. (6230/6170-6070 cal BC). Levels 8-7 can be dated at ca. 5300-5200 B.C. (6070-5970 cal BC).

Levels 6-1 are part of the more developed stage of the Pottery Neolithic, the Balikh III period, and are dated at around 5200/5150-5000 B.C. (5970/5960-5760 cal BC). At Sabi Abyad, the subphases Balikh IIIA and IIIB have been attested. The Balikh IIIA phase is marked by levels 6 to 4, representing the transitional stage between the earlier Pottery Neolithic and the upper Early Halaf. This phase is dated at ca. 5200/5150-5100 B.C. (5970/5960-5910 cal BC). Finally, the Balikh IIIB phase comprises the topmost Early Halaf levels 3-1, dated at ca. 5100-5000 B.C. (5910-5760 cal BC).

Significantly, the sequence at Tell Sabi Abyad has shown an uninterrupted development of Halaf out of local Neolithic traditions. Such a transitional sequence (i.e. Balikh IIIA) has as yet not been attested elsewhere; at all other investigated sites there was a hiatus in settlement between earlier Neolithic layers and Halaf depositions. Especially the new and exciting discoveries in the Transitional level 6 are important in this respect. The level 6 village was reduced to ashes by a violent fire, which has resulted in an extraordinary preservation of architecture and related finds (Akkermans and Verhoeven 1995;

Verhoeven and Kranendonk 1996:38-63). This so-called Burnt Village, excavated over an area of ca. 800 m², consisted of rectangular and circular buildings, ovens, bins, benches, etc. Some structures were still preserved to a height of ca. 1.5 m. Vast quantities of in-situ finds have been recovered from the various burnt structures, including ceramic and stone vessels, ground-stone implements, flint and obsidian tools, human and animal figurines of unbaked clay, labrets, axes, jewellery and hundreds of clay sealings with stamp-seal impressions.

3 Research Objectives

Our knowledge and understanding of the cultural processes at Tell Sabi Abyad had mainly been based upon the typo-chronology of the various artefact assemblages (pottery, flint, etc.). These typo-chronological sequences were founded upon the stratigraphical and architectural analysis of the various levels at Tell Sabi Abyad (Verhoeven and Kranendonk 1996). Among other things, this analysis has resulted in the assignment, per stratigraphical unit, of all excavated objects to spatial units: rooms, open areas, etc. Prior to the present study this information regarding the spatial distribution of objects had not been analysed in detail. It was felt, however, that an analysis of the spatial distribution of these objects could provide important new information about the prehistoric communities at Tell Sabi Abyad; a spatial analysis may result in an insight in the organization of activities, the use of space, the economy, and the social structure of those communities.[2] It will be clear that information about these topics may increase our understanding of cultural developments such as mentioned in the introduction. Thus, comparison of the range and nature of activities, the use of space, and socio-economic aspects within and between levels may be an important addition to the above-mentioned typo-chronological studies; not only the archaeological objects themselves are now being studied, but also, and foremost, their context.

When compared with excavation reports and typo-chronological studies, there are few detailed studies devoted to the spatial distribution of artefacts from tell excavations (but see e.g. Daviau 1993; Roaf 1989; Voigt 1983:295-321). This scarcity of spatial studies may be explained by the fact that often the three main prerequisites with regard to spatial analysis cannot, or have not, been met.

The first of these prerequisites is that the objects which are included in the study must have been analysed. This means that classifications of the objects (typological, techno-logical, etc.), have to be available. Consequently, the 'spatial analyst' is dependent on the results of studies of other researchers (e.g. the pottery specialist).

Secondly, in order to detect artefact patterning and to obtain statistically meaningful results, as many excavated objects as possible should be included in the analysis. Ideally, all the different assemblages (e.g. pottery, flint, etc.) should be dealt with. In the case of tell-settlements this means that (ten-) thousands of objects have to be included in the analysis. Not only the objects themselves (e.g. pots) can be studied, but also the different

---

[2] Commonly, the term *social* refers to relations between humans, *economic* refers to production, use and consumption. This distinction is useful as a heuristic device, but in fact the two cannot be separated, since 'economic' activities always have social dimensions as well. Therefore the phrase *socio-economic* will frequently be used.

characteristics of the objects (e.g. decorations on pots). Combinations and associations of data may be an interesting research topic as well. As one can well imagine, the sheer quantity of data, and the many possibilities for combining different sets of data may result in extremely time-consuming analyses.

Thirdly, an assessment has to be made of the suitability of the archaeological record for the proposed spatial analysis. In other words, the archaeologist should be aware of the 'formation processes' of the archaeological record. By this all processes are meant which "... create the evidence of past societies and environments that remains for the archaeologist to study" (Schiffer 1983:676). In many instances the patterning in the archaeological record reflects a process of the mixing of material over a considerable period of time. It should never automatically be assumed that we are dealing with little Pompeiis (Schiffer 1985:38). Therefore, identification of formation processes should be the first step in intra-site spatial investigations (see e.g. Schiffer 1985; Verhoeven 1990). If, for instance, the spatial investigation is aimed at the reconstruction of activity areas, then only those objects which have been left in the place of use should be included in the analysis. An understanding of formation processes, then, may reveal which objects are still in their place of use. Once this has been established, decisions can be made regarding the potential of a spatial analysis, and the proper techniques to be used in the research.

With regard to tell settlements in the Near East, in-situ distributions of objects are in many instances the result of catastrophic fires, quickly covering, and thereby fixing objects. Thus, if a settlement was suddenly destroyed by a fire, and if the site has subsequently not been heavily disturbed, the original distribution of prehistoric objects can be reconstructed.

It may be clear by now that spatial analysis of in-situ contexts can be a powerful tool for providing an understanding of the activities and, concomitantly, of the socio-economic structure of prehistoric communities. The level 6 Burnt Village with its rich in-situ finds presents a unique possibility for such an understanding. Moreover, as has been indicated above, the settlement allows an insight into a prehistoric period of which until recently virtually nothing was known. The present study, then, is designed to:

> reconstruct the socio-economic structure of the level 6 community at Sabi Abyad. This reconstruction is based upon a spatial analysis, a method which contextualizes the archaeological material, i.e. a method which is suited to provide an insight into the distribution of objects, the function of spatial units, the organization of activities, the use of space, and, concomitantly, the meaning of material culture and the socio-economic organization of a prehistoric community.

More specifically, answers to the following questions are sought:

1. What is the spatial distribution of the objects?
2. Which formation processes can be distinguished?
3. What was the function of spatial units?
4. What does the spatial distribution of objects and the function and meaning of spatial units tell us about the socio-economic structure of the level 6 people?

The study is thus synchronic in character, and centers around the level 6 village. The main diachronic aspect of this study is a comparison between levels 6 and 3: in order to come to a better understanding of the processes of abandonment in the level 6 settlement, a restricted analysis of artefact distributions within the Halafian level 3 settlement will be presented.[3]

Furthermore, this study is designed as a methodological contribution to spatial analysis of tell settlements; models for assessing formation processes and the function of spatial units are presented.

4 An Archaeological Ethnography

According to the *Shorter Oxford English Dictionary* (1973), ethnography is "the scientific description of nations or races of men, their customs, habits and differences". Ethnology is described as "the science which treats of races and peoples, their relations, their distinctive characteristics, etc." Anthropology is the science of man, or of mankind in its widest sense. Ethnography and ethnology are subdisciplines of anthropology. Ethnography refers to studies of individual cultures or cultural systems, often resulting in detailed descriptions. Ethnology, on the other hand, has a more general perspective; comparisons among ethnographic data are used to understand cultural processes (Sharer and Ashmore 1987:23).

The present study is meant to be an archaeological ethnography of a Neolithic community, i.e. of the level 6 society at Tell Sabi Abyad. I argue that the term ethnography is relevant for the present study, since (1) the focus of the study is on one village community, (2) the analysis is synchronic instead of diachronic (I am presenting 'a slice of history'); (3) the analysis aims at a detailed description of such a community; (4) I wish to deal with a social group, instead of with its material culture only; (5) the present study makes extensive use of ethnographic and anthropological literature. Naturally, material culture is at the basis of this study, but I particularly want to investigate and reconstruct socio-economic relations and classifications.

Of course, an archaeologist cannot obtain as detailed a picture as an ethnographer or anthropologist (mainly due to the impossibility of verbal communication), but basically both deal with unfamiliar and 'strange' communities. Both the archaeologist and anthropologist have to interpret their observations and data, and both produce one of many possible (re) constructions, since objective realities do not exist. Furthermore, both seek to investigate and analyse patterns rather than isolated events. The main difference between anthropology and archaeology is that archaeologists cannot directly observe and communicate with the people they study. What archaeologists can do, and anthropologists cannot, however, is to study long-term processes. However, since the present study is synchronic rather than diachronic, it is a prehistoric ethnography rather than an ordinary archaeological study.

---

[3] The present study is based on the 1986 to 1993 excavations at Tell Sabi Abyad (Akkermans 1993; Akkermans, ed., 1989, 1996). Level 3 was excavated in 1986 and 1988, level 6 is being excavated from 1991 onwards.

5 Etic and Emic Perspectives

It is stressed that the models such as presented in this study represent etic categories and classifications, which should not be uncritically applied to the archaeological data. The models should be used to structure one's thoughts about formation processes (chapter 4) and functions (chapter 5). They are general frameworks, to be used as eye-openers, and they guide further research. The models are a clear example of 'reasoning by analogy' (cf. chapter 5). Such reasoning, i.e. the use of etic perspectives ('our view'), is the only way to come to an understanding of emic classifications ('their view'). In this respect I want to mention that inductive and deductive approaches should not be opposed (contra Binford 1981), but used in a dialectic fashion, the information acquired through the one approach informing the other approach.

In the case at hand, i.e. in the spatial analysis (chapter 6), specific processes and areas have been designated, i.e. etic classifications have been constructed. On the basis of the patterns observed, reclassifications have subsequently been set up (chapter 7). For instance, within the general framework of activity areas, storage areas and discard areas (cf. chapter 4), more peculiar and specific areas such as chambers where broken objects were stored, have appeared (see chapter 6). Furthermore, specific formation processes could be isolated. For example, the large amount of refuse in in-situ contexts was unexpected, and had to be accounted for in an emic perspective. These reclassifications, then, are representations and reconstructions of the emic viewpoint (which can only be approached by reasoning from the etic perspective). The past can only be understood through the present.

6 This Book

This book consists of seven chapters, including the present one. Chapter 2 (*The Study of Space*) is a theoretical discussion of the archaeological study and interpretation of space. A model for assessing the function of spatial units is presented, a post-processual perspective is advocated, and Bourdieu's *Theory of Practice* is introduced as the theoretical framework. In this chapter the foundations are laid for the interpretations of the spatial analysis. In chapter 3 (*The Burnt Village*) the level 6 settlement and the related finds are introduced. In chapter 4 (*The Past into the Present*) formation processes are discussed. A series of models of such processes relevant for tells is presented. By delineating and discussing such processes and by using the models, the nature and structure of the assemblages, the process of abandonment and the formation of the archaeological record in tell settlements can be elucidated. In chapter 5 (*Reasoning by Analogy*) a series of models of the functions of prehistoric artefacts from Sabi Abyad is presented. In appendix 1 the extensive text upon which these models have been based can be found. Chapter 6 (*Putting the Pieces together: Spatial Analysis*) is the core of the present study, in which a detailed spatial and contextual analysis of level 6 and a general one of level 3 are presented. Functions of architectural units (e.g. buildings and rooms) are reconstructed, and both levels are compared. The final chapter 7 (*On a Dual Basis: Nomads and Residents at Tell Sabi Abyad*) is the synthesis: the results of the spatial analysis are interpreted, i.e. the meaning of the material culture is dealt with and a reconstruction of the socio-economic structure of the level 6 community is presented.

As has been indicated above, appendix 1 presents a detailed discussion of the functions of artefacts of the Burnt Village. In appendix 2 the excavation methodology and the recording and retrieval policies are explained in order to give an insight into the formation of the archaeological database which has been used. Appendix 3 is a list of numbers and percentages related to the visual inspection in chapter 6 (section 2). In appendix 4 an example of the functional assessment of spatial units is given.

CHAPTER 2

THE STUDY OF SPACE

1 Introduction

In this chapter the theoretical and methodological framework of the present study is presented. Section 2 is a short introduction about intra-site spatial analysis. In section 3 the relationship between human behaviour and material culture is discussed. This section is the introduction to section 4, which stresses the social dimension of space. In section 5 Bourdieu's *Theory of Practice*, which will be applied for interpreting the architectural structure of the Burnt Village and the distribution of objects in it, is presented. Finally, in section 6, I shall present a model especially constructed as a methodology for intra-site spatial analysis of tell settlements. In fact the present book is largely organized according to the various subsequent 'steps' taken in the model.

2 Intra-Site Spatial Analysis

The archaeological study of space and architecture has received much attention recently, as shown by the many publications about these topics (e.g. Allison 1995; Clarke 1977; Darvill and Thomas, eds. 1996; Fletcher 1989; Grön et al. eds., 1992; Heinz 1997; Hietala 1984; Kent, ed., 1990; Larsson and Saunders 1997; Locock, ed., 1994; Parker Pearson and Richards, eds., 1994; Richards 1996; Samson, ed., 1990; Verhoeven 1990, in press). Furthermore, in anthropology, sociology and human geography too a number of interesting studies about the meaning of space have recently appeared (e.g. Ardener, ed., 1993; Bahloul 1996; Blanton 1994; Bourdieu 1973; Carsten and Hugh-Jones, eds., 1995; Fox, ed., 1993; Gregory and Urry, eds., 1985; Hillier and Hanson 1984; King 1980; Lawrence 1987; Moore 1986; Nas and Prins, eds., 1991; Rapoport 1969, 1982; Unwin 1997).

The spatial organization of activities is, in a general sense, the subject of intra-site spatial analysis. The basic assumption is that the spatial distribution of material remains can provide information about the cultural or natural processes which produced these distributions.

Intra-site spatial analysis is an important field of study within the discipline of archaeology, and many (mathematical) techniques have been developed. Since it is not directly relevant for the present research objectives, however, it is not attempted here to present an overview of these many techniques. Moreover, in a number of useful studies such overviews have already been given, e.g. Blankholm 1991; Carr 1984, 1985; Hodder and

Orton 1976; Hietala 1984; Orton 1982; Verhoeven 1990; Voorrips 1987; Wandsnider 1996.

Within a tell-settlement space is divided and defined by structures: buildings and/or features like pits, burials, basins, platforms, etc.[4] These are the "fixed", immovable, elements of a site. These "structural" elements clearly demarcate and bound space within a settlement. Related to the structures (found within, around, under and above) are the "non-fixed", transportable elements:

1. artefacts (ceramics, flint tools and debitage, etc.);
2. osteological and botanical remains.

If we want to obtain an insight in the use of space, i.e. the spatial organization of activities, in an ancient settlement, then these constituent elements and their relationships should be studied in detail.

Here it should be pointed out that spatial analysis of tell settlements is a specific type of intra-site spatial analysis. The large majority of the techniques presented in the theoretical and methodological literature such as indicated above, are constructed as pattern recognition/searching approaches. These methods "... are used to determine whether or not a given spatial pattern deviates from randomness, and if it does, whether the deviation is towards a clustered pattern or towards a regular pattern" (Voorrips 1987:434).

These methods have in common that they search for patterns which are not directly obvious. In fact, what they do is to set up boundaries between different sets of objects; these techniques have been specially constructed for unbounded sites such as a Paleolithic site marked by a flint scatter. Such a site is mostly characterized by a concentration, or by a diffuse distribution, of lithic artefacts; structures are most often absent. By using pattern recognition techniques spatial patterning may be discovered and activities may be reconstructed (e.g. Whallon 1984).

At tells, on the other hand, as at many other sedentary sites, we deal with clearly demarcated/bounded spaces and structures. The objects found in a room, for instance, are spatially related to each other. Without having to use pattern recognition techniques (with all their problems and shortcomings, see e.g. Voorrips 1987) we, fortunately, have straightforward and clear analytical units (e.g. rooms)! Of course, the objects in a room are not necessarily functionally related, but since they are related one way or another, they should be studied together. After an assessment of the formation processes (see chapter 4) the function of a room may (or may not) be reconstructed by using the object inventory. Thus, spatial analysis of tells may start directly by analysing inventories of bounded spatial units. This is indeed the approach taken in the present study. At a later stage, of course, pattern recognition techniques can be used to investigate patterning at site level. For instance, in analysing the functions of rooms from Grasshopper Pueblo in Arizona, Ciolek-Torrello (1984, 1985) used multivariate techniques (pattern recognition

---

[4]   Spatial/functional studies of tells encountered during the research were: Al-Khalesi 1977, 1978; Austin 1995; Caldwell, ed., 1967:236-255; Chapman 1990; Chavalas 1988; Daviau 1993; Dollfus 1983; Durand 1987; Falconer 1995; Gallery 1976; Gnivecki 1983; Gullini 1970/71; Henrickson 1981, 1982; Malek-Shahmirzadi 1977; Marfoe et al. 1986:75-87; Nicholas 1980; Pfälzner 1996; Pollock, Pope and Coursey 1996; Redman 1986; Roaf 1989; Seeden 1982; Verhoeven 1990; Voigt 1983.

techniques that include more than one artefact type), i.e. factor analysis, principal components analysis and multi-dimensional scaling, to define clusters of rooms with similar
functions (see for similar approaches also Cowgill, Altschul and Sload 1984; Hill 1970;
Verhoeven 1990).

In a spatial analysis the various archaeological datasets are contextualized. Preceding
any spatial analysis of a tell settlement, the architecture, the artefacts and the osteological/
botanical data should have been studied in some detail, because first we must obtain basic
knowledge about the intrinsic properties of these data. In the spatial analysis the results of
these investigations have to be combined to produce context, i.e. to establish the distribution of artefacts within the settlement. In the end, the analysis of the architectural data
and the distributions should bring us 'from excavated space to prehistoric place'.

Before I proceed it should be pointed out that the present spatial analysis is an
integrated spatial analysis, i.e. a spatial analysis in which, as far as possible, the distribution of objects from various assemblages, and not just one, (i.e. 'small finds', ceramics,
flint/obsidian artefacts, animal bones) is studied.[5] I have tried to put together all the pieces
of the prehistoric puzzle that I was dealing with.

3 Archaeology and Material Culture

Archaeological research of space and the use of space and architecture is characterized
by two different approaches which correspond with different views of the relationship
between material culture and social practice. In the first, processual, approach, material
culture, such as architecture, is regarded as a reflection or as a passive by-product of
human action. It is assumed that the relationships between material culture and social
practice are straightforward and predictable. Narrol (1962), for instance, accepted a direct
relation between the floor surface of a house and the number of its inhabitants. A more
recent example of the processual approach is a study by Kent (1990). Based on a cross-
cultural ethnographic survey she postulates a direct relation between the segmentation of
architecture and the degree of socio-political complexity of its users. Kent states that as
groups become socially and politically more segmented (or complex), their use of space
and architecture becomes more segmented as well. Thus both Kent and Narrol propose
direct, one-to-one relationships between behaviour and material culture. And in
concordance with this view both investigators regard architecture as a passive factor: as a
reflection of social practice.

In the second, post-processual, approach, material culture is regarded as an active and
meaningful element of societies (e.g. Hodder 1986, 1992; Moore 1986).[6] It is maintained
that the production, use and perception of material objects is not only a physical, but also
a cognitive process. People consciously or unconsciously connect objects with abstract
concepts, values and emotions. The term meaning designates the cognitive effect that
objects have on persons (Fiske 1990). Meanings attributed to material culture actively
structure human behaviour. To put it differently: meaning mediates between social

---

[5] By small finds are meant artefacts other than pottery and flint and obsidian implements, e.g. pestles,
bone awls, spindle whorls, etc.
[6] In recent years post-processual archaeology has come to be known as 'interpretative' archaeology
(Hodder et al., eds. 1995; Tilley, ed. 1993).

practice and material culture. The differences of use of more or less similar material objects in different societies, for instance, can be explained by differences in the meanings attributed to objects. Within the post-processual approach, then, a direct functional and universal relation between behaviour and material culture is rejected.

Meaning does not reside in objects, but has to be added to them. Due to differences of context and apprehension multiple interpretations of similar objects are often possible. In this sense subjectivity is acknowledged, as opposed to objectivity, predictability and universality, i.e. dimensions which receive much attention within processual studies. Material culture is thus not only a functional element in societies: it is also a symbolic element. In this respect Miller (1987:105) stresses that the physicality of objects is not an 'ultimate constraint' or final determining factor, but that the manner in which mundane objects alter their presence is simultaneously material and symbolic or meaningful. It is often assumed that there is a direct efficient relationship between the physical shape of an object and its use, due to adaptation or deliberate design processes. Certainly there is form-function relation (see chapter 5), but, within a general shape repertoire, the functional constraint on an object is generally loose. The specific form given to material culture is not only related to functional requirements, but is also guarded by non-functional and symbolic dimensions. As Hodder (1994) has recently argued, however, the symbolic nature of material culture should not be overstressed. In this respect it may be useful to distinguish different types of material culture, i.e. material culture with primarily: (1) an emotional effect (e.g. a teddy bear); (2) an aesthetic effect (a painting); (3) a semiotic effect (a flag); (4) an utilitarian effect (a screwdriver). Clearly the flag has more symbolic meaning than the screwdriver.

Furthermore, in the post-processual approach it is argued that there is an active and dialectical relation between social practice and human behaviour, which depends on the actions of individuals in their culture-historical contexts. In this way material culture is not only the outcome, but also the medium, or even the trigger, of human action. Space and material culture structure everyday practice and as a corollary also influence societal rules and values. Among others, Hodder (1986, 1992) has argued that the particular relation between material culture and human behaviour is especially determined by the activities of individuals in their culture-historical context. In this respect the organization of space is a cultural representation, and it is through this representation that the individual and his/her image of the world is constructed (Moore 1986:120).

In general, the post-processual approach in archaeology has the following three basic assumptions (as defined in Hodder 1982a, 1995:12-16):

1. Material culture is meaningfully constituted, i.e. besides being functional things, objects have a symbolic dimension: "... there are ideas and concepts embedded in social life which influence the way material culture is used, embellished and discarded" (Hodder 1995:12).
2. Material culture and humans are active, not passive. Objects are not passive by-products of human behaviour. Human action is seen as creative and interpretive, and it is acknowledged that meanings can be actively constructed. There is a dialectical relation between material culture and social action: both are structured by each other.

3. Material culture should be studied contextually. The historical, spatial and associative context should be taken into account when studying objects. Within these contexts, furthermore, there is a multiplicity of messages, a network of different levels of meanings.

Floor areas and partitioned buildings are not fixed and unchangeable functional elements. Both can be interpreted and used in various ways by different groups within societies. If this were not so, it would be difficult to explain change!

Archaeology, then, can be defined as: "... the study of material culture as a manifestation of structured symbolic practices meaningfully constituted and situated in relation to the social" (Tilley 1989:188).

4 The Social Dimension of Space

Within the post-processual approach, space and architecture are regarded as 'social space' (e.g. Grön et al. eds., 1992; Parker Pearson and Richards, eds., 1994; Samson, ed., 1990). For instance: "The house is to be interpreted not only as a feature that protects one from the elements but also as a meaningful and complex artifact which can express physical, social, and symbolic aspects of the lives of the people who dwell in it" (Stevanović 1997:335). It is thus emphasized that buildings and other structures are literally social constructions, that buildings are much more than physical structures, and that they can play an active role in the structuring of activities and in the manipulation of social relations. Apart from being a functional container architecture is regarded as a system of what has been called 'non-verbal communication' (Fletcher 1989; Rapoport 1982). This implies that the built environment is meaningfully structured, and that it communicates information about appropriate behaviour. Especially in prehistoric pre-literate societies the communicative aspect of material culture will have been important.

Spatial structures may act as media through which social relations are produced and reproduced; buildings and places actively constitute society. Architecture helps not only to define, but also to actually create the social framework it embodies. Buildings and monuments are important agents for projecting and displaying social conventions and ideas. In this view buildings are consciously created arenas for social action that simultaneously impose constraints on action. Consequently, there is a dialectical relationship between spatial structures and social structures (e.g. Soja 1985). Architecture can be seen as a material element which connects functional, social and symbolic aspects of human societies: it interrelates buildings, peoples and ideas. Architecture is used to think about the world, and also as a place where cultural knowledge is obtained through activities (Bourdieu 1977). From physical structures one can begin to trace the ideas and social values of a society (Fox 1993:2). It is important to realize that the meaningfully constituted space may carry a multiplicity of messages. Symbolic, functional and material aspects of spatial structures are interrelated, and should not be studied as separate domains, but as different levels, i.e. as dynamic relationships and networks of meanings (Larsson and Saunders 1997:81).

5 Structuring Structures

In this section Bourdieu's *Theory of Practice* (1977, 1990) will be introduced and commented upon. Giddens's theory of structuration (1984) is highly comparable to Bourdieu's work; both investigators emphasize the dialectical relation between structural meanings and structured social practice. For convenience both theories will be called structuration theories. These theories ultimately derive from the structuralist approach as mainly defined by Lévi-Strauss (1963). He has argued that cultural forms and social relations are produced by classificatory systems or sets of homologies (such as left-right). His theory, classical structuralism, has been described very clearly by Moore (1986:2): "Classificatory systems may be understood as systems of meaning, where meaning is given through a series of structural contrasts or oppositions. Such systems are models of intelligibility which are projected onto the world and which, in their myriad forms - dress, cuisine, house space - are all transformations of the same underlying logic. The natural world is thus divided into cultural categories, and the same conceptions and categories are to be apprehended in the social structure of society, the organization of the cosmos, the layout of ritual, and a wide variety of cultural activities and representations from the symbolic to the quotidian. The use of conceptual and material boundaries to mark and maintain differences means that the same spatial 'grid' may underlie the social and symbolic distinctions between people, between those people and their world, and between their world and the world of their gods. Space is thus often analysed as a reflection of social categories and systems of classification".

In classical structuralism, as in processualism, material culture, including space, is regarded as a passive element in societies. Moore (ibid.:2-6) has distinguished four related problems with this aspect of structuralism (see also Hodder 1986:47-53):[7]

1. Contextuality. In principle structuralism is ahistorical, and it takes no account of the specific social conditions in which symbolic codes are used.
2. Reflection. It is assumed that cultural forms reflect society in a rather direct way. This takes no account of ideological formations or why some representations appear as closer 'reflections' of the historical conditions which generate them than others.
3. Social Change. As in processualism, social change is very difficult to conceptualize due to the inadequate linkage between structure and process.
4. Social Actors. No account is given of the activities of individual actors or the meanings they ascribe to those activities, since it is held that the underlying meanings are not directly perceptible by the social actors themselves: "Individuals play the game, but they cannot change the rules" (Moore 1986:5).

Both Bourdieu's and Giddens's work (and see also Miller 1987) overcome these problems by presenting theories in which the strategies and intentions of social actors are connected to the production and reproduction of institutional structures. Contextuality, reflexivity (or dialectic), social change and human agency are all taken into account in their theories. Structures are regarded as enabling rather than constraining social action; they are involved in the generation and negotiation of meanings and action. This meaning

---

[7]   In fact these criticisms are also relevant with regard to the processual paradigm.

is not intrinsic, but must be invoked through practice. Bourdieu and Giddens avoid the objectivism (in which social actors are ignorant) which is characteristic of processual studies and the subjectivism (social action is highly autonomous) which surfaces in some post-processual contributions. They do this by acknowledging the duality of structure, which is both medium and outcome of action. Material culture and individuals are both active elements, creating and continually changing society (Hodder 1989:74).

In the following Bourdieu's theory will be dealt with in somewhat more detail, as it will be the theoretical framework which is used to interpret the results of the spatial analysis (chapter 7).

In his *Theory of Practice* the French sociologist Pierre Bourdieu (1977, see also Bourdieu 1989, 1990, 1992) has dealt with the concept of non-verbal communication (see section 4) in detail. With regard to architecture Bourdieu uses the phrase *structuring structures*. This refers to a dialectical relationship: buildings and their inventories are structured by social practice which in its turn is structured by architecture and other material objects in people's surroundings (see also Giddens 1984). Architecture is therefore not only the outcome, but also the cause of human action or *practice* in Bourdieu's terminology. The organization of space is not just the reflection of a set of social and economic relations; rather it is a product, through practice, of individuals' images of those relations (Moore 1986:120). The cognitive framework which, largely unconsciously, is used for the interpretation and attribution of meaning to material objects is called *habitus* (from habitat) by Bourdieu. Habitus, as the word implies, refers to habits, customs and dispositions, which are at the basis of and which shape the above-mentioned framework. In other words, habitus is the set of ideas, values and experiences with which a person gives meaning to the environment. It appears that habitus is structured on the basis of interrelated sets of material and immaterial or conceptual oppositions. These structural oppositions are observed and experienced in the surroundings (for example front versus back, light versus dark, or male versus female). Habitus is embedded and shaped in the mundane world, in spatial and material objectifications such as works of art, dress, etc., but it is particularly the built environment and the use of space and objects which lead to an understanding of the habitus (Bourdieu 1973). As the word implies, habitus refers to the practice of dwelling in the house, which is the principal locus for the objectification (in a literal sense) of generative schemes (Bourdieu 1977:89). By living in the house the occupants are constantly and largely unconsciously picking up clues about societal principles, which are thus inculcated and reinforced (e.g. Robben 1989). In principle *structure* is perceived unconsciously and habitus is unconscious: the symbolic codes are generally beyond the grasp of consciousness (Bourdieu 1977:94): fortunately, people do not have to think about each action they carry out. Generally we fail to notice that which is most immediate, or commonplace: "The schemes of the habitus, the primary forms of classification, owe their specific efficacy to the fact that they function below the level of consciousness and language, beyond the reach of introspective scrutiny or control by will" (Bourdieu 1984:466). Moving in ordered space, the body 'reads' its surroundings, particularly the house, which serves as a mnemonic for the embodied person. Particularly in non-literate societies the house acts as an instrument of thought, as a microcosm (ibid. 1977:89; Carsten and Hugh-Jones 1995:5, 22). Houses are as much in us as we are in them (Bachelard 1964:xxxiii), they are structures for remembering. Houses and other buildings serve as points of reference in our understanding of the world. Habitus is produced and reproduced in the house, which "through the intermediary of the divisions and hierarchies it sets up between things, persons and practices ... continuously inculcates and reinforces the taxonomic principles underlying all the arbitrary provisions of this culture" (Bourdieu 1977:89). It should be taken into account that habitus does not simply reproduce structural principles, but that it is an expression of a strategy of household members (Blanton 1994; Miller 1987). Reproduction (through *practice*) is almost always imperfect, and therefore changes in *structure* and habitus take place over time.

The frameworks in which the habitus is shaped are termed *fields*, for instance the field (or 'world') of art, sports, etc. Fields are relatively autonomous social arenas, or 'worlds', which each have their own rules and their own logic. The different fields are connected by various structural and functional homologies. A society as a whole consists of fields and the relations between fields. Persons act on various fields, each having its own logic and rules, i.e its own ideology. Because of these interactions field ideologies are transmitted and linked, i.e. the fields are tied up and the society is made coherent. Social relations in fields are in principle asymmetrical, and in each field there is a continuous struggle for influence and power through manipulation of relations, i.e. by changing the *structure* and thus the habitus. The field as structured space structures the habitus, which in its turn structures the field (Bourdieu 1989:66).

Among other things, the material world acts a symbol system which presents clues as to appropriate behaviour: it contains non-verbal messages, *structure* is encoded in it (e.g. one normally does not shout in a library, which in the first place presents itself as a building with lots of books on shelves). From this it may seem that artefacts contain rather straightforward clues, and are much less multi-interpretable than the other main symbolic system in human societies: language. Miller (1987:106-107) has however argued that material culture is also ambiguous in its message content: objects, just like words, may evoke variable responses and a variety of interpretations. Artefacts are extremely visible and at the same time extremely invisible: "... its physical and external presence belies its actual flexibility as a symbol" (ibid.:106). Thus the response to objects is context-dependent (e.g. within different fields objects may be perceived differently).

Bourdieu acknowledges the existence of basic sets of oppositions, but he points out that in itself the isolation of such structures (as in many classical structuralist studies) is inadequate, since such a procedure ignores the social and economic conditions which produced them. Bourdieu, like Giddens, stresses the dialectical relationship between social meaning and social practice. Space acquires meaning through *practice*, which in its turn is informed by a set of conceptual schemes which are represented in the order of space. The active individual is re-introduced in these structural analyses. In this way a structuring structure, for example a house, is both medium and outcome of social practice.

The potential of Bourdieu's theory for the analysis of architecture and the use of space is considerable. His ideas have succesfully been applied in a number of ethnoar–chaeological and archaeological sudies (e.g. Barret 1981; Blanton 1994; Braithwaite 1982; Davis 1984; Donley-Reid 1990; Hingley 1990; Moore 1982, 1986; Rippengal 1993). Furthermore many other structuralist analyses have indicated the importance of the symbolic organization of space (e.g. Cunningham 1973; Lawrence 1987; Lévi-Strauss 1963:132-163; Tuan 1977).

	SPACE	
	Spatial unit	
Steps		
1	**Visual inspection**	
2	**Architectural analysis**	Description of spatial unit: Dimensions, dooropenings, architectural features
3	**Depositional analysis**	Determination of formation processes: stratigraphy, condition of objects, contextual information
4	**Determination of object context'**	- Depositional - Interactive - Discard - Transformed
5	**Determination of general function**	- Activity area - Storage area - Discard area
6	**Functional analysis**	Delineation of activity categories: - Subsistence activities - Manufacturing/maintenance activities - Administration - Social life - Other Delineation of specific activities on the basis of analogy: architecture, features, artefacts
7	**Synthesis and interpretation**	*Structure, habitus* and *practice*
	PLACE	

Figure 2.1. From space to place: model for assessing the function of prehistoric spatial units.

It is exciting to realize that mundane objects, excavated by archaeologists, can play an active role in the structuring of society, in the constitution of the habitus of persons and groups (Hodder 1986:73). By taking the role of material culture in societies seriously many non-material aspects, such as ideology and cosmology, may become part of archaeological inquiry. The meaningful association of specific objects in specific areas of buildings, for instance, would be preserved in the archaeological record if buildings were suddenly abandoned and subsequently quickly covered by debris. Many associations are of course of a functional character, but, as was shown in the example above, functional and symbolic aspects are highly interwoven.

Practice can be analysed by reconstructing architecture, the use of space and the organization of activities, i.e. by contextualizing material culture (chapter 6). *Structure* and
habitus can be reconstructed by delineating patterning in architectural layout and material
culture. Patterning in material culture may especially be detected by looking for
associations and oppositions (chapter 7).

6 The Model: From Space to Place

Space is largely a geographical and physical concept; it is a three-dimensional
surrounding, an undefined dimension upon which humans may act, something which they
may shape, or give meaning to. By this shaping, by this working on the landscape, space
is transformed into place: nature is transformed into culture. Order and meaning are
created in a seemingly chaotic (i.e. natural) world, and randomness is transformed into
predictable patterns (Bargatzky 1994:9). Place is thus much more specific than space; it
is a dimension with meaning. Place is not necessarily created by erecting visible boundaries; it is well-known that in many cultures natural features in the landscape, such as
mountains, rocks, waterfalls or caves are sacred places, which have special meanings
(Carmichael et al., eds., 1997; Hirsch 1995:4). Place, therefore, should not be defined as
something created by humans, but as something that "... owes its character to the experiences it affords to those who spend time there" (Ingold 1994:155). Gramsch (1996)
contends that there is a dialectic between space and place, place structuring space, and
space containing place. The meanings of place are largely culture-specific, i.e. they
depend on the worldview of its users, which in its turn is shaped by place (see also Hirsch
1995:4).

In the following paragraphs the various 'steps' which have been taken in the spatial/
contextual analysis of level 6 will be described. In other words, the methodology of this
analysis will be presented. In figure 2.1 the method advocated is presented as a model
entitled 'from space to place'.

1. The first step in the model consists of a general *visual inspection* of the distribution
of objects in the settlement. To make distribution maps (and for all other steps in the
spatial analysis) first computerfiles indicating the horizontal and vertical (i.e. stratigraphical) distribution of objects in level 6 at Sabi Abyad were constructed. For the
construction of these databases the computer packages DBASE and EXCEL were used.
For the same object various properties of this distributional information were constructed,
e.g. stratum, level, coordinate (X, Y and Z), room, building, etc. These records were
linked to records of the same objects in which intrinsic properties of these objects (e.g.
designation, shape, dimensions, etc.) were recorded. All this resulted in elaborate
databases in which as many data as available (i.e. the 'small finds', the flint and obsidian
artefacts, the pottery and the animal bones) were included.[8] This extended database
(28,241 records) forms the basis of the present spatial analysis.

---

[8] Botanical remains have been sampled, but not in a systematic fashion; therefore the distribution of
plant remains has not been analysed in the present study.

In order to make distribution maps, the level 6 plan was digitized, so that the objects recorded in the databases could be linked to this computer map. For digitizing and editing the subsequent drawings the program AUTOCAD was used. Simply put, AUTOCAD is a computer program in which drawings can be made and manipulated. The AUTOCAD drawings and the DBASE/EXCEL databases have been linked by the program MAPINFO. MAPINFO is a simple GIS (Geographical Information System). A GIS is "a computerized set of tools for collecting, storing, retrieving at will, transforming, and displaying spatial data from the real world for a particular set of purposes" (Burrough 1986:6, see also: Aldenfelder and Maschner, eds., 1996; Allen, Green and Zubrow, eds., 1990; Llobera 1996). A GIS consists of two major components: "One component stores the spatial information (database management), whereas the other serves to manipulate and display these data. The data, transformed or not, are usually displayed cartographically. Both the database management capacity and the possibilities for transformation and display are crucial to the performance of a GIS" (Wansleeben 1988:435-436). Commonly GIS is used in archaeology for investigating patterns in large-scale site distributions (large areas with many - different - sites, e.g. Wansleeben 1988), i.e. in inter-site or regional spatial analysis. Most often these analyses are set up to investigate the relation between site distributions and landscape, i.e. site location. GIS may also be used, however, for intra-site spatial analysis, especially in the case of complex sites with many structures and artefacts (e.g. Quesada et al. 1995). By using MAPINFO, maps can be made of all the characteristics of objects which have been recorded in the databases. Within seconds maps can be made not only of e.g. all stone objects versus all bone objects in a settlement, but also of e.g. all objects between 5 and 10 cm in length which were found on floors.

In general, the visual inspection proceeded from analysis of general patterns to the study of specific distributions. For example: first the distribution of all bone objects was plotted, then that of specific categories of bone tools (e.g. awls, spatulas, etc.).

In this way the visual inspection has been a general exploratory and qualitative approach. This first act of the spatial analysis is as it were a first acquaintance with the 'contextualized settlement'. On the basis of the patterns observed (or the lack of patterns) first tentative hypotheses, guiding further research, may be formulated.

2. The second and following steps (all based on the databases introduced above) consist of assessing the functions of spaces. The second step is the *architectural analysis*. The spatial unit (e.g. a room or an oven) of which the function is to be assessed is described in detail; the dimensions, construction techniques (e.g. absence or presence of plaster) and architectural features, if present, are described. In the present study the architecture is described in chapter 3.

3. Then follows a *depositional analysis* (cf. chapter 4). This study consists of a determination of the formation processes which have resulted in the distributional patterns observed in the spatial unit which is being studied (e.g. Schiffer 1987). As is argued in chapter 4, an insight in formation processes is important if one wishes to come to a reliable functional assessment. By looking at the stratigraphy, the condition of objects (are they broken or complete?) and contextual information (e.g. the possibility of use of an object in the area analysed) and by using the models of the characteristics of formation processes

as presented in chapter 4 the depositional history of the object and deposit in which it is imbedded may be established.

A distinction is made between floor and fill deposits. Floor deposits consist of depositions and objects on or directly above (up to ca. 15 cm) floors. In the level 6 village at Sabi Abyad the floor deposits in the burnt structures and areas are marked by a ca. 10 cm thick black ashy layer containing many objects (see chapter 3, fig. 3.4 and fig. 2.10 in Verhoeven and Kranendonk 1996). Objects found on and directly above floors or surfaces (including roofs) may actually have been used in the space where they have been recovered. When a conflagration has trapped such artefacts the relation of floors and artefacts is particularly clear. However, it should be noted that fired roofs cave in at their weakest point or where the fire is most intense, thus causing roof goods to slide onto the floor below, to become mixed with objects there before being sealed by the rest of the roof. In practice, this contamination of a systemic inventory on room floors (see chapter 4) will be very difficult or impossible to detect, even by detailed stratigraphical analysis. Therefore, it should be taken into account that each floor deposit may contain non-floor objects, and in each instance the archaeologist should asses the impact of the disturbance (see chapter 4 and appendix 4).

Fill deposits, covering floor deposits, represent dump or erosion material; in general it is unclear how objects in these depositions relate to the function of spaces in which they are found. Therefore, with a few exceptions, only objects on or directly above floors/ surfaces have been taken into account in the functional assessments.

4. After and on the basis of the depositional analysis the *object context* should be established. Were the objects in an area in an interactive, depositional, discard, or transformed context (see chapter 4)? An arrowhead found in a small room, for instance, is not likely to have been used there. Therefore it is probably not in an interactive context. If it is complete and not totally worn down it is much more likely that the arrowhead is in a depositional context, i.e. that it is stored in that room, awaiting further use.

In this fourth step use is made of tables and contingency tables in which, per artefact assemblage, the quantity of objects has been expressed. These tables have been made with the statistical computer program SPSS, which has been linked to the DBASE or EXCEL files.

5. Now, with the help of the model as presented in chapter 4 (i.e. table 4.5), it may be possible to *determine the general function* of the area under study: is it an activity area, a storage area, or a discard area? It is argued that these three kind of areas are the most basic areas which can be distinguished in prehistoric tell settlements.

An activity area is simply defined as an area where activities have been carried out, or as the locus at which a particular human event occurred (Kent 1984:1). A somewhat more detailed definition would be that an activity area is a spatially demarcated area where a specific task or a series of related or unrelated activities have been carried out. Usually activity areas are marked by specific structures or features, tools, debris and raw materials (e.g. Flannery and Winter 1976, see also O'Connell 1987). Activities have been divided into a number of classes, i.e. subsistence activities (including storage), manufacturing/ maintenance activities, administration and social life (see chapter 5). With regard to these activities it has to be acknowledged that one object may be used for several activities, and that it may be used in different contexts. For instance, objects or activities may have symbolic/ritual connotations without being used in a ritual context only.

A storage area is an area where objects and products were stored, i.e. these elements are temporally and spatially displaced from contexts of use and consumption (e.g. Schiffer 1972:158). A distinction between long-term and short-term storage can be made. With long-term storage is meant the storage of objects or products for extended periods of time, i.e. weeks and longer periods. Short-term storage refers to storage less than about a week.

A discard area is an area where objects and goods are deliberately discarded, i.e. an area where refuse or elements that no longer participate in a behavioral system are dumped (ibid.:159).

How does one recognize such areas in the archaeological record? An activity area must of course simply be large enough for carrying out activities: many spaces are of such restricted dimensions that they are unsuited for conducting activities. Secondly, activity areas are often marked by architectural features or installations, e.g. ovens, benches, or mortars sunk into the ground. Third, the artefacts found in an area which is to be designated as an activity area must indicate activities that actually can have been carried out in that area, i.e. the object context must be interactive. Through ethnographic studies of Near Eastern settlements (e.g. Krafeld-Daugherty 1994; Kramer 1982; Watson 1979) we know that activity areas (i.e. living rooms, kitchens and reception rooms) are generally well-constructed and well-accessible. Furthermore, typical objects of daily use are present in them.

A storage area is often small and less accessible than activity areas. Secondly, architectural features are generally (but not always!) absent. Third, items in storage areas often occur clustered in (relatively) large numbers. Fourth, typical stored objects or products occur, such as staple products (grain in bulk), large storage jars, and sealed containers (e.g. Christensen 1967; Horne 1980:23; Krafeld-Daugherty 1994:132; Kramer 1982:105).

A discard area is marked by refuse, i.e. broken or heavily damaged objects that can no longer be used, animal bones and other organic debris, etc. Most often a variety of different materials is found together in such areas. The refuse can simply be dumped in open areas, in pits, in deteriorating buildings or other architectural structures.

In practice it is often difficult to assess if a certain area was an activity area, storage area, or discard area. Especially the distinction between activity areas and storage areas is problematic. For instance, very small activity areas without architectural features may have existed, and, vice versa, large storage areas without characteristic items. Storage and activities, moreover, may have been combined in one and the same area, most probably the storage being short term, but long-term storage cannot be excluded either. It has been tried to tackle this problem by designating different kinds of activity, storage and discard areas (see chapter 4; table 4.5).

6. The subsequent *functional analysis* is based upon analogy (cf. chapter 5). On the basis of the various models of the use of artefacts, ceramic containers and flint/obsidian artefacts, the function of an area is specified. First, it is reconstructed which activity categories, i.e. subsistence activities, manufacturing/maintenance activities, administration, social life and 'other', may be distinguished. Then, if possible, specific activities are isolated.

7. The final step consists of a - functional - comparison of the various buildings and other features. A synthesis is given of the results of the preceding analysis. First the activities, the *practice,* will be dealt with (chapter 6), next it will be tried to reconstruct *structure* and *habitus* (chapter 7), this according to the theoretical framework presented in section 5.

CHAPTER 3

THE BURNT VILLAGE

1 Introduction

Since the level 6 Burnt Village is the focus of the present study, this settlement and its finds will first be introduced.[9] First the architecture (section 2), and then the finds (sections 3 to 9) are dealt with. The architecture is discussed in detail, since it represents the context of the spatial analysis (chapter 6). The finds are dealt with in a more general manner; detailed reports can be found in Akkermans, ed., 1996. Moreover, the function of the various artefacts is discussed in chapter 5.

As indicated in chapter 1, level 6 is the earliest of the so-called Transitional or Balikh IIIA levels, and is dated at around 5200/5150 B.C. (5970/5960 cal BC).

2 Architecture

Up to 1996, the Burnt Village remains, partially standing to a height of 1.40 m, have been excavated over an area of about 800 m² (fig. 3.1). The settlement was built in terraces: part of the mound had been dug away along the slope, and the floors, walls, etc., of the structures low on the slope were founded about two metres below those of the buildings somewhat higher on the mound. Consequently, it appears that the floors of the upper buildings must have been more or less on the same level as the roofs of the lower-situated buildings; one could easily walk onto these roofs (figs. 2.14 and 2.15 in Verhoeven and Kranendonk 1996). Actually, we have some evidence that this was indeed the case and that various kinds of activities were carried out on the roof.

So far, the Burnt Village is represented by eight rectangular, multi-roomed structures (buildings I-V, X-XII) and four circular ones (the so-called tholoi; buildings VI-IX). In addition, ten ovens were unearthed in and between the house remains. The dimensions of the rectangular buildings seem to have varied between ca. 90 m² and 120 m². Generally speaking, the buildings of the Burnt Village were originally built along very regular lines and were closely attached to each other (cf. fig. 3.1)[10], although all kinds of renovations and reconstructions took place in the course of time. Some buildings seem to have had

---

[9] This presentation of level 6 mainly deals with the results of the 1991 to 1993 excavations (Akkermans, ed., 1996; Akkermans and Verhoeven 1995). The general results of the 1996 excavations have been added. The discussion is largely based upon Akkermans and Verhoeven 1995 and Verhoeven and Kranendonk 1996.

[10] In the colour plate at the beginning of this book building II has been completed (cf. fig. 3.3) and a tholoi recovered in 1996 (east of building IV) has been added.

Figure 3.1. Plan of level 6 architecture (1991-1993).

more than one floor (each consisting of hard-tamped loam layers ca. 1-3 cm thick). Basically, the oblong structures seem to have been divided into three rows or wings, each of which consisted of a series of small rooms (or cells). Some of these buildings had 15 or more rooms, all very small and varying in size between about 3 and 5 m².

The generally 40 cm wide walls of the level 6 buildings were simply founded on earth and were all built of pisé. Some rooms had a doorway at floor level (ca. 50 cm wide), but many rooms gave no evidence of doorways, and it seems that these rooms must have been accessible from a higher level. A number of rooms in buildings IV and V had doorways of such restricted size (diameter ca. 50 cm) that one would have to crawl through them on hands and knees. These 'portholes'[11], earlier reported from sites like Bouqras, Umm

Dabaghiyah, Beidha, Abu Hureyra and Ganj Dareh (see e.g. P.A. Akkermans et al. 1983; Smith 1990), were all situated at a somewhat higher level in the wall and had a rounded, almost 'arched' superstructure (cf. fig. 2.8 in Verhoeven and Kranendonk 1996).

Apart from some tholoi, all level 6 structures were heavily affected by an intense fire which penetrated the walls throughout and which caused a considerable accumulation of orange to brown, crumbly loam, wall fragments, dark ashes and charred wood in the buildings (fig. 3.2). The lowest, ca. 10 cm thick part of these deposits, directly situated on the floors, virtually always consisted of fine and powdery, black ashes; most likely, these ashes were the burnt residue of the roof cover (reed mats). The common occurrence of charred beams and hard-burnt loam fragments with impressions of reeds and circular wooden poles in the various buildings reveals that the roofs were all made in the same way: wooden rafters were placed at regular intervals and covered with reed mats, in their turn covered by a thick mud layer. The section through buildings IV and V in trench P15 on figure 3.4 (respectively walls AM, G and H and walls I and R) may serve as an example of typical level 6 roomfills (here represented by strata 5C and 5A).[12]

As yet, the various buildings have not given evidence for the presence of a second storey, and there are various indications that such storeys were absent. First, no floors or other architectural features (e.g. hearths) were found high up in fill deposits. Second, no mud staircases, such as found in Early Neolithic Çayönü (Schirmer 1990:372) and Bouqras (Banning 1997:46, but see P.A. Akkermans et al. 1983:348), were noted at Sabi Abyad. Third, the walls of the level 6 buildings are only ca. 40 cm wide, i.e. they are rather thin for having served as foundations for a second storey. Generally the walls of two-storey mud-brick buildings in the Near East are at least 60 cm wide, and ideally they are ca. 1 m wide (cf. Kramer 1982:132). Furthermore, in her ethnographic research in Aliabad, Kramer has noted that when thin (ca. 50 to 60 cm thick) walls are used as bases for a second storey, an abutting second wall is erected to reinforce the second storey's support. Two-storey houses in Aliabad were also marked by wide and deep foundation trenches (ibid.:132). All these features, which reinforce the strength of a wall so that a second storey can be build (i.e. stone foundations, buttresses along walls, abutting walls and foundation trenches) are absent from the Burnt Village.[13]

[11] Portholes were present between rooms 1 and 2, 2 and 6, 5 and 12, 10 and 11 in building IV, and in the west wall of room 10 of building V; see fig. 3.1.

[12] In square Q15 and trench P15 level 6 is represented by strata 5D-5A. Building IV was erected at the very beginning of the level 6 period; this structure has been ascribed to stratum 5D. The next stratum 5C contained a ca. 50 cm thick accumulation of debris south of building IV. This debris consisted of grey-brown loam intermingled with reddish-brown pisé fragments. Stratum 5C debris was also encountered in room 10 of building IV; here ca. 30 cm of grey-brown loamy layers accumulated. Apparently room 10 had been abandoned when the other rooms were still in use. In the area south of room 10, pit AC had been sunk into the upper part of the stratum 5C debris. Upon this pit and upon stratum 5C debris, another multi-roomed building, V, had been erected, which represented stratum 5B. The characteristic burnt debris (consisting of a lower black and ashy layer upon floors, followed by a crumbly orange/reddish burnt loam) has been attributed to stratum 5A. This material had accumulated directly upon the floors of all rooms of building V and most rooms of building IV. In the open area or, perhaps, courtyard east of building V, the burnt debris was present along the walls of the building only. In trench P15, on top of the stratum 5A remains, a wall (S) had been raised, which was partly hidden in the north section (Verhoeven and Kranendonk 1996:41).

[13] The level 3 building I at Sabi Abyad was marked by a number of these features (i.e. thick walls, buttresses along the walls, stone foundations); therefore this building has been reconstructed as a structure with two storeys; see chapter 6, section 9.

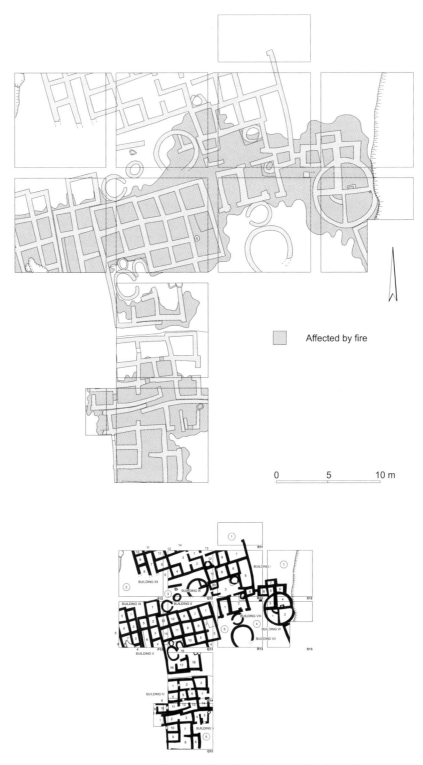

Figure 3.2. The area affected by fire (dotted) in the level 6 village.

Fourth, as has been indicated above, pieces of clay with impressions of circular wooden beams were regularly retrieved from roomfills. In case of collapse of a two-storeyed building (the floor of the second storey, like the roof, most likely having been made of wooden poles, covered by reed and mud, cf. Krafeld-Daugherty 1994 Abb. 137) one would expect two superimposed layers of these fragments, since logically first the material of the floor of the second storey would cave in, which would then be covered by wall debris and the roof material. Room stratigraphies, however, have never shown such a succession, and therefore the clay fragments with impressions most likely stem from the roof, and not from a second storey floor and the roof. In this respect it should also be mentioned that no supports for wooden beams were found in the level 6 walls, sometimes preserved to a height of 1.40 m, i.e. heights where a second floor could be postulated.

It could be argued that the level 6 buildings consisted of low basement cubicles and second storey living/working areas. However, a number of architectural features suggest a 'normal' height for the level 6 buildings: (1) doorways at floorlevel in various buildings and a door-socket in room 1 of building II, suggesting that some rooms could be closed by doors; (2) a platform in room 2 of building II; (3) ovens in rooms 2 and 12 of building I; (4) in a number of instances complete ceramic vessels were standing on floors; these could not have fallen from a second storey, and must have been stored or used. These features suggest activities and upright human postures in at least some of the rooms of the rectilinear buildings, which therefore must have had a height of at least 1.60 m.

Building I. Building I was oriented NE-SW, at least 12.50 m long and 9.50 m wide (fig. 3.1). Excavation has so far shown that it consisted of 13 rooms constructed around a large, central room or, perhaps, courtyard (area 3). The southernmost series of rooms (1-4 and 10-11) was heavily affected by fire and filled with ashes, but the other rooms had been left undisturbed. These rooms may have been already out of use when the fire started; there is some, admittedly poor, evidence that an accumulation of grey to brown loam (undoubtedly wall debris) found in these rooms preceded the deposition of ashes and other burnt materials. On the other hand, it may also be the case that the fire stopped for one reason or another, and that these rooms were simply left to the elements.

Doorways were recognized in the case of rooms 1-2 (marked by small buttresses) and rooms 12-13. No passages were found in the other areas, probably largely due to the rather poor state of wall preservation (some walls stood to a height of only 20 cm). Room 1 was accessible only from the open area or courtyard south of building I, whereas room 2 could be entered both from the south and the north.

A rounded, beehive-shaped oven (T) was found in the northeastern corner of room 2. The oven wall stood upon a low mud platform. A circular pit was found immediately southwest of the oven T. Another oven (M), similar to the one in room 2, was uncovered in the court area 3 of building I. An opening or ventilation hole was present in the southeast. The oven stood upon a low mud platform. A third oven (AT) was found in the southwestern corner of room 12. This oven was horseshoe-shaped and had an opening on its northern side. The opening was blocked with a rectangular piece of grey compact loam. The oven wall was constructed of grey pisé. The interior wall face was covered with red-burnt mud plaster, whereas the floor was black-coloured and thoroughly fired.

The stone construction found in the northern part of room 5 was rather curious and is as yet unexplained. It consisted of cobbles and gypsum boulders, carefully placed in line with the surrounding mud walls. The stones seemed to constitute a kind of platform, ca. 3.50 m² large and about 30 cm high.[14]

A series of rooms directly northwest of building I has been termed building XII, but it is not excluded that these rooms were originally part of building I; if so, building I was an extremely large, L-shaped structure. It was noticed that the southern wall of room 2 of this structure is connected to building I; in addition, room 1 of building XII and room 13 of building I were connected by a small doorway. On the other hand, it appeared that the rooms of building XII were all of more or less the same shape and dimensions, whereas the various rooms of building I were less coherent in this respect. Moreover, household features have been found to be wholly absent from building XII so far, whereas three ovens were present in building I. At present, it is suggested that buildings I and XII were separate structures.

Building II. Originally, buildings II and III were treated as separate units (Akkermans and Verhoeven 1995; Verhoeven and Kranendonk 1996, fig. 2.7). Subsequent analysis and especially the 1996 excavations, however, have revealed that buildings II and III are part of the same large structure. This is indicated by the western façade of the building: 'buildings' II and III are bonded. This can clearly be seen on fig. 3.3 (and the colour plate), where the rooms found during the 1996 excavations have been added. One immediately recognizes a large central building with an open area marked by ovens. Building II (as the central building will be called; building III is discarded, as it is part of building II) is reminiscent of building I in this respect. Building I, however, is smaller and less regular.

Building II, like building I oriented NE-SW, has been completely excavated. It was 12 m long and 12 m wide, representing the largest excavated level 6 building (fig. 3.1). In the north the structure consisted of 14 small and square rooms arranged in three rows. The rooms were virtually identical in size, measuring ca. 1.75 x 1.75 m. In the south the building was marked by a less regular row of 5 rooms, separated from the northern ones by an open area (area no. 15) with various ovens in the west and a passage leading to this area. This passage, giving access to the whole building, could be reached from the east by a narrow doorway. To the east of room 20 (cf. fig. 3.3) two very small rooms were probably added to building II.

Domestic installations in the rooms were represented by a low platform in room 2 and a 40 cm high bench along the southern wall of room 20 (fig. 3.3). In room 19, moreover, a mortar had been sunk into the floor (as in room 5 of building V). Evidence for interior doorways has been found only between rooms 2 and 3. A small pivot-hole was hollowed out in a rounded loam boulder, suggesting that this doorway was once closed by a wooden door. In all other instances the walls of the northern rooms (which stood to a limited height only) were uninterrupted, and apparently these rooms were accessible either through a passage situated at a higher and now eroded level in the wall (i.e. portholes) or from the roof of the building. Apart from room 20 the southern rooms were open to the north.

---

[14] Excavations in 1996 have indicated an east-west running wall in the middle of room 5; this room thus actually consisted of two separate chambers.

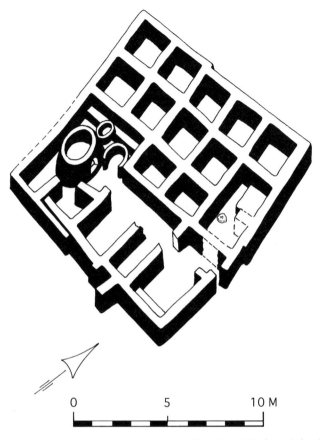

Figure 3.3. Axonometric reconstruction of building II of level 6 (1996).

The ovens, concentrated in the area north of rooms 16 and 17, were not conceived and raised at a single point in time but built in at least three stages. Apparently, when an oven went out of use for some reason or other, the feature was levelled and a new installation constructed. The largest oven (S), however, seems to have been used continuously for a considerable period of time. This impressive, beehive-shaped feature stood to a height of almost 1.50 m, with its vaulted roof still partly intact (cf. fig. 5 in Akkermans and Verhoeven 1995). The oven was oval in shape and had a maximum diameter of 2.90 m. Its wall was ca. 35 cm wide and constructed of layers of orange-brown clay, heavily tempered with straw. The interior was considerably sooted. The oven was accessible through a tapering opening in the east.

Two other types of ovens were constructed immediately to the north and the northeast of oven S. A horseshoe-shaped oven or hearthplace (CS) was, like oven S, used intensively: seven floor levels were recognized. When this hearthplace went out of use, a circular *tannur*-like oven (BQ) was built upon its remains. Another *tannur* (CR) stood to the west. North of ovens CR and CS, against the façade of rooms 11 and 14, a narrow bench was present.

A small keyhole-shaped oven (DG) was present in the north of room 18, built against its eastern wall.

Building IV. The remains of the east-west oriented building IV, partially standing to a height of 1.40 m, were (with building V) among the best preserved of the Burnt Village. Building IV measured at least 11 x 7 m and consisted of at least 14 rooms (fig. 3.1). As in building II, these rooms were arranged in three rows. The building was heavily affected by fire except in the northernmost series of rooms 1 to 4. Whereas most rooms were entirely filled with ashes and other burnt building debris, the latter areas gave evidence of loam only. Some of the walls in these northernmost rooms leaned over severely to the north and east, probably due to the pressure of the collapsed upper walls and roof covering.

Most rooms seem to have had normal doorways but portholes were found as well. Some rooms did not have a passage at floor level at all and were apparently accessible from the roof.

Building V. Building V measured at least 10.50 x 8 m and consisted of at least ten rooms of varying dimensions (fig. 3.1). The structure suffered severely from the fire that swept over the village and all rooms were filled in with ashes and other burnt building debris. The walls were generally preserved to a height of about 1 m but stood much lower in the heavily eroded westernmost room 1. Fragments of charred wooden poles, up to 1.50 m long (undoubtedly part of the roof cover), were found in various rooms, as were burnt impressions of reeds. Roof remains were most clearly recognized in rooms 2 and 3: these oblong chambers were originally covered by north-south oriented timbers laid at regular intervals and a thick layer of reeds oriented east-west. The reeds were subsequently covered with a thick layer of mud.

Basically, building V seems to have consisted of room 2 and rooms 7 to 10. The other parts (rooms 3 to 6) were added to the main structure at a somewhat later stage. The walls of this added part were not bonded with those of the original structure but abutted them instead. Moreover, some rooms (3 to 5) were raised upon debris which had accumulated in the open area or court in front of both buildings IV and V and blocked the southern doorway of the former structure. The various rooms had either normal doorways (chambers 2-8) or portholes (room 10). Direct access to room 2 was blocked when the neighbouring rooms 3 to 5 were constructed; apparently, access to this room 2 (and to the newly built room 3) now shifted to the roof. Room 9 in the centre of building V yielded no passage at floor level either and must also have been accessible from the roof only.

No domestic installations were found in building V except in room 5. Here a small *tannur*-like oven (EN) appeared. The oven was somewhat peculiarly located in what originally may have been a passage from this area 5 to either the courtyard or another room situated further east. In addition to this oven, room 5 contained a large limestone mortar partly sunk into the floor. By means of a lining of stones, this mortar was solidly secured into the floor.

Building X. Building X was a rather narrow, NW-SE oriented structure measuring ca. 10 x 5 m (fig. 3.1). It was composed of two rows of rectangular rooms. Unfortunately, the building was preserved to a limited height only (i.e. 30 cm at the most). There can be little doubt that building X was raised at a somewhat later time than its neighbours; it seems to have been constructed in what originally must have been an open area or, perhaps, alley between buildings II and XI. The walls of building X were positioned against the exterior façades of the other structures.

In the case of rooms 2 and 3, it appears that the interior division walls stood against the outer façade of building II; apparently, the exterior wall of building II served to bound building X as well, thus suggesting a close relationship between both structures.

Room 1 was a relatively large square area, measuring ca. 2.30 x 2.30 m. Room 6 was rectangular in plan and measured ca. 3.50 x 1.50 m. The other rooms all represented small rectangular compartments, each measuring ca. 1.80 x 1.50 m. All rooms except room 1 had been affected by the fire. An unroofed area was present north of room 5, whereas traces of yet another room seem to be present north of room 1, largely hidden in the section baulk. So far, no evidence of this additional room has been found in the neighbouring square P12 (perhaps due to slope erosion).

A low and narrow bench had been constructed along the western wall of room 6.

Building XI. The NW-SE oriented building XI measured at least 9 x 5 m and consisted of at least three rows of rectangular rooms (nos. 1-9; fig. 3.1). The eastern row consisted of four rooms (nos. 1-4) which each measured ca. 2 x 1.25 m, whereas the central series of rooms (5-7) each measured about 2 x 1.75 m. The central row of rooms protruded slightly to the north when compared with the eastern wing. Similar protrusions were found along the north façade of building II (north of room 12), the west façade of building I (west of room 10) and the north façade of building IV (north of room 3).

Building XI seems to have been less heavily affected by the fire than the other level 6 structures. The northern rooms 1 and 5 were not touched at all; their walls were unburnt and they were filled with grey-brown loam instead of the orange-brown burnt material.

Building XII. Building XII was a large, NE-SW oriented structure, very regular in layout and measuring at least 13 x 8 m (fig. 3.1). The building consisted of at least 14 small rooms. The southwestern part of the building had been heavily affected by slope erosion. The building seems to have been closely associated with the neighbouring structure I; actually, it is not excluded that building XII was originally incorporated within building I (see above).

Three types of rooms could be distinguished: (a) small square rooms, each measuring ca. 1.50 x 1.50 m (nos. 4, 9 and 12), (b) larger rectangular rooms, each measuring ca. 2/ 2.30 x 1.50/1.75 m (nos. 1-2, 5-6) and (c) a narrow rectangular compartment measuring ca. 4 x 1 m (no. 7). Room 3 is L-shaped but most likely it originally consisted of three separate, rectangular chambers (the walls in this area were very eroded). The eastern rooms 1-3 were connected by small doorways at floor level. No other entrances have been found, most likely due to the rather poor state of preservation of building XII.

Building XII seems to have been abandoned before the start of the devastating fire (similar to the northern part of building I). The wall stubbs, which stood to a height of maximally 60 cm, were mainly surrounded by grey-brown wall debris instead of the burnt materials.

A series of pits (AC, AM, AP and AS) were sunk in the open area or courtyard of building XII, shortly after its abandonment.

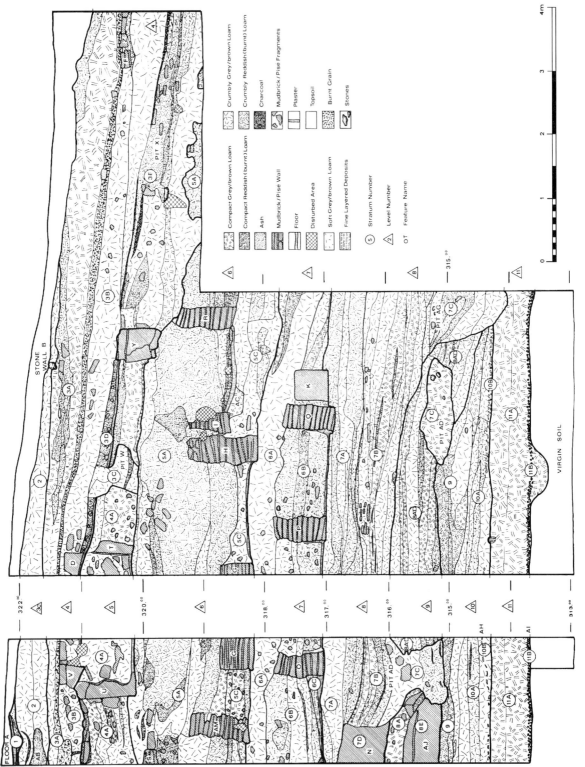

Figure 3.4. North and east sections of trench P15.

The circular buildings VI-IX. In addition to the rectangular structures, four circular buildings or tholoi were found (fig. 3.1). The largest one was building VI, situated to the east of building I. It had an interior diameter of about 6.75 m and was divided into a series of smaller compartments. Some very small rectangular rooms (5 to 8) seem to have been added to the circular chamber at a somewhat later date. Another rectangular structure could be found immediately south of the tholos. It had been partly destroyed by a large pit sunk from a late second millennium B.C. layer of occupation. The main entrance to tholos VI was found in the northeast corner of compartment 2. This doorway had a low, clay threshold and contained a stone door-socket, indicating that the passage was originally closed by a wooden door. No other doorways were recognized, perhaps due to the fact that the walls stood to a very limited height only, i.e. about 20 cm. A small and low, rounded bin made of clay slabs was found in the northeast corner of compartment 1, and a rectangular bin, sunk into the floor, was found along the wall in room 4.

Two other circular structures (VII-VIII) were found in the open area to the southwest of tholos VI. Building VII had an interior diameter of about 4.50 m and stood to a height of ca. 70 cm. Its wall, thickly white-plastered on the exterior façade, curved slightly inwards already at floor level, thus suggesting a domed superstructure. The southeastern part of the tholos was disturbed by later building activities.

In 1996 two more tholoi were recovered underneath tholos VII. Apparently, the earliest tholos had been rebuilt twice (tholos VII being the upper one). Stratigraphic analysis indicates that the lower two tholoi are related to an early, unburnt phase of the level 6 settlement. During this phase the rectangular buildings were already present (and in use).[15] So far these tholoi are among the largest circular structures at Sabi Abyad; they had an interior diameter of ca. 6 m. The walls of the tholoi were ca. 50 cm wide and built of grey-brown pisé. In the south a ca. 1 m wide opening was present. In two of these super-imposed tholoi a large horeshoe-shaped hearth or fireplace was built against the wall in the northeast.

Tholos VIII stood immediately to the north of building VII but was much smaller in size: it had an interior diameter of ca. 2.50 m only. The structure was preserved to a height of ca. 80 cm and was accessible through a doorway in the eastern façade. This small tholos seems to have been in use for a considerable period of time: at least five superimposed floor levels of tamped loam were found, each separated from the other by a series of thin and compact grey-brown layers of loam. The lower floors all sharply inclined towards the west, i.e. to the entrance of the building; the topmost surface, however, had been levelled. Evidently, the construction of a new floor in the tholos was related to a heightening of the open area or courtyard around the structure.

The fourth tholos (IX) stood in the courtyard between buildings I, II and XII. The tholos had an interior diameter of ca. 3.50 m and still stood to a height of ca. 50 cm. The entrance was hidden in the section baulk between squares Q12 and Q13. Tholos IX seems to have been divided into two compartments, the smallest measuring ca. 100 x 80 cm. The steeply southwards sloping floor was made of tamped loam, covered by a thin white plaster. The walls of tholos IX were straight; most likely the structure had a flat roof instead of a

---

[15] In 1996, in both the early unburnt and later burnt level 6 phases, other multi-roomed structures consisting of small square and rectangular chambers, some with ovens, were recovered to the east of the area excavated in previous campaigns (i.e. in squares S14, T13 and T14).

beehive-shaped superstructure. Fragments of hard-burnt loam with impressions of reeds and circular wooden poles made it clear that this roof was made of wooden rafters covered by reed mats, in their turn covered by a thick mud layer (cf. the rectangular buildings). Actually, it seems that these flat roofs made of reeds and timber (i.e. highly inflammable materials) accounted for the burning of some of the level 6 tholoi; the unburnt circular structures all seem to have had a beehive-shaped, clay superstructure. Only the eastern half of the tholos was affected by the fire; here the wall was burnt throughout and surrounded by burnt debris.

Apart from the two tholoi underneath tholos VII, two other tholoi were recovered in the 1996 campaign: in square R14, east of building IV (cf. the colour plate) a smaller tholos (ca. 4.90 m in diameter) with a small central hearth and a small compartment; in square S13 a tholos with a diameter of ca. 3.50 m was found underneath tholos VI of the Burnt Village. This tholos had a floor which was partially constructed of Coarse Ware sherds.

Open areas. The various open areas or courtyards between and around the various buildings have been numbered 1 to 6 (see the encircled numbers on fig. 3.1).[16] Apart from area 3 none of the areas have been wholly exposed. Area 1, measuring at least 9 x 9 m, is located to the north and east of buildings I and VI, in squares R11, R12 and S12. No architectural features were found in this locality. Area 2 was found in squares P12 and P13, north of buildings X and XI and west of building XII.[17] Architectural features were absent from area 2. The area measured at least 13 x 9 m.

Area 3 (measuring 11 x 7.50 m) is represented by the central open area (in squares Q12 and Q13) surrounded by buildings I, II, X and XII. Apart from tholos IX, two circular ovens were found in area 3. Both features were unaffected by the level 6 fire. The beehive-shaped oven DA stood in the southwestern corner of the courtyard. It had an interior diameter of ca. 2 m and stood to a height of ca. 40 cm. Its wall was constructed of compact grey pisé. The exterior façade was covered by a mud plaster. The interior gave evidence of a twice-renewed, hard-burnt mud plaster on both floor and wall. The oven was accesible from the east through a narrow opening at floor level.

Oven AR stood next to oven DA. The oven was beehive-shaped and built of compact, grey-brown pisé layers. The oven wall stood upon a low platform of compact brown loam. The interior oven face carried a brown mud plaster. An opening has not been found, due to the poor state of preservation. A low bench abutted oven AR to the east. Immediately north of oven AR a small, oval and quite shallow pit was present.

A NE-SW oriented wall was found north of oven AR. Perhaps this wall served as the boundary of an activity area around ovens AD and AR, and, at the same time, it may have enclosed the courtyard. Area 3 could be entered from three places: (1) from open area 3 of building I; (2) from open area 5 through a narrow passage between buildings I and II; (3) from open area 2 through a narrow passage between building XII and a long wall, which was probably connected to building X. All these entrances were very narrow; for some reason area 3 was a more or less secluded space, perhaps with a special significance.

---

[16] Two other open areas are situated *within* buildings I and II; these areas (resp. nos. 3 and 15) have already been dealt with in the above discussion of these buildings.

[17] The area south of rooms 6 to 8 of building XII was heavily eroded; it cannot be excluded that more rooms of building XII were present in this place.

The area between tholoi VI, VII and VIII has been designated area 4, which measured at least 8 x 6 m. Architectural features were absent from this area. Area 5 is represented by the open space between building II and tholoi VII and VIII. Area 5 measured ca. 9 x 3 m. Architectural features are absent.[18] Area 6, finally, is represented by the open space east of rooms 5, 6 and 7 of building V. Area 6 measured at least 8 x 4 m.

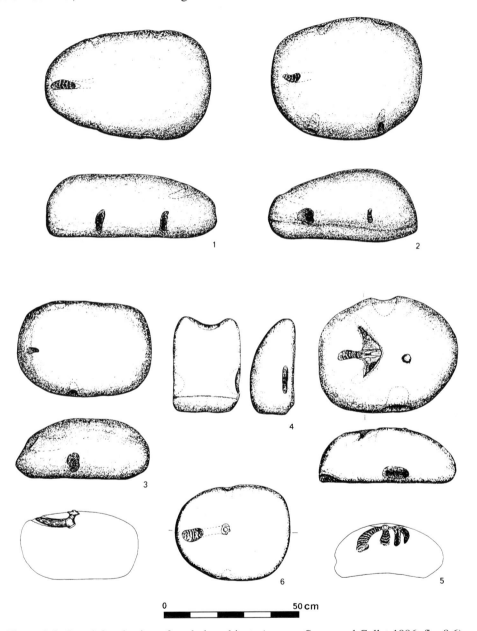

Figure 3.5. Level 6: selection of oval clay objects (source: Spoor and Collet 1996, fig. 8.6).

[18] During the 1996 excavations it appeared that 'hearth' ED (Verhoeven and Kranendonk 1996:52), south of room 4 of building II was in fact part of an eroded wall, i.e the western wall of room 20 of building II.

3 Burials

Three burials, one of a child and two of infants, have been recovered from the level 6 village. First, attention is drawn to a well-preserved child inhumation (SAB92-B1; Aten 1996) found just along the northern wall and below the floor of building II, room 10 (actually, the floor was renewed after interment). The dead child was lying on its right side in a tightly flexed position in a shallow pit ca. 45 cm in diameter and 22 cm deep (cf. fig. 2.11 in Verhoeven and Kranendonk 1996). The body was oriented east-west (atlas to sacrum), with the head facing south. No burial gifts were found.

The find of an infant burial (SAB91-B1; Aten 1996) in oven T in room 2 in building I, unfortunately rather poorly preserved, was surprising. The dead child was lying upon a ca. 10 cm thick layer of brown loam; a similar deposit, in its turn followed by burnt debris, was found on top of the skeletal remains. The dead infant was oriented NNE-SSW and was lying on its back, with the head towards the southwest, the legs spread and the right arm in a flexed position. A small bowl seems to have been placed at the feet of the child, perhaps as a burial gift.

Between the level 6 ovens CR and CS a right femur measuring 82.7 mm ± 1 mm was found as well as four metacarpals, nine phalanges, a large part of a fibula and two skull fragments, all belonging to a newly-born child. Most likely, this child's grave was largely destroyed when the ovens were constructed.

4 Oval Clay Objects

Eleven large, oval and loaf-shaped clay objects (fig. 3.5) were found in the fill of building V (in addition, one was found in the upper fill of room 11 of building IV), sometimes high above the floor and between charred roof beams and impressions of reed mats. In view of their position, these heavy objects must originally have stood on the (flat) roof and fallen down when the building collapsed.

Interestingly, in the fill of room 7 and amidst some of these oval clay objects, which have been interpreted as 'animals' (Akkermans and Verhoeven 1995:16), the skeletal remains were found of two adults, their bones completely crushed and burnt. These persons, too, must have fallen down from the roof. It is stressed that in none of the other buildings human skeletal remains were uncovered which could be associated with the fire; apparently, most of the inhabitants of Sabi Abyad escaped from the disaster in time. In chapter 7, section 5, the dead persons and the clay 'animals' have been interpreted as the remains of a mortuary ritual.

5 The Pottery

A large amount of sherds and a number of complete or, at least, reconstructable vessels have been recovered from the Burnt Village (Le Mière and Nieuwenhuyse 1996). The majority of the pottery (up to 85%) consists of plant-tempered and often burnished Mineral Coarse Ware and so-called Standard Ware (figs. 9 and 10, nos. 1-13, 16, 19 in

Akkermans and Verhoeven 1995). Occasionally these Coarse Ware ceramics carry a red slip or have incised or impressed patterns of crosshatching, oblique lines and herringbone. Others are decorated by means of bands of dark-red paint, sometimes in combination with incision, or, very rarely, have nobs in *appliqué*.

The Coarse Ware pottery shows a restricted variety in shape and mainly consists of simple, plain-rim bowls with a rounded or occasionally straight vessel wall, hole-mouth pots, and jars with flaring or straight necks. Some of these jars are of considerable size, i.e. up to 1 m in height. Many of the bowls have a distinct oval shape. Flattened, oval-shaped discs, simply made of sun-dried clay, are often found in association with these vessels and may have served as lids. The oval shape is not restricted to pottery only but is also recognizable in vessels made of stone.

In addition to the Coarse Ware, small quantities of locally manufactured Grey-Black Ware and imported Dark-Faced Burnished Ware were found. The Grey-Black Ware (figs. 3.18-3.20 in Le Mière and Nieuwenhuyse 1996) has a very fine paste, is mainly mineral-tempered and is purposefully blackened. Usually these ceramics are overall burnished but crosshatched pattern-burnishing occurs as well. In addition they sometimes carry incised patterns of crosshatching or herringbone. Shapes are simple and consist mainly of small bowls and angle-necked jars. The Dark-Faced Burnished Ware (ibid., figs. 3.14, 3.16-3.17) differs from the Grey-Black Ware in both technological and typological respect, and clay analyses have made it clear that these vessels are import products from western Syria or southeastern Turkey. The pottery has a reddish-brown to greyish or black paste and surface colour, contains mineral inclusions of rather large size and is carefully burnished (an unburnished variety occurs occasionally as well). Some vessels are incised or have broad bands of red paint. Shapes mainly consist of rather large angle-necked jars.

Special attention is drawn to the small sample of so-called Fine Ware, comprising around 6% of the ceramic bulk (and consisting of Standard Fine Ware, Orange Fine Ware and Fine Painted Ware, cf. Le Mière and Nieuwenhuyse 1996). So far this pottery is absent from the lower levels (11 to 7) of occupation at Sabi Abyad and seems to represent a true innovation in local ceramic production. The finely textured and mineral-tempered pottery consists of various kinds of bowls and small jars of the angle-neck type (fig. 10, nos. 17-18, 20-21 in Akkermans and Verhoeven 1995), which have a brown to orange or buff surface colour and are often burnished. The majority of these Fine Ware ceramics (about 66%) is decorated, either painted or, less commonly, incised or painted and incised. The paint is matt, reddish-brown to black in colour. The emphasis of decoration is on the vessel's neck and upper body and mainly consists of horizontal bands enclosing geometric designs in narrow zones (crosshatching, chevrons, zigzags, herringbone, etc.). Naturalistic designs, showing horned animals, are found in very small numbers.

In general terms, the pottery found in the Burnt Village at Sabi Abyad has its best counterparts in western regions such as coastal Syria and southeastern Turkey. The busily painted Fine Ware, however, shows close parallels with that of the Samarra and, perhaps, Hassuna cultures of northern Mesopotamia.

6 The Lithic Industry

The level 6 flint material (studied by Copeland, 1996) is characterized by being in a very bad condition due to burning, while this does not appear to have affected the obsidian element. In certain areas large numbers of flints are blackened, calcined and fractured or shattered into tiny fragments and dust.

In the flint samples the debitage consists mainly of flakes, with blades being markedly fewer and almost always broken into sections. The cores are worked down and, judging by the numerous refreshment elements, often re-shaped. This suggests that the tools were fashioned and repaired on the site. It is already clear that the tool kit consists of 'domestic' types such as scrapers, borers or drills, burins, notches and denticulates, as well as composites of the same; these together form almost half of the tools. In the scraper group the tabular scrapers or 'Tile Knives' are notable for their extreme thinness (3 mm on average). 'Agricultural' types such as picks are scarce, and lustred sickle-blade elements are neither common nor well-made. Weapons are virtually absent but a javelin fragment, expertly bifacially pressure-flaked, is worthy of note.

The obsidian industry consists mainly of numerous irregularly retouched or unretouched small blades, most often deliberately broken into sections. In contrast, certain types are present which, as we now begin to recognize, are northern Fertile Crescent specialities. Side-blow blade flakes (SBBF) and the 'cores' for the same form almost a third of the obsidian tools. SBBFs appear to characterize Pottery Neolithic assemblages in northeastern Syria and northern Iraqi regions, e.g. at Kashkashok (mainly on the surface), Bouqras and Umm Dabaghiyah (upper levels). At Sabi Abyad SBBFs begin before level 6 and continue to be plentiful in the upper Transitional layers, levels 5 and 4. Another early type occurs, although very rarely: the corner-thinned blade (CTB), as defined by Nishiaki (1990).

7 Sealings and Tokens

Among the most exciting finds in the buildings of the Burnt Village were the clay sealings with stamp-seal impressions, of which 300 were found, and the small tokens (n= 182) which may have served as calculi. The earliest sealings in clay known so far stem from the final stage of the Halaf period, i.e. from the early fifth millennium B.C., and have been found at very few sites only (see Akkermans and Duistermaat 1997:18-19).[19]

The sealings at Sabi Abyad (Duistermaat 1996, figs. 5.7-5.22) consisted of lumps of clay originally placed on the fastening of various kinds of containers or covering their opening entirely. The reverse shows that mainly ceramics and baskets were sealed, but stone bowls, mats and sacks originally carried sealings as well. The vast majority of the sealings was subsequently provided with one or more stamp-seal impressions. In this manner they secured the containers against unauthorized opening, whereas at the same time they may have carried information on the contents, destination or ownership of the containers.

---

[19] In Akkermans and Verhoeven 1995 it has been suggested that the sealings figured in an extensive exchange network. Subsequent analysis, however, has made it clear that the sealings functioned on a local level and most likely they served for the securing of products of a nomadic population related to Sabi Abyad (Akkermans and Duistermaat 1997, and see chapter 7 for a detailed discussion of this viewpoint).

The sealings showed a wide variety of designs. Until now, 26 different motifs have been recognized, most of which were geometric (zigzag lines, triangles, concentric circles, diamonds, crosshatching, etc.) but naturalistic representations (animals and plants) were found as well. Some designs occurred only once or twice, others were found in considerable numbers. Very common were the billy-goats or gazelles, depicted in a lively manner with long, curving horns and with great attention to detail (ibid., figs. 5.7 and 5.8).

It appears that at least 61 different stamp seals were used for sealing purposes, this on the basis of the size, shape and design of the various impressions. Remarkably enough, however, not a single stamp seal was found in the buildings of the Burnt Village. Perhaps the actual seals should be seen as precious items carried on the body of the owners, who consequently had left the site at the time of its destruction. It may also be the case that the seals were made of perishable materials such as bone or wood.

The numerous clay tokens found in association with the sealings seem to support this view. The tokens (Spoor and Collet 1996, figs. 8.4 and 8.5) were all very small and had simple geometric shapes, e.g. balls, cylinders, discs and cones. Most likely they acted as counting devices expressing the quantities of objects exchanged or otherwise deployed (Schmandt-Besserat 1992).

8 Human and Animal Figurines

Dozens of very schematically rendered human figurines were found, together with some animal representations (fig. 3.6 and see Collet 1996, figs. 6.1-6.4). Interestingly enough, in the case of the human figurines it is only the lower body that has been preserved. The upper part and the head seem to have been intentionally broken off, perhaps for ritual purposes. So far, only one head has been recovered, long-drawn and conical in shape with oblong eyes set obliquely (ibid., fig. 6.3, no. 2). Some figurines had a hole in the neck, suggesting that the head was separately added and fitted onto the body by means of a dowel.

The human figurines were all made of sun-dried clay and invariably depicted women. Virtually all showed a thin vertical incision near the base, indicating the vulva, and in some cases the breasts were clearly marked. Most figurines had a narrow waist, whereas the hips and belly were of a considerably exaggerated size. The limbs were rarely indicated. In one case the right arm was rendered which seems to support the breasts (ibid., fig. 6.1, no. 5). Horizontally incised bands and other impressions perhaps served to indicate clothes. So far, the closest parallels for these figurines have been found at the Hassunan site of Yarim Tepe I in northern Iraq (Merpert and Munchaev 1969).

In addition to the anthropomorphic representations, crudely-shaped animal figurines, all made of sun-dried clay, were found in small numbers (Collet 1996, fig. 6.4). All figurines were considerably damaged but the heads, legs and tails are still clearly recognizable. Some of the animals seem to have had horns and perhaps represent bulls.

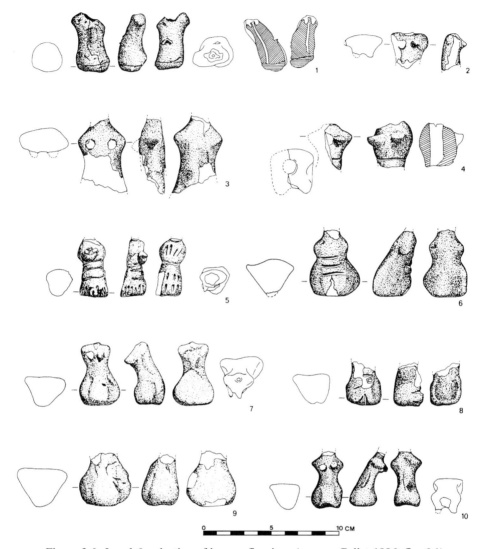

Figure 3.6. Level 6: selection of human figurines (source: Collet 1996, fig. 6.1).

9 Other Finds

Small objects other than figurines, sealings and tokens were found in rather large quantities in the buildings of the Burnt Village. Hundreds of stone objects of all kinds appeared, made of a wide variety of raw materials, some of which were locally available (e.g. limestone, gypsum, sandstone, quartzite), whereas others must have come from regions far away, such as the Euphrates valley or the Turkish piedmont near Urfa (e.g. basalt, dolerite, granodiorite, serpentinite, chlorite, granite and steatite). Apart from basalt, which was commonly used for the production of ground-stone tools, the non-indigenous materials were each found in minute quantities and seem to be largely restricted to 'luxury items' like small, carefully finished bowls and personal adornments (beads, pendants and labrets). Carving or manufacturing debris was entirely lacking, thus suggesting that these items reached the site in finished form (perhaps as part of the exchange network) or that production was restricted to specific and as yet unexcavated parts of the site.

Grinding implements, made of basalt and to a lesser extent limestone, granodiorite, flint and sandstone, were among the most commonly found items in the Burnt Village (Collet and Spoor 1996, figs. 7.1-7.7). The grinding slabs had an oval or rounded shape, a convex base and a flat or slightly concave working surface. Pestles and handgrinders came in many different dimensions and shapes. Most of them were conical in shape but cylindrical or spherical ones were found as well. Mortars appeared in two varieties, one portable and consisting of simple bowls with thick and straight, flaring sides and flat bases, the other more irregularly shaped and sunk into the floor (e.g. in room 5 of building V). These grinding tools often show very smoothed and polished surfaces due to intensive use. In addition to these worked items, a number of unworked but very smooth stone objects were found, which, according to their use traces, seem to have served as grinding slabs or working platforms. At least one of these was used as a palette, as shown by the traces of red ochre and dark-blue pigment on its surface. Actually, the common occurrence of ochre traces on pestles, mortars and grinding slabs and its presence in small pieces on the floor in some rooms suggests that this kind of pigment was widely used at Sabi Abyad, perhaps for body embellishment or the decoration of ceramics and other artefacts.

Shallow to deep and carefully finished stone bowls were found in rather small quantities and were made of a wide variety of raw materials. The various vessels showed flat or concave bases and flaring straight or rounded walls (occasionally slightly curving inwards at the rim), with plain, flat or bead rims.

Also small axes (ibid., fig. 7.7, nos. 9-11) were recovered from the various buildings, most of them made of fine-grained grey-black dolerite, the remainder of quartzite, serpentinite and limestone. The celts were wedge-shaped and heavily polished, with the sharp cutting edge in all but two cases seriously damaged due to domestic use. The two exceptions were each of very small size (each hardly 3 cm high), carefully finished and in perfect condition, perhaps suggesting that these objects served in luxury or ritual contexts.

Some stone objects occurred only once or twice, such as the carefully finished gypsum macehead or the palm-sized and re-used grinding-slab fragment of fine-grained basalt carrying a rather wide but shallow groove, on both sides surrounded by finely incised linear and arrow-like patterns (ibid.: fig. 7.7, nos. 16 and 17).

Implements made of bone consisted of simple awls and, to a lesser extent, spatulas and burins (Spoor and Collet 1996, figs. 8.11-8.13). The awls were all made of caprid metapodia, the other tools of the ribs or tibia of sheep, goat or cattle. A small piece of worked bone from building IV shows cut marks of varying length at small intervals and perhaps served as a rattle or counting device (tally; ibid., fig. 8.13, nos. 4-8).

Objects serving for personal adornment, such as beads, pendants and labrets, were often found in isolation and in rather limited numbers; most likely most of these items were worn on the body and left the site at the time of destruction together with their owners. The small beads, disc-like or cylindrical in shape, were all made of fine-grained stones. Labrets, most likely serving as lip or ear ornaments (Hole et al. 1996:235-236), were made of both stone and sun-dried clay. Some of these labrets have a squat appearance with a flat, slightly protruding top and flattened base, whereas others are taller in shape with a somewhat protruding flat or convex head and pointed base.

Other small finds include oval-shaped discs or lids, biconical sling missiles and either conical or, more often, biconical spindle whorls, all made of sun-dried clay and found in considerable quantities (Spoor and Collet 1996, fig. 8.1). Very common, too, were the small perforated discs (3 or 4 cm in diameter) made of chipped sherds or occasionally stone (ibid., fig. 8.2).

CHAPTER 4

THE PAST INTO THE PRESENT

1 Introduction

In this chapter I am going to discuss formation processes, i.e. those processes which affect the way in which archaeological materials came to be buried, and which affect the subsequent history afterwards. I do not intend to give the subject an exhaustive treatment, since studies especially dealing with formation processes abound (especially within processual archaeology formation processes have been given ample attention). The foremost author dealing with formation processes has been Schiffer (1972, 1976, 1983, 1987), but many others have also made valuable contributions (e.g. Ascher 1968; Cameron and Tomka, eds., 1993; Goldberg et al. eds., 1993; Hayden and Cannon 1983; Lange and Rydberg 1972; Rathje and Murphy 1992; South 1979; Stevenson 1982; Tani 1995; Verhoeven 1990; Villa 1982; Wood and Johnson 1978). In the following paragraphs the concept of in-situ contexts is discussed (section 2), the study of formation processes is introduced (section 3), the various formation processes relevant for Tell Sabi Abyad are delineated (section 3.1), the process of abandonment of level 6 is dealt with (section 4), two scenarios for this abandonment are suggested (section 4.1), and in the final section 5 models for designating formation processes are presented.

2 In-Situ Contexts

"Only if activities were suddenly stopped because of some catastrophe and materials remained 'frozen' where they fell could archaeologists reliably reconstruct the behaviour of the ongoing community without reference to abandonment and postabandonment processess" (Joyce and Johannessen 1993:150). With this remark Joyce and Johannessen, in their ethnoarchaeological study of abandonment of household compounds in rural Mexico, refer to Pompeii-like situations (see Binford 1981; Schiffer 1985a, 1985b). These situations seemingly represent the most favourable of archaeological circumstances, i.e. fully-functioning sites which are rapidly abandoned, where all is left behind, and where disturbances have been absent or minimal. In that case the archaeological context is an almost exact copy of the systemic context (leaving the living organisms aside).

In archaeological reports such contexts are often referred to as in-situ contexts, i.e. Pompeii-like assemblages of de facto refuse. It is most often assumed that these assemblages have been undisturbed. In the following I wish to comment upon the generally uncritical use of the concept of in-situ contexts.

First of all, the concept in situ is potentially misleading, as it refers to objects 'in a place'. Now every object recovered during excavations is in a place, whether it is found in a dump, a pit, or on the floor of a room. In situ refers to the place where the object was found; strictly speaking, it does not refer to the place where it was used. Obviously, the distinction between contexts of use (or interactive contexts) and other contexts (i.e. storage and discard) is important, especially when it is attempted to reconstruct the function of spaces.

In the present study with in situ and in-situ (or primary) contexts deposits and finds are meant which were deposited in their place of use. Often such contexts denote the material results of abandonment due to catastrophes such as fires. Secondary contexts denote finds outside their places of (primary) use.

Furthermore, a distinction should be made between undisturbed and disturbed in-situ contexts. Undisturbed in-situ contexts are very rare. Material which is really undisturbed may have been found in the recently discovered cave of Chauvet in France (Chauvet et al. 1996). Here prehistoric footprints (ca. 27,000 B.C.) were found in the soft and dusty soil upon the floor of the cave! The artefacts which were found may have been lying there completely undisturbed for thousands of years. The exceptional state of preservation in the cave may allow a very detailed reconstruction of the exact location and range of various activities at the site.

In most instances, however, in-situ contexts are somewhat disturbed (see e.g. Rosen 1986:92). In fact, the very catastrophes creating the contexts may have caused disturbance. In Pompeii, for instance, the lava which preserved a large amount of things also disturbed the locations of many objects; one can well imagine the force and the related disturbance of the lava which came upon the town. These disturbed contexts, however, may provide valuable information too. For example: in Near Eastern Bronze Age sites cuneiform tablets may be found scattered on the floor of a room. In contrast with the Chauvet Cave, there is no 1:1 relation between the systemic and the archaeological patterning. But still it can be ascertained that clay tablets were stored in a particular space, and, depending on the degree of disturbance, it may even be possible to reconstruct the precise place of storage (e.g. in Ebla, western Syria, the location of Bronze Age cuneiform tablets, among other things, strongly suggest that they were stored on wooden shelves along the walls; Archi 1993:108-111).

It is often assumed that in-situ contexts represent frozen moments in time, which are much less problematic for spatial analysis than other (secondary) contexts. The main difference is the nature of the abandonment process. With in-situ contexts we deal with settlements, or parts thereof, that are unintentionally abandoned. Therefore, things are left behind which normally would have been taken away: the systemic inventories are largely left in tact. This *de facto refuse* may give a direct indication of room use. Secondary contexts, on the other hand, are the result of planned abandonment, leading to a depletion of systemic inventories. After abandonment both primary and secondary contexts may be affected by various formation processes.

The fact that objects in in-situ contexts may have been left in their last place of use, makes these contexts most suitable for spatial and functional analyses, i.e. for the reconstruction of the function of spaces. Moreover, systemic inventories are less depleted as in secondary contexts. Spatial analysis of secondary contexts, however, may yield valuable insights into discard practices and abandonment behaviour, both of which are important cultural practices.

Something should also be said about the temporal dimension of the use of space. What we encounter in the archaeological record are the final products of ancient communities. What we are in effect dealing with is a static picture of dynamic practices. Room functions, for instance, may have changed in the course of time. Similarly the footprints in the Chauvet Cave only represent the latest prehistoric pedestrians. Stratigraphical studies may reveal these changes, but often rooms and other spaces are kept clean, the deposits inside them only representing secondary fill after abandonment. Moreover, stratigraphy does not necessarily indicate room functions; fine sediments rather than objects may have been allowed to accumulate. To complicate matters even further, different spaces may have gone through different trajectories of use. Clearly, this affects comparisons between such spaces.

3 Formation Processes

Formation processes can be divided in cultural and natural formation processes. Cultural formation processes include the deliberate or accidental activities of humans; natural formation processes refer to natural or environmental events which result in, and have effect upon, the archaeological record.

Many formation processes may have had an effect upon the objects which were left behind by ancient people. Especially in spatial and functional analyses, analysis of formation processes is of the utmost importance since it may not be assumed that archaeological deposits and assemblages are equivalent to prehistoric (or emic) inventories and deposits.

In a recent spatial analysis of Middle and Late Bronze Age sites in Palestine, Daviau (1993) has attempted to localize activity areas and to reconstruct the organization of domestic space. While acknowledging the importance of assessing formation processes (ibid.:26, 52, 63) no proper account has been taken of these processes (mainly due to the fact that her study is based on published data in which formation processes have not been given attention!). When reading the book it becomes clear that Daviau uses the implicit and in each case untested assumption that use location equals discard location. Moreover, she does not indicate whether the artefacts in the spatial analysis are complete or broken. Clearly, the preservation of objects is an important element in a spatial analysis, and it should be indicated if an object is complete or broken. Therefore, due to the general neglect of formation processes alone, Daviau's reconstruction of room functions remains highly speculative.

The same holds for Voigt's analysis of the location of activities in sixth millennium B.C. Hajji Firuz Tepe in Iran (Voigt 1983:295-321). Voigt also acknowledges the importance of formation processes (ibid.:295), but these processes are not further dealt with. Implicitly it is, again, assumed that use location equals discard location. Therefore this reconstruction, too, is speculative at best. These analyses fail to consider the contributions of formation processes to object distributions.

Discard process	Archaeological indication
Primary refuse	Object: small, broken/damaged Context: floors/surfaces Scattered distribution of objects, or distribution around features in corners and along walls; In sedentary sites like tell settlements regular maintenance activities will not allow primary refuse to accumulate in large numbers; The relation space-refuse should be 'logical' (e.g. fragments of unbaked sling missiles bear no relationship to the use of ovens)
Secondary refuse	Object: broken/damaged Context: foremost fill deposit, but also floors/surfaces It is expected that secondary refuse appears as clusters of broken/damaged objects, i.e. dumps; Secondary refuse is mainly expected in abandoned structures, natural depressions, pits, and in 'out of the way' corners in open areas; Often the relation space-refuse is not logical
Abandonment stage refuse	Object: broken/damaged, but (largely) reconstructable Context: floors/surfaces Various different objects on floors; Probably clustered; The relation space-refuse should be 'logical'
Provisional refuse	Object: broken/damaged but (largely) reconstructable Context: floors/surfaces Cluster of various objects; In 'out of the way' places in rooms, open areas; Clustered appearance of objects; The relation space-refuse should be 'logical'

Table 4.1. Model of cultural formation processes at Tell Sabi Abyad: discard processes and their archaeological indications.

3.1 Formation Processes at Sabi Abyad

In this section the various formation processes relevant for Tell Sabi Abyad, especially for level 6, are discussed. The various processes have been grouped into four main classes, i.e. discard processes, disposal modes, reclamation processes and disturbance processes. Discard processes refer to objects that are regarded as refuse. By disposal modes are meant the other processes that result in the deposition of objects. Reclamation processes are processes in which deposited objects are reclaimed. Disturbance processes, finally, are processes that disturb objects which were laid down due to discard processes and disposal modes.

DISCARD PROCESSES

Primary refuse

Primary refuse represents objects discarded at their location of use. It will be clear that primary deposition may result in artefact patterning which indicates the use of an area. Of the discard processes, primary refuse is not expected to have been deposited in architectural structures, since these will have been kept clean. In open areas primary refuse is more likely to occur, but on the whole primary refuse is not expected to have been a regular feature in tell settlements. At Tell Sabi Abyad primary refuse may be represented by small objects which were found on floors in rooms and open areas (these were not removed by maintenance processes), but generally speaking primary refuse is not expected in sedentary tell-settlements such as Sabi Abyad.

Secondary refuse

Secondary refuse represents objects that were discarded outside their location of use. When swept up and deposited elsewhere, primary refuse may become secondary refuse. A main characteristic of secondary refuse is the clustered occurrence of discarded objects, e.g. trash-heaps or dumps. Secondary refuse may be expected in areas and structures which are no longer used for activities and storage, and in 'out of the way' corners in open areas. Especially abandoned rooms and pits may have served as refuse receptacles. At Sabi Abyad the large pit X of level 3 (Verhoeven and Kranendonk 1996: fig. 2.3), located at the periphery of the settlement, had probably been a quarry pit. From this pit large amounts of trash were recovered, most likely dumped when the pit was no longer used for extracting clay.

Apart from the northern rooms 1 to 3 of building IV, only limited amounts of artefacts were recovered from the structures which were not affected by the fire (see fig. 3.2). In other words, there seems to be a clear difference between primary and secondary contexts. For instance: from room 8 of building I, which was partly covered by the burnt debris, a complete Dark-Faced Burnished jar, two complete Standard Ware bowls, and fragments of respectively a disc, a grinding slab and a mortar were found. From the neighbouring room 9, not covered by the burnt debris, just one token was recovered. The other unburnt areas from building I (i.e. rooms 5, 6, 7, 12 and 13) also gave evidence of few objects only. Likewise, building XII contained relatively few artefacts. What could be the reasons for this? It has already been said that we have some evidence that building XII was out of use and perhaps deteriorating, or even partly taken down, when the fire started. In that case the highly inflammable timber and reeds of the roof construction would have been removed, which would explain why building XII was untouched by the fire. Furthermore, the systemic inventories of the various rooms of the building would have been depleted and the rooms were most likely used as dumps for secondary refuse. Perhaps the same holds for the unburnt areas of building I.

On the other hand, it may be the case that building XII and parts of building I were complete and in full use when the fire started, but for some reason simply were not affected by the fire. After the fire, these structures with their roofs partly consisting of valuable timbers may have been stripped of their contents. Such a practice would also explain the scarcity of finds in buildings XII and the northern rooms of building I. Of the other buildings the valuable timbers had wholly burnt, and their contents were covered with a thick layer of burnt debris; these would have been much less attractive for looting. Whatever the case, building XII and parts of building I, like tholoi VII and VIII, lack the obvious in-situ contexts of the other areas. This is indicated by the stratigraphy, and by the scarcity of finds. Most likely, the objects recovered from these areas represent a mixture of abandonment stage refuse (see below) and secondary refuse. If indeed the buildings were looted after the fire we would be dealing with depleted de facto assemblages (reclamation processes). It is, however, felt that the first option is more realistic. In this respect it should be noted that the large majority of small finds of building XII (i.e. 13 out of 15) were broken. Therefore it is suggested that we are dealing with secondary refuse. Nevertheless, a functional assessment of building XII and the northern rooms of building I, will be attempted on the basis of the architectural information. Furthermore, the artefact inventories will be analysed in two ways, i.e. as secondary refuse, and as abandonment stage refuse. On the basis of treatment as abandonment stage refuse, just as an option, tentative hypotheses of activities carried out in a room are presented.

Abandonment stage refuse

In anticipation of abandonment of a structure or settlement in the immediate future standards of cleanliness and maintenance activities may be relaxed or not be carried out at all. In these instances refuse may be discarded in areas which were previously kept clean. Such refuse is termed abandonment stage refuse, which may be primary or secondary (Schiffer 1987:98; Stevenson 1982).

Provisional refuse

Provisional refuse represents debris which is provisionally discarded (not necessarily in its place of use), awaiting final discard in another place. Our rubbish bins, for instance, are containers for provisional refuse, which becomes secondary refuse when finally dumped. Considerable amounts of provisional refuse may accumulate in workshops or activity areas. It should be taken into account that objects in provisional refuse are susceptible to other formation processes, such as child's play and scavenging, and reclamation processes.

As will be argued below (section 4), provisional refuse may have been quite common in the level 6 settlement; in a large number of areas broken or damaged artefacts were recovered from in-situ contexts. The possibility of abandonment stage refuse will also be discussed below.

DISPOSAL MODES

De facto refuse

Deposition of de facto refuse is an important process related to the abandonment of places. De facto refuse "... consists of tools, facilities, structures, and other cultural materials that, although still useable (or reuseable), are left behind when an activity area is abandoned" (Schiffer 1987:89). De facto refuse is not necessarily deposited in its place of use; only in the case of what may be termed primary de facto refuse the objects are left behind in their place of use. Primary de facto refuse thus refers to objects in in-situ or primary contexts.

Disposal mode	Archaeological indication
De facto refuse	Object: complete/reconstructable Context: on floors or, less likely, in fill deposits Usually large, heavy or bulky items with low replacement costs
Loss	Object: complete, small, mobile, highly valued, long uselife. Context: floors/surfaces, fill deposits 'Artefact traps' (e.g. dark corners, pits, etc.)
Secondary use	Object: broken/damaged Context: floors/surfaces, but especially fill deposits Peculiar associations of deposits, structures, features and objects, e.g. hearths in fill deposits
Ritual cache	Object: mostly complete/reconstructable, but broken/damaged items may also occur Context: on or sunk into floors/surfaces Discrete concentration of artefacts, usually not in a secondary refuse deposit; Isolated and 'non-functional' concentration of objects with a possible ritual function, e.g. burials, caches

Table 4.2. Model of cultural formation processes at Tell Sabi Abyad: various disposal modes and their archaeological indications.

At Tell Sabi Abyad a large quantity of primary de facto refuse was recovered from the level 6 village, at least part of which was quickly abandoned, leaving behind complete artefacts, due to a fire. Of the disposal modes primary de facto refuse seems to be represented by the complete objects in in-situ deposits.

Loss

There are five main variables that influence loss and retrieval rates (Schiffer 1987:77). First the object size is of main importance. Loss and retrieval rates are inversely related to object size or mass, i.e. small objects (e.g. beads) are more easily lost (and not found back) than large objects. Secondly, the formal properties of objects, such as colour, shape and texture, have to be taken into account, e.g. brightly coloured artefacts are retrieved more easily than objects without bright colours. Third, the character of the surface on which objects are used is of importance. Whether the soil is loose or hard, the presence or absence of vegetation, wetness or dryness all affect retrieval rates. Fourth, the value or replacement cost of an object affects the effort put into searching. Fifth, the mobility of an object influences loss rates, mobile objects being more susceptible to loss than immobile elements. At Sabi Abyad especially the beads which were recovered seem to represent lost items, since all of them were found in isolation, and not as parts of necklaces or bracelets. Moreover, these objects were potentially valuable, and would not have been left behind. Also other items of personal adornment, such as labrets, may have been lost rather than left behind.

Secondary use

After abandonment sites and structures may be used for a variety of purposes; this I call secondary use. In the Near East, for instance, shepherds often used abandoned structures as shelters. In such instances traces may be left which should not be confused with primary uses. The traces consist on the one hand of features such as hearths and pits, and on the other hand of artefacts and refuse. The artefacts may have been deposited as caches of de facto refuse, which were to be used on future occasions. Such 'site furniture' (Binford 1979:264) may consist of a teapot, cooking utensils, glasses, etc. It can be expected that the traces of secondary use occur especially in the fills in and around structures.

At Tell Sabi Abyad a number of hearths and a basin in the fill in and around building II of level 5 seem to represent secondary use. These features were constructed at a time when the building had already been deserted and debris had accumulated in it to a considerable extent. The hearths were situated among the wall remains of building II. Most likely, the wall stubs protected these hearths against the wind (Verhoeven and Kranendonk 1996:72, fig. 2.17).

Ritual caches

Objects may also be ritually deposited as a cache. Well known examples of this are foundation deposits, dedicated to the construction of a site or structure. Other kinds of ritual deposits consist of e.g. votives, hoards and deposits at sacred places (e.g. Carmichael et al., eds., 1997). Also burials may be seen as ritual caches. An interesting instance of a ritual deposit in the Balikh valley is represented by the skull of an ox across the threshold in one room of a Neolithic building (possibly a shrine) at Tell Assouad, not far from Tell Sabi Abyad (Mallowan 1946:124).

At Sabi Abyad itself a number of child burials have been recovered beneath floors, some with gifts such as pottery vessels (see the contribution of Nico Aten in Verhoeven and Kranendonk 1996:114-116). Other clear ritual depositions have as yet not been uncovered at Sabi Abyad. However, the strange clay objects and human skeletons on the roof of level 6 building V (Akkermans and Verhoeven 1995:16) most likely represent a ritual deposition (cf. chapter 7, section 5).

RECLAMATION PROCESSES

The main reclamation processes are reincorporation and salvage, scavenging, collecting and pothunting, and, last but not least, archaeological excavations.

Reincorporation and salvage

Reincorporation and salvage refer to the incorporation of objects, e.g. de facto refuse, when a settlement is reoccupied. Especially at tells, which were continually reoccupied, these processes are of importance. Stones and wooden building materials, for instance, are often reincorporated in new settlements (e.g. Lange and Rydberg 1972). Reincorporation/ salvage is not only related to de facto refuse, but also to other kinds of refuse. Sherds, for instance may be reused to provide temper for new ceramics. At Sabi Abyad a large stone wall was found in the level 3 settlement. Presumably the wall served as a kind of retaining wall supporting a terrace (Akkermans 1993:57). The wall was well-preserved, but at some locations stones seemed to be missing. Most likely, when the wall was out of use, these stones were reincorporated in new structures somewhere at the tell. Furthermore it can be assumed that wooden rafters supporting roofs were taken away at abandonment and subsequently reused.

Reclamation process	Archaeological indication
Reincorporation/ salvage scavenging	Object: complete (incorporation of de facto refuse), but also broken/damaged (e.g. reuse of sherds for temper) Context: floors/surfaces Depletion of de facto refuse, perhaps indicated by deviation from composition of comparable de facto refuse assemblage(s); Reuse of building materials, particularly stone and wood: missing parts in structures
Collecting Pothunting	Object: complete (incorporation of de facto refuse), 'valuable' objects Context: floors/surfaces, burials Depletion of de facto refuse. Perhaps indicated by deviation from composition of similar de facto refuse assemblage(s); Few intact items at site; Reuse of building materials, particularly stone and wood: missing parts in structures; Large pits ('potholes') and fill depositions that have been disturbed by digging, and spoilheaps may be indicative of pothunters, especially if the de facto refuse assemblage in the disturbed spatial unit seems to be depleted; Small objects seem to be missing but large and heavy objects remain

Table 4.3. Model of cultural formation processes at Tell Sabi Abyad: reclamation processes and their archaeological indications.

Scavenging

By scavenging is meant the exploitation of previously deposited artefacts in a settlement by that settlement's inhabitants (Schiffer 1987:106). Scavenged materials are reincorporated in systemic inventories. Scavenging thus leads to depletion of inventories, whereas reincorporation leads to growth of such inventories. Apart from artefacts deposits may also be scavenged and reclaimed. In the Near East, for instance, clay for making mud-bricks is often quarried in areas around the village which are littered with all kinds of garbage (e.g. Watson 1979:119). In this way the walls of buildings and other structures may contain sherds, flint artefacts, etc. (e.g. Kirkby and Kirkby 1976:230; Kohlmeyer 1981:57; Rosen 1986:75-91; Schiffer 1987 fig. 5.4). For example, in a spatial analysis of the Late Bronze Age 'palace' at Tell Hammam et-Turkman in the Balikh valley, I have noticed that up to 35.5% of the sherds in the various roomfills of the building dated to the Middle Bronze Age. Most likely, these sherds were included in the bricks of the walls, which were constructed of mud from the tell, containing Middle Bronze Age materials (Verhoeven 1990:46 and table 2). When analysing roomfills, this and similar processes (such as the deliberate filling of rooms to create foundations) leading to reversed stratigraphies, should be taken into account.

Collecting and pothunting

Scavenging is related to the inhabitants of the settlement from which materials are reclaimed. Collecting and pothunting, however, indicate processes that result in the transfer of objects from an archaeological site to an occupied settlement elsewhere. "Collecting processes are those that involve the disturbance, removal, and transport of surface materials; pothunting refers to the disturbance, reclamation, and transport of subsurface materials" (Schiffer 1987:114). As Schiffer notes, it is important to distinguish between intra-site reclamation processes (scavenging) and intersite processes (collecting and pothunting), since collecting and pothunting affect the formation processes of two sites, and thus have other effects than processes on one site. Secondly, intra-site processes operate with hardly or no transport restraints.

Especially valuable artefacts, such as fine ceramics and jewelry, are looted by collectors and pothunters. In the Near East many archaeological sites and regions have been subjected to extensive looting. In the Balikh valley, for instance, a large number of Islamic sites have been systematically 'excavated' by robbers in the 1960s, leaving tell surfaces which look like moon landscapes. Another well-known example is Luristan, where large numbers of graves have been plundered of their rich contents (i.e. bronze objects) by grave robbers to be sold at the international art markets.

At Tell Sabi Abyad an example of prehistoric collecting may be indicated by the find of a typical Pre-Pottery Neolithic B figurine (such as found at Tell Sabi Abyad II and Tell Assouad; Verhoeven and Akkermans, eds., in prep.; Cauvin 1972a) in the Pottery Neolithic level 6 settlement at Tell Sabi Abyad I. Moreover in the Middle Assyrian (ca. 1250 B.C.) fortified settlement at the top of Sabi Abyad, a complete Halaf jar was found on the floor of a room, amidst a Middle Assyrian domestic assemblage (Nieuwenhuyse 1997:233). Most likely this vessel was kept as a curiosity.

Apart from valuables, more ordinary artefacts may be collected, e.g. sherds for temper, and building materials such as wood and stones.

In the case of the level 6 village, it has already been remarked that "... some categories of material, particularly luxury items such as personal adornments, seals, and finely painted pottery, are present in much smaller quantities than one would expect in the case of a fully preserved domestic context" (Akkermans and Verhoeven 1995:31). Here the Middle Assyrian settlement at Sabi Abyad may also be mentioned. This settlement was partly destroyed by a fire, and large amounts of complete objects were recovered, including luxury items such as finely made pottery, necklaces, beads, etc. (Akkermans et al. 1993). In this instance, the in-situ context appeared as one would expect it: all or most categories of objects used within domestic contexts seem to have been left behind due to the fire and immediate abandonment.

In the level 6 village, however, people had perhaps been able to remove - valuable - objects before the destruction caused by the fire. Or maybe they returned after the disaster to search for their possessions. In that case, however, pits would have been expected in the burnt layer; these have, however, not been found in the burnt debris.

Archaeological excavations

It is a well-known fact that archaeological excavations are paradoxical instances of scientific practice: by excavating, archaeologists actually destroy their evidence. The mobile objects (e.g. sherds, lithic implements) are retrieved, but their spatial context and relations are destroyed. Furthermore, architectural features are often destroyed. What we are left with is decontextualized material and a record on paper and in computerfiles. In an intra-site spatial analysis it is tried to re-contextualize the material on the basis of this record. A number of factors related to excavation and post-excavation practices may influence such an analysis, the most important being: (1) the techniques of excavation (do the procedures allow contextual reconstructions?); (2) the area covered by the excavation (e.g. is the area large enough for a spatial analysis?; are buildings completely or incompletely recovered?); (3) methodological procedures (are the - statistical - techniques used in concordance with the archaeological data?); (4) the theoretical framework ("Theoretical orientations condition how and what is seen, how it is interpreted, how interpretations are explained, and how understanding is achieved" (Kent 1987:513)); (5) the researcher (is he/she qualified?).

DISTURBANCE PROCESSES

Earth-moving

Earth-moving activities disturb previously deposited objects and layers. One major earth-moving activity of relevance for archaeologists is the digging of pits in antiquity, which occur in a large variety of functions, shapes and dimensions. Examples abound, e.g. digging for the construction of houses, digging of foundation trenches, construction of graves, quarrying mud for mud-bricks, mining. It is evident that earth-moving may result in the upward migration of artefacts, and consequently in 'reverse' stratigraphies.

At Sabi Abyad, prehistoric earth-moving activities have been of great influence. In the central squares P13 to S13 the tell had been levelled, and levels 5 and 4 had been removed, to enable the construction of a large level 3 building (Verhoeven and Kranendonk 1996:

63). Certainly, this levelling will have affected the level 6 settlement, especially buildings I, II, VI, VII, VIII, X and XI. In this respect it should be noticed that buildings IV and V, located outside the terrain of levelling, were preserved higher (i.e. to ca. 1.20 m) than the buildings higher up the mound (preserved to 1 m at the most). Earth-moving will especially have affected the upper part of the level 6 structures; the objects on floors were in the majority of cases undisturbed, as indicated by the fact that they were surrounded by (low) walls. Of course, it cannot be excluded that due to the levelling mixing took place, but, most likely, this will have affected the upper parts of deposits only.

Trampling

Trampling (by people and animals) is an important disturbance process, since it disturbs previously deposited artefacts on floors and surfaces. Objects may break due to trampling and it may result in false stratigraphical associations, due to vertical movement of objects in soils (Villa 1982; Villa and Courtin 1983). Trampling can cause mixing of materials belonging to two separate levels. By refitting the vertical displacement of materials (e.g. flint, pottery, bone) trampling may be assessed. It should be noted, however, that trampling results in the fixation of small objects, such as primary refuse. Therefore, trampled material can often be used to assess the function of spaces, since in many cases the sub-surface material is directly related to use.

At Sabi Abyad trampling will have resulted in the fixation of at least some small objects (< 4 cm; Rosen 1986) in and beneath floors and floor levels. As has been argued, floors were apparently not kept meticulously clean inside the domestic structures of the level 6 village. Therefore, small objects may have indeed been trampled into floors of domestic spaces. Furthermore, trampled objects may be expected in the open areas or courtyards.[20]

Erosion/deterioration

During and especially after occupation (when maintenance stops), erosion affects sites, structures and objects, resulting in their deterioration. Apart from chemical, physical and biological degeneration, structures and objects at tell settlements are influenced by the two main forms of erosion: eolian and hydrological processes. Mud-brick structures undergo deflation by strong winds, and artefacts may erode out of the walls (cf. fig. 9.3 in Schiffer 1987). Depending on the force of the wind, objects may also be transported by the wind. Wind-blown sediments may of course also cover structures and objects. At Sabi Abyad, for instance, due to aeolic-fluviatile accumulations, the present-day field level is, ca. 4 m above the natural soil reached in trench P15 (Verhoeven and Kranendonk 1996:25). On the basis of the chronology, this deposition must have taken place at ca. 5700-5200 B.C. (6460-5970 cal BC). Archaeologically, this accumulation of deposits (which has also been attested at nearby Tell Hammam et-Turkman) has serious consequences, since small settlements and other features (e.g. burial sites, field systems, etc.) may have been buried.

Undoubtedly, erosion/deterioration processes have affected the level 6 remains. In the first place, of course, the devastating fire should be mentioned: apart from some tholoi (nos. VII and VIII) and one rectangular structure (building XII), all level 6 buildings were

---

[20] No systematic sieving of floors has taken place; see appendix 2, therefore the effect of trampling, as indicated by small objects in floors, cannot be verified.

heavily affected by an intense fire which penetrated the walls throughout and which caused a considerable accumulation of orange-brown to brown crumbly loam, wall fragments, dark ashes and charred wood in the buildings.

Disturbance process	Archaeological indication
Earth-moving	Object: deposits, structures and objects Context: fill deposits Construction of pits, basins and foundation trenches; Levelling activities (these may be indicated by horizontal surfaces and by architectural features of which the upper part seems to have been 'cut off'); Earth-moving can result in deposits whose order of deposition is not matched by the order of manufacture or use of the constituent artefacts; Artefact displaced by earth-moving may result in 'reverse stratigraphies'; Objects in different spatial units which can be refitted
Trampling	Object: small objects or parts of larger objects (generally < 4 cm) Context: penetrable floors/surfaces In rooms and especially in open areas; In places with considerable deposition (e.g. secondary refuse areas) low intensity of trampling can be expected
Graviturbation	Object: deposits, structures and objects Context: deposits, structures and objects on slopes Mixed sediments and artefacts
Erosion/deterioration: eolian and hydrological processes	Object: complete and broken/damaged Context: floors/surfaces, fill deposits Eolian and hydrological processes: Breaking down of building material: building debris (depositions of loam, collapsed and eroded walls); Smoothening of tell surface topography Erosion gullies; Deflation; Abrasion of objects; Burial of features beneath deposits; Chemical, physical and biological degeneration of objects; Eolian processes: Deposits of fine grained sediment; Deflation of mud-brick structures; 'Sandblasting' of objects; Hydrological process: Undercutting of walls
Faunalturbation and floralturbation	Object: complete, broken/damaged Context: fill deposits Burrowing animals: tunnels, burrows, krotovinas; Root action: root casts; 'Reverse stratification' due to krotovinas and root casts; Tree throws; Gnaw marks on bones

Table 4.4. Model of cultural and natural/environmental formation processes at Tell Sabi Abyad: disturbance processes and their archaeological indications.

Erosion/deterioration processes (eolian as well as hydrological), i.e. the breaking down of building material, degeneration of objects, wind erosion, and undercutting of walls, will also have affected the settlement prior to and after abandonment. However, due to the fire, most parts of the settlement will have been covered rapidly by burnt debris. Therefore, the effects of post-abandonment erosion are likely to have been limited to those parts of the structures which protruded from the burnt debris. The floors of burnt structures covered with objects in primary context were generally untouched by erosion processes. Likewise, it is not expected that faunal and botanical agents, such as burrowing animals and plant roots, have resulted in large disturbances; during excavation such 'bioturbations' were not attested.

Faunalturbation and floralturbation

Animals and plants also have effects on deposits and the objects these contain, and together the disturbance of animals and plants may be called bioturbation. Burrowing animals (e.g. rats), larger mammals (e.g. the fox), and even earthworms, may mix various deposits and thereby displace objects while digging their holes. Earthworms may even utterly blur the boundaries between various strata (Schiffer 1987:208). Especially bones are susceptible to animal disturbance. Domesticated dogs may account for the differential preservation of skeletal parts: bones can be spatially and physically disturbed by them. Chewing marks may indicate the disturbance, but, as Kent (1984:179) has noted, dogs do not necessarily leave marks on the bones from which they have chewed meat. At Sabi Abyad traces of carnivore gnawing on a number of bones from the prehistoric levels can probably be attributed to dogs, the presence of which has been confirmed at the tell (Cavallo 1997:97). Not only predators or dogs consume bones: other animals may eat bones as a source of calcium (Yellen 1977).

The principal form of floralturbation is root action. Roots (especially large ones) can move buried artefacts aside, and if a root decays roots casts are left behind. Furthermore vegetation growth may obscure sites or large parts of them. 'Tree throw' (= the toppling of trees) often causes pits which at first sight look like human-made depressions. Furthermore tree throws can result in the redeposition and mixing of materials (ibid.:212).

Prehistoric animal holes are present at Sabi Abyad, as appeared from the various sections (e.g. Verhoeven and Kranendonk fig. 2.5). Recent animal disturbances are represented by holes which birds (e.g. bee-eaters) make in the section walls when they are exposed after the excavations. Even foxes dig their holes in the sections or structures after the excavation team has left. Also shrubby plants and grasses gradually colonize abandoned excavation squares. In their turn sheep may consume this vegetation.

Graviturbation

Graviturbation or downslope movement leads to the mixing of sediments and artefacts (Wood and Johnson 1978:346). Especially on high and steep tells graviturbation can be an important formation process. Due to erosion buildings collapse and their remains may move downwards. This way the covered deposits below may be obscured. At Sabi Abyad, for instance, the remains of the Late Bronze Age buildings at the top of the tell have covered the prehistoric remains lower on the tell.

4 What happened to the Level 6 Village?

It has already been indicated that broken objects represent a major component of the level 6 artefact assemblage as a whole. Of the small finds assemblage 61.5% of the objects were broken. For example only two complete basalt grinding slabs were found. Within the group of ceramics only a handful of complete vessels was recovered. Of the flint/ obsidian industry, however, only ca. 25% of the tools were broken (debitage makes up ca. 82% of all flint/obsidian artefacts). How do we account for this? The case of the ceramics is most intriguing; in primary contexts, would one not expect large numbers of complete/ reconstructable vessels on floors? The scarcity of such vessels in the level 6 settlement may be due to a number of - interrelated - factors: (1) as yet no systematic programme of refitting sherds has taken place: it is expected that such an activity would yield a number of additional ceramic vessels[21]; (2) despite the fact that there is no clear evidence of recla- mation processes such as reincorporation/salvage, scavenging, collecting and pothunting, it may be the case that after the fire people came back to collect their pots, or ceramics may have been collected by non-inhabitants; (3) perhaps people were able to remove some pottery before the destruction caused by the fire; (4) maybe only a limited number of vessels was actually in use when the fire started.

Especially the last hypothesis may be significant; a number of complete vessels have been recovered from the settlement, e.g. in room 8 of building I three complete vessels were present, and in room 6 of building IV eight complete potstands were found (Akkermans and Verhoeven 1995:12, 15). At least these complete vessels were easily recognizable, and it is expected that if large numbers of complete vessels had really been present many more of them would have been found. During excavation great care was taken to sample all the sherds of spatial units, and during subsequent analysis these sherds have been kept together. If a broken but complete vessel had been amongst those sherds, this would in most instances immediately have been noticed. Therefore, it seems that only a limited number of vessels was in use at the time the fire broke out. How, then, to explain the large number of sherds (ca. 22,000)?

Clearly, one or more of the formation processes distinguished above must have caused the presence of all these sherds. Especially the disturbance and the discard processes come to mind. Of the former processes it has already been argued that earth-moving has disturbed the settlement, but floor deposits seem to have been left largely intact.[22] Erosion may have been of effect, but this does not explain the scarcity of complete vessels; vessels may have been destroyed by e.g. collapsing walls, but even then the debris would have covered these vessels and they would have appeared during excavation as broken but reconstructable. However, it has to be taken into account that vessels (and other objects) may have been used and deposited on the roofs of buildings. Especially buildings IV and V, the roofs of which were at the same height as the floorlevel of neighbouring level 6 buildings, gave evidence of objects high in their fills. Most likely, these objects had been used and deposited on the roofs (Verhoeven and Kranendonk 1996: fig. 2.15). When, due

---

[21] It was attempted, however, to refit most of the Fine Ware sherds, without results.

[22] In this respect room 6 of building II should be mentioned; although the walls surrounding this room (like all walls of building II) were preserved to a height of ca. 30 cm at the most, this room gave evidence of a large number of objects (Verhoeven and Kranendonk 1996: fig. 2.10), undoubtedly all in primary context (actually, these objects may have been stored on shelves along the walls).

to the fire, these roofs collapsed, the vessels upon them may have been smashed to pieces and their sherds could easily have been deposited in various spatial units (i.e. in various rooms).[23] Also at the other buildings, vessels present on roofs may have been destroyed and their sherds may have been scattered among the upper fills of different spatial units due to the levelling activities when the level 3 building was constructed.

Consequently, disturbance processes must be taken into account when dealing with the virtual absence of complete vessels, but it is felt that, first and foremost, discard processes have resulted in the fragmented nature of the ceramic assemblage.[24] It has already been indicated that of the discard processes, provisional refuse may represent a regular formation process at work in the Burnt Village. It may be suggested that the presence of so many sherds in the level 6 village is to a large extent due to the circumstance that vessels that were used and broken in spatial units (e.g. rooms) were (temporarily) left there awaiting final discard. A problem with this interpretation, however, is that in that case reconstructable vessels would be expected, and this is in the majority of instances not so. Perhaps, when a vessel broke, some parts were removed, whereas other parts were left. In this respect it should be taken into account that the interior of the small rooms in the various buildings was probably dim, or perhaps even dark; if broken objects were no direct obstacle they could easily be overseen.[25] Broken objects may also have been stored temporarily, not necessarily in their place of use, in anticipation of future re-use. As will be argued in chapter 6, a number of rooms in buildings II and V seem to have functioned as such temporal storage places.

There is also another possibility which would explain the apparently limited number of complete vessels and the large quantity of sherds (and the many other broken objects). It may be the case that at the time when the fire started the level 6 village was already being abandoned. In that case the broken objects in spatial units may represent abandonment stage refuse and especially secondary refuse, i.e. material that had been broken and which was subsequently dumped in abandoned structures such as rooms. In that case, however, it would be expected that wall and roof debris had accumulated upon the floor. Such debris depositions were however only encountered in the unburnt and secondary contexts (e.g. building XII). If the settlement was already in its 'abandonment phase', then it must have been at the very beginning of this phase.

4.1 Two Scenarios

Taking the above problems seriously two 'scenarios' are proposed. In both scenarios the complete objects are regarded as representing in-situ artefacts, i.e. as primary de facto refuse. Furthermore, both scenarios only take floor/surface deposits into account; as has been said above, objects in fill deposits are generally unrelated to the use of a space.

---

[23] Actually we do have some evidence that this was indeed the case: in building V fitting sherds (all belonging to a large Orange Fine Ware jar-neck) were found high in the fill of at least three separate rooms.
[24] Above it has been argued that reclamation processes and disposal modes were not major formation processes.
[25] Olivier Nieuwenhuyse has observed that compared with the other, unburnt, levels the sherds in the Burnt Village are generally larger and less eroded. Perhaps this indicates that in the Burnt Village the sherds were not subject to as much post-depositional disturbance as the sherds in the other levels were.

In the first scenario all artefacts, broken as well as complete, are taken into account. In this scenario the broken artefacts are regarded as provisional or abandonment stage refuse: like the complete objects, these objects can give clues about the function of the space in which they were found (since it is assumed that they were actually used in that space). In the second scenario, only the complete artefacts are taken into account, which are, as in the first scenario, regarded as objects in primary context. The broken objects are in this case regarded as (secondary) refuse which bears no relationship to the function of its findspot; therefore they are omitted from this scenario. The difference between both scenarios is thus represented by the broken artefacts. The second scenario can be regarded as a 'stripped' version of the first scenario; the broken objects of the first scenario are now omitted.

As can be observed in modern villages in the Near East, abandoned structures (e.g. rooms) are often used as trash receptacles for the waste of the occupants of other still functioning buildings; this may also have been the case in the Burnt Village. As will be clear, the second scenario is on the safe side; it is more likely that complete objects in insitu contexts were used or stored at their find locations than broken objects, which represent refuse (primary or secondary). The first scenario should be regarded as the *maximal* option, i.e. it is based on all the material (broken and complete). The second scenario is the *minimal* option: it is based on part of the material only (the complete objects).[26] It may be argued that this minimum option is the most reliable, since it rigorously omits the 'unreliable' broken objects. On the other hand, as has been argued above, these broken objects are perhaps remnants of objects actually used or stored in the place where they were found. We have evidence that broken objects were actually stored in the Burnt Village (see chapter 6), and also that they were used secondarily. In room 5 of building IV, for instance, sawn-off necks of Dark-Faced Burnished Ware jars were reused as pottery stands. In other instances large sherds of Coarse Ware jars were used in the construction of ovens. Both scenarios, therefore, are not optimal scenarios, but should be related to one another.

The first scenario suggests that the village was still largely functioning when the fire broke out; provisional refuse implies a use, and not a discard context. The second scenario, however, suggests that large parts of the village were being abandoned or had already been deserted when the fire started; secondary refuse implies abandonment and disuse. The complete objects, then, were for some reason left behind.

Due to the absence of clear evidence we can only speculate about the cause of the fire. It may be suggested that an accident, possibly with one of the ovens, resulted in the conflagration. Furthermore it can be proposed that the destruction of the level 6 village by fire was the result of hostilities between the inhabitants of the level 6 settlement and other groups. In this respect Bernbeck (1995:36) has argued that in the Samarra culture (with which Sabi Abyad level 6 is more or less contemporaneous, e.g. Campbell 1998) warfare may have been a regular feature, as suggested by features such as the large wall around the level IIIA settlement at Tell es-Sawwan, which has been interpreted as a defensive wall (e.g. Yasin 1970). However, no bodies were found inside the buildings of level 6 (apart from the two skeletons on the roof of building V, which were most likely not victims, but part of a mortuary ritual, cf. chapter 7, section 5); if indeed hostilities resulted in the fire, people seem to have escaped from the village.

---

[26] Consequently, in this scenario 99.9% of the ceramic assemblage (i.e. the sherds) are omitted!

Third, it may be suggested that the settlement, or part of it, was purposely set on fire by its inhabitants. Several authors have pointed out that the burning of buildings may take place as a ritual associated with abandonment. This seems especially to be the case for pueblos in the southwest of North America, e.g. Grasshopper Pueblo, Chodistaas Pueblo and the Duckfoot site. Many of these sites contain large numbers of de facto refuse.

A number of studies have pointed out that pueblo earthen structures are difficult to ignite and burn very slowly (e.g. Glennie and Lipe 1984; Wilshusen 1986). Intentional, instead of accidental, burning at the time of abandonment, therefore, is more likely. In Chodistaas Pueblo (Arizona), it appears that soon after the burning of the pueblo many of the rooms were filled with sherds (other formation processes are unlikely). Montgomery (1993:161) interprets this abandonment behaviour as a ritual act associated with the 'death' of the pueblo and the subsequent burial of household belongings. At the Duckfoot site (Colorado) Lightfoot (1993:169) found that at the time of abandonment seven individuals were placed on the floors of pit structures. Very likely these interments were deliberate, and not accidental catastrophes. Also a study of Wilshusen (1986) may be mentioned here. He reports significant associations between the occurrence of 'ritual features' in structures and both the deliberate destruction of roofs and human interments in Anasazi pit structures. Besides these archaeological examples there are many ethnographic accounts from the American Southwest of the burning of one or more structures after the death of persons (e.g. Allen 1891; Beals 1934; Kent 1984).

Recently Stevanović (1997) has discussed the social dynamics of house destruction at settlements of the later Neolithic Vinča culture in the central Balkans (ca. 5500-3800 B.C.), especially at the Opovo site. So far, this region has not yielded a single later Neolithic site with architectural remains that are completely unburnt. On the basis of ample evidence and a most detailed technical analysis of the burnt wattle-and-daub houses, Stevanović forcefully argues that they were intentionally set on fire.[27] She suggests that this was done in order to secure "... its postutilitarian visibility in order to show social and material continuity of the Neolithic society" (ibid.:334).

In the Near East the renewed excavations at Neolithic Çatal Hüyük in Anatolia (Hodder, ed., 1996) gave possible evidence of deliberate burning of structures as abandonment rituals. Detailed analysis of the stratigraphy of some buildings at the site has revealed that in a number of instances there was a strictly controlled use of fire within discrete buildings. 'Ritual cleansing' or 'closing-off' of these buildings has been suggested (Matthews et al. 1996). Other types of possible abandonment rituals at Çatal Hüyük which are mentioned are apparently deliberate acts of depositing mineral sediments, figurines and refuse. Also at the Halafian sites Arpachiyah and Yarim Tepe II in northern Iraq there are indications for ritual destruction of buildings, i.e. in both sites structures seem to have been intentionally set on fire and, moreover, these buildings were associated with what seems to be deliberately smashed, and luxurious, objects (Campbell 1992).

---

[27] Intentional house conflagration, instead of accidental fire or fire due to warfare, has been argued on the basis of: (1) the temperatures of house burning, which were much too high to have resulted from the burning of the construction wood; (2) the fire path, indicating ignition points on the floors; (3) the stratigraphy of the houses, indicating that they were pulled down so as to bring them to a closure; (4) the fact that the houses were not used secondarily; (5) the absence of bodies in the houses, and (6) the fact that complete house inventories were recovered (Stevanović 1997:363-364).

It can therefore not be excluded that at Sabi Abyad similar abandonment rituals were carried out, and interestingly, a number of Stevanović's arguments for deliberate house burning in Vinča sites seem to fit the Sabi Abyad evidence regarding the conflagration of the level 6 village. First, as in Opovo, the temperatures of burning at Sabi Abyad (up to 800 °C: Aten 1996) seem to have been much too high to have been achieved only by the fire of construction materials (i.e. mud, wood and reeds). Second, (due to this high temperature) the walls in the level 6 village were completely and heavily burnt. Of the buildings, the wooden poles and the reeds represented the inflammable materials. In the case of an exterior fire (due to an accident or conflict), only the roofs would have burnt, since "... the nature of fire is such that it always burns upwards either the entire roof or some portions of it would be caught in fire but the burn would not go under the roof" (Stevanović 1997:378). Taking the heavily burnt walls into account, this would indicate that the fire started within the buildings. Given the question of how the fire spread unaided from one room to the other (which were in most instances not connected by doorways), it can be suggested that several buildings were burnt due to intentionally started interior fires. Third, as already mentioned, no bodies (i.e. victims of hostilities) were found inside the buildings (apart from the two skeletons, see above). Fourth (as in Arpachiyah and Yarim tepe II), inventories were not systematically taken out of the fired buildings, which were moreover not used secondarily.

As yet, we have to find more contemporary settlements in order to verify whether intentional house conflagatrion was, as in the Vinča culture, indeed common practice in the Late Neolithic of Syria and surrounding areas. The purpose of such a practice may have been, as Stevanović has suggested, to 'seal the house off' from possible future utilitarian purposes. "At this point of their use-life the houses might have acquired a new, nonutilitarian function, such as ensuring the continuation of ancestral line in one place" (ibid.:385). The fire destroyed the buildings, but at the same time transformed it in baked clay and thereby preserved it as foundations for new (i.e. level 5) buildings. In a sense, the fire preserved buildings, and thereby secured social and material continuity in society.

If indeed the fire in the level 6 village was due to abandonment rituals, it might be argued that this practice was associated with the mortuary ritual centred around the skeletons on the roof of building V. Like at the some of the pueblo sites discussed above, then, death of persons and 'death' of the village may have been associated.

Summarizing, it has been argued that the scarcity of complete/reconstructable pottery vessels and other artefacts may be due to a number of interrelated factors. Disturbance processes, i.e. earth-moving and erosion account for at least part of the phenomenon. It is felt, however, that relatively few complete vessels were in use at the time the fire destroyed the village. Furthermore it has been argued that the sherds and the broken objects may be interpreted in two ways: (1) as provisional and abandonment stage refuse, i.e. as parts which were simply left at the place of breakage; (2) as secondary refuse. Both scenarios imply different modes of abandonment: in the first option the settlement caught fire while still (largely) in use; in the second option the settlement was already (partly) abandoned (as perhaps indicated by the many broken objects).[28]

---

[28] Both scenarios include possible intentional firing.

Problems related to the first scenario are: (1) why would one leave refuse inside buildings instead of removing it?;[29] (2) why are there so few complete vessels and relatively few complete objects? Problems with the second scenario are: (1) why doesn't the stratigraphy indicate that (apart from building XII and the north of building I) the buildings were in disuse and abandoned? (i.e. lack of unburnt debris beneath burnt layer); (2) in spite of the relatively large number of broken objects, there is a significant number of complete objects of all kinds; why were these left behind?; (3) why did a supposedly largely dismantled village (roof beams) burn so fiercely?

It has already been indicated that the open areas were on the whole unaffected by the fire. In-situ contexts, therefore, seem to be largely absent from these spaces. As opposed to the many small rooms within the buildings, the open areas were well-accessible and they would have been traversed regularly. Furthermore the architectural features (e.g. ovens) and the finds indicate that many activities were carried out in these areas. It may therefore be expected that various formation processes have affected the open areas. As has been argued, even in in-situ contexts the designation of specific processes is a difficult task. When dealing with secondary contexts in areas that may have been affected by many different formation processes, the task becomes even more problematic. Most likely, the situation as observed in the archaeological context was due to an amalgam of formation processes. Therefore, the isolation of specific formation processes will not be attempted in the case of open areas. It will simply be indicated that discard processes, disposal modes, reclamation processes and disturbance processes have been of influence. Most likely, the majority of the objects recovered from the open areas have to be regarded as primary and secondary refuse.

Despite the absence of in-situ contexts an attempt will be made to determine the function of the various open areas. These designations are based on artefact distributions, potential characteristics of open areas (e.g. dimensions, accessibility) and presence of architectural features (e.g. ovens). In the spatial analysis of the Burnt Village both introduced scenarios will be used.

The level 3 Halafian settlement at Sabi Abyad represents an unburnt village which was gradually and 'normally' abandoned. In chapter 6 (section 9) aspects of the distribution of selected artefacts in this settlement will be studied in order to shed some light on the problems related to the mode of abandonment in the Burnt Village.

5 Models of Formation Processes

In order to asses the formation processes at Sabi Abyad a number of models or schemes of the different processes and the archaeological indications which may occur in tell-settlements have been made. It is stressed here that these models (largely based on Schiffer 1987) are *general* frameworks. Only the most important characteristics of the various formation processes have been enumerated. In practice, the models should be used in combination with Schiffer's more exhausitive studies on the identification of formation processes (Schiffer 1983, 1987:263-302). In these studies he distinguishes simple properties of artefacts (e.g. size, density, shape, etc.), complex properties of

[29] In the case of intentional firing, this would not be a problem, but indeed a characteristic of such a practice.

artefacts (e.g. quantity, diversity, etc.), and properties of deposits (e.g. intrusive materials and morphology).

In chapter 6 the models have been applied in the spatial analysis. In tables 4.1 to 4.4 the possible archaeological indications for the various formation processes discussed have been listed.

Object context Spatial context	INTER-ACTIVE Activity	DEPOSITIONAL		DISCARD		TRANS-FORMED Disturbance processes
		Deposi-tion	Storage	Primary discard	Secondary discard	
ACTIVITY AREA	A	B	C	D	E	?
STORAGE AREA	F	G	H	I	J	?
DISCARD AREA	-	-	-	-	K	?

Table 4.5. Model of the different object contexts and spatial contexts.

Table 4.5 represents a model of the different contexts in which objects play a role (the object contexts and spatial contexts). First I will deal with the object contexts, which indicate whether an object was used, placed, stored, discarded or otherwise deposited in the area where it was found (see also Sullivan 1978). The interactive context refers to a context of use or activity. In a depositional context objects are deposited between periods of use, or they are stored. Objects in a discard context represent refuse, i.e. damaged or broken objects which are no longer of use, and which are therefore discarded (primarily or secondarily). Objects in the transformed context, finally, are objects that are not deposited due to use, deposition or discard. These objects have been moved from their original depositional location as a result of disturbance and reclamation processes such as earth-moving and scavenging.

The spatial context refers to the three functional classes of spatial units in which objects can be present:

1. activity areas
2. storage areas
3. discard areas

Definitions of these areas have been given in chapter 2, section 6. As can be seen in table 4.5, the combination of object context and these three spatial contexts results in a distinction of 11 different types of spatial units, indicated by capitals. An area A, for instance, is a 'normal' activity area. Area B, however, designates an activity area where objects were also deposited, i.e. stored for a short time. In area F some activity or activities took place in an area which was mainly used for storage. By indicating the relations between spatial and object contexts it has been tried to take into account the complexity of the use of space.

The criteria for assigning an object to an area (and so for designating the function(s) of an area) consist of a reconstruction of: findspot (floor/surface or fill), object condition (broken or complete), object function (subsistence activities, manufacturing/maintenance activities, administration, social life), spatial distribution (random or clustered), relational context (can an object actually have been used, deposited or discarded in this area?), formation processes and object context (interactive, depositional, discard, transformed).

CHAPTER 5

REASONING BY ANALOGY

1 Introduction

"...there is no way to form hypotheses about ancient cultural assemblages except by analogical reasoning..." (Watson 1979:4).

Traditional typologies are based on criteria of form or morphology. In these typologies speculative assumptions about the function of artefact classes are most often present (e.g. scraper, grinder, etc.). Morphological and functional features are thus combined. This results in nomenclatures that are based upon assumed, and not proved, uses of artefacts. It is generally accepted, for instance, that grinders were used for the grinding of raw materials, spindle whorls for spinning, and so on. In many instances these form-function correlations seem obvious and logical; they have been based upon tradition, the archaeologist's 'common sense', and his/her general knowledge of the functional aspects of material culture. It should be taken into account, however, that such functional 'common sense' categories may have made less sense, or worse, no sense at all, for the people who used and produced these artefacts. Furthermore, many artefacts such as 'pierced discs', and 'notched bones' cannot be explained so readily.

Therefore, typological classifications to be used in a spatial analysis cannot be taken for granted. Each class of artefacts, and ideally each artefact, should be assigned a function only after specific research aimed at functional assessment. In the present chapter this has been achieved by 'reasoning by analogy'.

Analogy can be defined as "... selective transportation of information from source to subject on the basis of a comparison that, fully developed, specifies how the terms compared are similar, different, or of unknown likeness" (Wylie 1985:28). The use of analogy in archaeology, in particular ethnoarchaeology, has been discussed by various researchers (e.g. Ascher 1961; Aurenche 1981; Gould 1980; Hodder 1982c:11-27; Kramer 1979, 1982; Skibo 1992; Watson 1980; Wylie 1985). In 1968 Spaulding already remarked that: "... the past can be understood only through the present. All studies of the past are conducted by taking present objects (or present memories) as relics of the past and drawing inferences as to past events from them. The premises by means of which the inferences are drawn are based on observations of present things, events, and relation-ships" (Spaulding 1968:37, see also Hodder 1982c; Noll 1996).

This does not mean that "... every trait that existed in the past must have an analog in the present" (Watson 1979:1). However, it is felt that in many cases there is enough conti-nuity to use ethnographic analogy in archaeological interpretation (ibid.:1). An analogy is a hypothesis that must be tested against the archaeological record (Watson 1980:56).

'Reasoning by analogy' is in fact at the core of all archaeological research; without comparisons, juxtapositions, differentiations and analogies, interpretive frameworks cannot be established (see also Hodder 1982c:11-27). In this process of comparison, however, one runs the risk of transposing one's own cultural categories to the object of study (e.g. Shanks and Tilley 1987:7-28). In relation to this point, it has been argued that if we interpret the past by analogy to the present, we can never find out about forms of society and culture which do not exist today (Dalton 1981). Furthermore, deterministic 'uniformitarianism' (cf. the study of Clark, 1954) must be avoided; it may not be assumed that societies and cultures similar in some aspects are uniformly similar (Hodder 1982c:13). Another often heard criticism against the use of analogy is that analogies can never be checked or proved, because alternative analogies which fit the data from the past equally well, can always be found.

Reacting to these criticisms, Hodder (1982c), among others (e.g. Gramsch and Reinhold 1996), advocates the use of relational rather than formal analogies (Hodder 1982c:16-24). Formal analogies are those that are based upon "... a point for point assessment of similarities or differences in the properties of source and subject" (Wylie 1985:94). Relational analogies "... seek to determine some natural or cultural link between the different aspects in the analogy" (Hodder 1982c:16). According to Wylie (1985:95) they are based upon "... knowledge about underlying 'principles of connection' that structure source and subject and that assume, on the basis, the existence of specific further similarities between them".

Formal analogies are associated with the 'direct historical approach' which holds that when continuity between the past and the present can be assumed, many formal similarities between the information being compared may be acknowledged (see e.g. Watson 1980:56). There are two main problems related to this approach. First, problems related to comparison of proper contexts may arise (e.g. Noll 1996:246, 248). For instance, two groups in the same area may have produced very different artefacts. Cultural change, furthermore, may have led to drastic changes in all kinds of respects.

Second, frequently direct analogies situated in the region or period of interest are not available. For the Balikh valley, the regional context of Sabi Abyad, for instance, only one short and rather general ethnographic account is available, in which the use of material culture is not mentioned at all (i.e. Lewis 1988). Furthermore, detailed archaeological studies of the Later Neolithic period (the chronological context of Sabi Abyad), are rather scarce.

In relational analogies not only the attributes of artefacts, but also their cultural context are taken into account. The relevance of the association of two variables needs to be examined. Hodder states that: "We can avoid the charges of unreliability and anti-science by increasing the number and range of points of comparison between past and present, but also by identifying the relevance of the comparisons" (Hodder 1982c:23). The context of the object of study, including functional but also ideological aspects, should be taken into account. Often cultural contexts are unique, but there are at least two ways in which to deal with this 'problem'. First, general principles of meaning and symbolism do exist; generalizations can indeed in many cases be used. Second, the relevance of the analogy can be assessed by illuminating the empirical value of the analogy; it should be tried to show that the negative or uncertain likenesses are of limited consequence.

In this chapter, then, the results of a survey of the functions of the material culture of Tell Sabi Abyad are presented. When possible, relational analogies have been used to investigate the use and the variability of use of artefacts, the process of manufacture, possible symbolic connotations, etc. The analogies have been sought in the archaeological, ethnoarchaeological, historical and ethnographical literature. Moreover, studies which describe experiments for assessing the function of artefacts have occasionally been used. The focus has been upon publications related to the Near East, but literature about other areas, and more general sources have also been taken into account.

Ideally, in a spatial/functional analysis, each category of artefacts, or each artefact, should be investigated on microwear. This would result in the most reliable set of data on use of artefacts.[30] In the present study, a selection of flint implements has indeed been investigated on microwear. For the other artefacts, however, such studies were not available. In these cases analogy has been used to assess the general function of artefacts. It was not attempted to designate the specific function, for in that case microwear analyses would be needed. For example: it is determined that the general function of pestles is grinding/pounding food and pigments. The specific function (e.g. the kind of food and pigments ground) can only be hypothesized for most of the artefacts. In a number of instances, however, contextual information provided information about specific functions (e.g. distinct traces of use such as ochre on a pestle, or grain in a primary context in a room).

1.1 Editorial Remarks

For the Near East, a number of useful ethnoarchaeological studies which deal with the use of material culture are available (e.g. Aurenche et al. 1997; Kadour and Seeden 1983b; Krafeld-Daugherty 1994; Kramer 1982; Seeden 1985; Seeden and Wilson 1988; Watson 1979, 1980). Furthermore, there are a number of Near Eastern ethnographical accounts which are useful for assessing the function of architecture and artefacts, and for obtaining insight into the socio-economic and ideological practices in Near Eastern villages (e.g. Antoun 1972; Canaan 1932; Christensen 1967; Dalman 1964; Horne 1980, 1983, 1994; Kadour and Seeden 1983a; Lutfiyya 1966; Makal 1987; Nippa 1991; Sweet 1960; Thoumin 1932; Van Der Kooij 1976).

In the following, the results of the analysis of the function of artefacts such as have been found in level 6 at Sabi Abyad are presented as a series of 'models'.[31] In the models related to the small finds (section 2) and the flint/obsidian objects (section 4) the various activities attested have been ordered (thereby largely following Voigt's classification: Voigt 1983: tables 51, 53 and 54) in the following (etic) categories: (1) subsistence activities, including food procurement, food preparation, food serving (or consumption), food storage and animal keeping; (2) manufacturing/maintenance activities (of artefacts of

---

[30] However, it should be kept in mind that, like most archaeological methodologies, microwear analysis has its limitations. Due to many uncertain factors, direct identification of use is often not possible. Therefore the results of microwear analyses, like most archaeological research, should be regarded as interpretations and not as identifications (e.g. Van Gijn 1990:21).

[31] The *meaning* of artefacts is dealt with in chapter 7.

different materials); (3) administration (i.e. record keeping and sealing practices), and (4) 'social life' (cf. table 5.1). Social life refers to all kinds of processes of human interaction which are not directly related to subsistence, manufacturing/maintenance and administration. Whereas it is acknowledged that all the mentioned activities may have important social dimensions, it is felt that the category of social life is a valid distinction.

SUBSISTENCE ACTIVITIES	ADMINISTRATION
Food procurement: Working the fields Harvesting Hunting Food preparation: Butchering (including cutting) Plant-processing (working and cutting) Grinding Food deposition (temporary storage during processing) Sun drying Fermenting Baking Cooking Food serving **Food storage:** Dry Liquid **Animal keeping**	Record keeping Sealing
MANUFACTURING/MAINTENANCE ACTIVITIES	SOCIAL LIFE
Chipped stone artefacts Ground stone artefacts Bone artefacts Wooden artefacts Shell artefacts Unbaked clay artefacts Baked clay artefacts Pottery Pigments Textiles Basketry Skin-processing Leather items Construction of buildings	Eating and drinking Entertaining Meeting Assembling Recreation Music Celebrations Play Decoration/ornaments Sleeping Ritual Funeral Conflict/fighting
OTHER	
Storage of non-food products	

Table 5.1. Activity categories and activities.

The models have been based upon an extensive study of relevant literature. This study has been included as appendix 1. The microwear analysis of a selection of flint artefacts is included in this chapter (section 4).

The large group of pottery artefacts has its own specific problems and has therefore been dealt with separately in section 3.

In the final section 5 a general overview of the activities in the Burnt Village is presented.

2 Models of Functions of Small Finds of the Burnt Village (tables 5.2-5.4)

In the following tables 5.2 to 5.4 the functions of the various small finds such as recovered from the level 6 settlement have been depicted according to the categories listed in table 5.1. In tables 5.2 to 5.4 numbers and percentages have been added in order to indicate frequency of activities in the Burnt Village. The numbers indicate the total numbers of artefacts ascribed to a specific activity (e.g. butchering), the percentages express the relative frequency of these activities per activity category (e.g. subsistence activities). In table 5.4 percentages have not been added, since here three different activity categories (administration, social life and other) are dealt with. It should be noted that the same objects have often been assigned to different categories and activities, since according to the analysis presented in appendix 1, most artefacts were multifunctional. Consequently, the same artefacts have in many cases been counted several times (see e.g. the stone vessels in table 5.2). Furthermore, complete as well as broken artefacts have been counted. Using the numbers, it also should be noted that some artefacts were used regularly and had a long use-life (e.g. stone vessels), whereas other objects may have been used once and for a short period of time only (e.g. sling missiles). Taking these difficulties (which are also relevant for the flint and obsidian artefacts, cf. tables 5.25-5.27) into account, it becomes clear that the numbers should be applied with care; they can only present a rough indication of the frequency of activities at Sabi Abyad. Nevertheless, a general comparison may be interesting (in chapter 6 activities are compared in detail). Here the activity frequencies as indicated by the small finds are discussed briefly; in section 5 all the evidence (small finds, flint/obsidian tools and pottery) will be taken into account.

In table 5.2 the small finds which were most likely used for subsistence activities have been presented. Especially food storage (dry and liquid) and grinding are well-attested. Hunting is relatively well-represented, but it should be noted that this activity is over-represented due to relatively large number of sling missiles, which by their very nature can only be used once. Food deposition, sun drying, fermenting, baking and cooking are not well represented, since these activities are mainly related to pottery (section 3).

Of the manufacturing/maintenance activities in table 5.3 the production of textiles and of leather items seem to have been the two most popular activities. The production (grinding) of pigments accounts for ca. 8% of the manufacturing/maintenance activities. The production of wooden and shell artefacts is not well-attested, but the remaining activities may have been executed quite regularly, considering their 6 to 7% 'share'.

Activity	Small finds	N total	%
Food procurement:			
Working the fields	Adze Perforated stone	20	1.7
Harvesting	-		
Hunting	Sling missile	104	8.9
Food preparation:			
Butchering	Axe and adze Gouge	18	1.5
Grinding	Grinding slab Mortar Grinder Pestle Stone ball/hammerstone	204	17.5
Food deposition	Stone vessel	35	3
Sun drying	Stone vessel	35	3
Fermenting	Stone vessel	35	3
Baking	-		
Cooking	Cooking stone	2	0.2
Food serving	-		
Food storage:			
Dry	Sealing Stone vessel Stone disc Stand? White ware Basketry Cloth	358	30.7
Liquid	Sealing Stone vessel Stone disc Stand? White ware Bitumen	356	30.5
Animal keeping	-		

Table 5.2. Small finds: subsistence activities.

Production of:	Small finds	N total	%
Chipped stone artefacts	Stone spatula? Whetstone Working platform Stone ball/hammerstone	59	6.4
Ground-stone artefacts	Stone disc Whetstone Working platform Stone ball/hammerstone	63	6.8
Bone artefacts	Unworked stone/flat stone/pebble Working platform Arrowshaft straightener	56	6.1
Wooden artefacts	Axe, adze, chisel Working platform Arrowshaft straightener	26	2.8
Shell artefacts	Working platform	10	1.1
Unbaked clay artefacts	Unworked stone/flat stone/pebble Stone spatula? Working platform	57	6.2
Baked clay artefacts	Unworked stone/flat stone/pebble Stone spatula? Spatula Loamer	60	6.5
Pottery	Unworked stone/flat stone/pebble Stone spatula? Spatula Loamer	60	6.5
Pigments	Mortar and pestle Stone disc Palette Working platform Stone ball/hammerstone	71	7.7
Textiles	Perforated stone Spindle whorl/pierced disc/loom Weight Awl/needle	176	19.1
Basketry	Awl/needle	55	6
Skin-processing	Unworked stone/flat stone/pebble Stone spatula? Spatula Gouge Prime knife	56	6.1

Table 5.3. Small finds: manufacturing/maintenance activities.

Production of:	Small finds	N total	%
Leather items	Unworked stone/flat stone/pebble Stone spatula? Awl/needle Spatula Gouge	110	11.9
Construction of buildings	Axe, adze, chisel Unworked stone/flat stone/pebble Door-socket Bitumen	62	6.7

Table 5.3 (continued). Small finds: manufacturing/maintenance activities.

Of the remaining activities in table 5.4 administration undoubtedly was an important activity as indicated by the 560 'administrative items', including 300 sealings. The storage of non-food products (n= 342) is probably over-represented, because the sealings have been included. The figurines (n= 50) were most likely used for ritual purposes (see chapter 7, section 6); when excluded from 'play' this activity is not attested. Decoration/ornaments is relatively well-represented by 62 items, especially labrets (n= 31).

3 Models of Functions of Ceramic Wares and Vessels of the Burnt Village (tables 5.5-5.19)

With regard to pottery Rice (1987:207-242) distinguishes five functional classes: storage vessels, cooking pots, food preparation (without heat), serving, transport. Pfälzner, in his study of the function of Mittani and Middle Assyrian ceramics from Sheikh Hamad in eastern Syria, has made a more specific model (Pfälzner 1995:22-30 and fig. 10). Like the model of Rice, this model is based upon the form-function hypothesis.

On the basis of the above models, ethnographic accounts and use-related properties (i.e. shape, colour, material, firing technique, temper, wall thickness, diameter, surface treatment and decoration), the various shape classes distinguished at Sabi Abyad have been assigned to the above-mentioned five functional categories. The shape categories bases, restricted forms (i.e. pots and jars) and unrestricted forms (bowls) and their sub-categories will be related to broad functional categories (e.g. storage, cooking, food preparation without heat, serving and transport).

Each of the ware groups from the Burnt Village (i.e. Dark-Faced Burnished Ware, Grey-Black Ware, Standard Ware, Standard Fine Ware, Orange Fine Ware, Fine Painted Ware) has been distinguished by a specific set of technological traits (Le Mière and Nieuwenhuyse 1996:121). Most likely, these traits have an important functional significance; each ware may have had a specific function (Nieuwenhuyse, pers. comm.). In appendix 1, section 2.1, the association of ware and vessel shape has been determined, and it appeared that there are significant (but not very strong) associations between vessel shapes and wares. This also seems to indicate functional differences between the various wares.

Activities	Small finds	N total
Administration:		
Record keeping	Sealing Pierced disc? Token Incised bone?	560
Sealing	Sealing	300
Social life:		
Eating and drinking	Stone vessel	35
Entertaining	-	
Meeting	-	
Assembling	-	
Recreation	-	
Music	Whistle Incised bone	3
Celebrations	-	
Play	Figurine	50
Decoration/ornaments	Bead and pendant Labret Copper ore Ochre	62
Sleeping	-	
Ritual	Figurine Axe and adze (miniature) Oval clay object	64
Funeral	-	
Conflict/fighting	Macehead Perforated stone Sling missile	110
Other:	-	
Storage of non-food products	Sealing Stone vessel Stone disc Basketry	342

Table 5.4. Small finds: administration, social life.

DARK-FACED BURNISHED WARE: functional categories		
Shape	Functional category	Remarks
Open orifice	Food preparation; serving	-
Vertical orifice	Food preparation; serving	-
Closed orifice	Cooking pots	-
Vertical neck	Transport/storage vessel	Dry goods and liquids?
Concave neck	Transport/storage vessel	Dry goods and liquids?
Closed neck	Transport/storage vessel	Liquids?
Flat base	Transport/storage vessel; food preparation; serving	-
Rounded base	Cooking pots?	-

Table 5.5. Dark-Faced Burnished Ware: functional categories.

DARK-FACED BURNISHED WARE: function	
General function of ware	Transport; special luxury ware; imported as jars; secondarily used for cooking/food preparation
Motivation	Imported vessels; carefully made; very suitable for cooking/food preparation (necks were often sawn off the bodies, perhaps indicating reuse of jar for cooking); strong, heat-resistant; mineral temper; traces of soot; repair holes and reuse suggest highly valued ceramics

Table 5.6. Dark-Faced Burnished Ware: function.

GREY-BLACK WARE: functional categories		
Shape	Functional category	Remarks
Open orifice	Food preparation; serving	-
Vertical orifice	Food preparation; serving	-
Closed orifice	Storage vessel	-
Vertical neck	Storage vessel	Dry goods and liquids?
Concave neck	Storage vessel	Dry goods and liquids?
Closed neck	Storage vessel	Liquids?

Table 5.7. Grey-Black Ware: functional categories.

GREY-BLACK WARE: functional categories		
Shape	Functional category	Remarks
Flat base	Storage vessel; food preparation; serving	-
Rounded base	Cooking pots?	-

Table 5.7 (continued). Grey-Black Ware: functional categories.

GREY-BLACK WARE: function	
General function of ware	Serving/display/'tableware'
Motivation	Usually decorated; fine fabric; well-made; carefully finished; thin walls; shape and size comparable to other 'fine wares'

Table 5.8. Grey-Black Ware: function.

STANDARD WARE: functional categories		
Shape	Functional category	Remarks
Open orifice	Food preparation; serving	-
Vertical orifice	Food preparation; serving	-
Closed orifice	Cooking pots	-
Vertical neck	Storage vessel	Dry goods and liquids?
Concave neck	Storage vessel	Dry goods and liquids?
Closed neck	Storage vessel	Liquids?
Flat base	Storage vessel; food preparation; serving	-
Rounded base	Cooking pots?	-
Husking Tray	Food preparation	-
Handle	Attached to cooking pots?	-

Table 5.9. Standard Ware: functional categories.

STANDARD WARE: function	
General function of ware	Storage/cooking pots/food preparation/ serving
Motivation	Varied assemblage; various functions

Table 5.10. Standard Ware: function.

STANDARD FINE WARE: unrestricted forms		
Shape	Functional category	Remarks
Simple rounded bowl: hemispherical	Serving; preparation	-
Simple rounded bowl: deep bowl	Serving; preparation	-
Simple rounded bowl: flaring	Serving; preparation	-
Simple rounded bowl: closed profile	Serving; preparation	-
Flat-based straight-sided bowl	Serving; preparation	-
Carinated closed bowl	Serving; preparation	-
Low carinated bowl	Serving; preparation	-
Carinated vertical bowl	Serving; preparation	-
Cream bowl	Serving; preparation	-
Short-collared bowl	Serving; preparation	-
Squat s-shaped bowl	Serving; preparation	-
Open s-shaped bowl	Serving; preparation	-
Closed s-shaped bowl	Serving; preparation	-

Table 5.11. Standard Fine Ware: unrestricted forms.

STANDARD FINE WARE: restricted forms		
Shape	Functional category	Remarks
Holemouth pot	Storage vessel; serving	Pots extremely rare
Straight and flaring angle-necked jar	Storage vessel; serving?	Small jars; storage of dry goods?
Angle-necked jar with vertical neck	Storage vessel; serving?	Small jars; short-term storage of liquids?
Angle-necked jar with concave neck	Storage vessel; serving?	Small jars; short-term storage of dry goods?

Table 5.12. Standard Fine Ware: restricted forms.

STANDARD FINE WARE: function	
General function of ware	Serving/display/'tableware'
Motivation	Well-made; majority painted, suggesting display function; repair holes and reuse suggest highly valued ceramics; pots very rare: probably no cooking; thin vessel walls make food preparation unlikely

Table 5.13. Standard Fine Ware: function.

ORANGE FINE WARE: unrestricted forms		
Shape	Functional category	Remarks
Rounded bowl	Serving; preparation	-
Low carinated bowl	Serving; preparation	-
S-shaped bowl	Serving; preparation	-
Flat-based straight-sided bowl	Serving; preparation	-

Table 5.14. Orange Fine Ware: unrestricted forms.

ORANGE FINE WARE: restricted forms		
Shape	Functional category	Remarks
Straight and flaring angle-necked jar	Storage vessel; serving?	Dry goods?
Angle-necked jar with vertical neck	Storage vessel; serving?	Liquids?
Angle-necked jar with concave neck	Storage vessel; serving?	Dry goods?

Table 5.15. Orange Fine Ware: restricted forms.

ORANGE FINE WARE: function	
General function of ware	Serving/display/'tableware'
Motivation	Well-made and finished; fine ware; majority painted, suggesting display function; repair holes and reuse suggest highly valued ceramics; pots absent: probably no cooking; thin vessel walls make food preparation unlikely; if import: highly valued?

Table 5.16. Orange Fine Ware: function.

FINE PAINTED WARE: unrestricted forms		
Shape	Functional category	Remarks
S-shaped bowl	Serving; preparation?	-

Table 5.17. Fine Painted Ware: unrestricted forms.

FINE PAINTED WARE: restricted forms		
Shape	Functional category	Remarks
Flaring angle-necked jar	Storage vessel; serving?	Small jars; short-term storage of dry goods?

Table 5.18. Fine Painted Ware: restricted forms.

FINE PAINTED WARE: function	
General function of ware	Serving/display/'tableware'
Motivation	Well-made and finished; fine ware; carefully painted decorations, suggesting serving/display function; pots absent; probably no cooking; thin vessel walls and absence of pots make food preparation unlikely

Table 5.19. Fine Painted Ware: function.

The characteristics of Mineral Coarse Ware (which due to the limited sample size regrettably could not be taken into account in the spatial analysis), for instance, strongly suggest that this group was used for cooking and other kinds of food preparation. Le Mière and Picon (1991) have analysed the suitability for cooking of early Neolithic pottery from Bouqras (dating from the first half of the sixth millennium B.C.). They suggest that one of the wares closely resembling the Mineral Coarse Ware from Sabi Abyad was probably used for cooking, because of (1) the large number of black traces observed on this pottery; (2) the presence of large mineral inclusions; (3) thick walls; (4) closed shapes; (5) the frequent occurrence of ledge-handles. Furthermore chemical analysis, measurement of firing temperatures and thermal expansion coefficients all pointed out that this group was much more suitable for cooking than the other Bouqras wares (ibid.:68).

Standard Ware accounts for 89.5% of all level 6 ceramics, the finer wares making up the rest (cf. table 6.25). As appears from tables 5.9 to 5.13 this 'Coarse Ware' was probably mainly used for storage and the preparation of food. Occasionally it may have been used for serving. The high percentage related to this ware indicates, not surprisingly, that storage and the preparation of food played a major role in the subsistence economy of the level 6 inhabitants.

The imported Dark-Faced Burnished Ware (tables 5.5 and 5.6) may have been a special luxury ware, and seems to have been secondarily used for food preparation.

The other finer wares (Grey-Black Ware, Standard Fine Ware, Orange Fine Ware and Fine Painted Ware) are carefully made and occasionally decorated, suggesting a serving and display function (cf. tables 5.7-5.8, 5.11-5.19). The rather limited amounts of the more luxurious finer wares (cf. table 6.25), which probably mainly functioned as serving and display vessels, suggests that these wares were used less frequently. It has to be taken into account, however, that the well-made and presumably highly valued finer wares were treated more carefully, and consequently had a longer use-life than the coarse Standard Ware.

4 Microwear Analysis of Flint Artefacts

4.1 Introduction

A total number of 151 flint artefacts from the prehistoric levels at Sabi Abyad has been submitted for microwear analysis to the 'Lithic Laboratory' at the University of Leiden.[32] These artefacts represent a selection of the most important typo-chronological classifications of Copeland's typelist of the Sabi Abyad lithic industry (Copeland 1996). Goal of the analysis was to assess the function of various artefact classes at Sabi Abyad in order to use these results, together with the ethnographic and archaeological clues, as a generalized framework for assessing the use of the flint artefacts at Sabi Abyad. The microwear analysis is based on the fact that using an implement causes wear-traces on its surface: these include edge-removals (frequently called use-retouch), edge-rounding, polish and striations, all of which can be examined microscopically. Experimental research has demonstrated that the configuration and appearance of these traces varies according to contact material and motion (see e.g. Keeley 1980; Odell 1977).

In the microwear analysis the wear-traces on working surfaces were studied; other aspects such as traces due to hafting, edge-angles, etc. are not dealt with in the present study.

Of the following 18 artefact categories samples were studied: pressure-flaked pieces (n= 4); lustred sickle elements (n= 13); shape-defined sickle elements (n= 12); truncated pieces (n= 10); backed knifes (n= 6); racloirs (n= 4); raclettes (n= 2); 'Tile Knives' (tabular scrapers, n= 6); end-scrapers (n= 6); steep-scrapers (n= 6); burins (n= 11); borers (n= 10); notches (n= 10); denticulates (n= 10); pieces with fine retouch (n= 10); pieces with abrupt retouch (n= 10); blades (n= 10); flakes (n= 11).

---

[32] The 'Lithic Laboratory' is under the direction of A.L. van Gijn, under whose supervision the flint implements were studied by M.C. Schallig. The results of the microwear analysis were presented as a list in which per artefact the presence or absence of traces, the material on which the tool was used, and the kind of movement was indicated. The subsequent use and interpretation of this list is wholly my responsibility.

Since the laboratory lacks the expertise for analysing usetraces on obsidian artefacts, these implements were not submitted for analysis.

4.2 Results

The outcome of the microwear analysis has been summarized in tables 5.20 to 5.24. The study of the flints has revealed various different activities at prehistoric Sabi Abyad (table 5.20): (1) plant-processing; (2) wood-processing; (3) butchering; (4) skin-processing; (5) plant and animal-processing; (6) boring; (7) carving; (8) working of an unknown soft material. In the following, these various activity categories will be discussed.

Plant-processing

The processing of plants has been divided into a number of activities: (1) working of plants; (2) cutting of plants; (3) cutting of grain; (4) working of silicious plants; (5) working of hard silicious plants; (6) cutting of hard silicious plants.[33] In general, plant-processing was the best-represented activity in the microwear analysis: 20.5% (n= 31) of the studied artefacts were used for this purpose. Especially the cutting of grain is well-attested: 15 flint implements were used for this activity. As was expected, all the analysed lustred sickle elements (n= 13), marked by a clear sickle-gloss (fig. 5.1 B, fig. 5.5 B), were used for cutting grain. More surprising, however, was the find that of the 'shape-defined sickle elements' ("... unlustred but apparently utilised pieces with appropriate shape and retouch type"; Copeland 1996:293) only one had been used for cutting grain. So it seems that these implements may not be regarded as sickle elements.[34] Another unexpected outcome was that one end-scraper had been used for cutting grain, although these artefacts are generally associated with the working of skin.

For working of plants in general various tools were used: shape-defined sickle elements (n= 2) and especially pieces with fine retouch (n= 5) and blades (n= 4) (see fig. 5.1). Apart from grain, silicious plants and hard silicious plants have been attested. The working and cutting of hard silicious plants was done by using end-scrapers (n= 3) and racloirs (n= 1). On fig. 5.1 A and D streaks of plant-polish due to working/cutting of hard silicious plants on, respectively, an end-scraper and a racloir are depicted (and see fig. 5.5 A and C). Most likely hard silicious plant is represented by the so-called Giant reed (*Arundo donax*), a hard and thick reed which today still grows along the banks of the Balikh (pers. comm. S. Bottema). The reed is used for roof covering; it is deposited upon the wooden (poplar) rafters and subsequently covered by mud. As was argued in chapter 3, the prehistoric roofs at Tell Sabi Abyad were probably built in the same way. Therefore it is here suggested that end-scrapers and racloirs were, among their other uses, employed for the cutting of thick reeds along the banks of the Balikh or other wadis, which were then used for roof covering. One piece with abrupt retouch was used for the working of 'normal' (i.e. not hard) silicious plants. It is also possible that the silicious plants were used for making basketry; as has been indicated in chapter 3, baskets must have been used in large numbers in the level 6 village, and probably also in the other prehistoric settlements at Sabi Abyad.

---

[33] In the microwear analysis a difference has been made between 'working' and 'cutting' of plants; cutting of plants refers to the carving of stems and leaves, whereas working indicates all other uses of implements on plants, e.g. scraping, peeling, chopping, etc. Also for the other 'Functional Categories' (i.e. wood-processing, skin-processing, plant- and animal-processing and working of an unknown material) it holds that 'working' is a general designation.

[34] However, it cannot be excluded that they represent freshly made and still unused sickles.

Functional categories	Tool typology	N
PLANT-PROCESSING		
Working of plants	Shape-defined sickle element (n= 2) Piece with fine retouch (n= 3) Blade (n= 4)	9
Cutting of plants	Piece with fine retouch (n= 2)	2
Cutting of grain	Lustred sickle element (n= 13) Shape-defined sickle element (n= 1) End-scraper (n= 1)	15
Working of silicious plants	Piece with abrupt retouch (n= 1)	1
Working of hard silicious plants	End-scraper (n= 3)	3
Cutting of hard silicious plants	Racloir (n= 1)	1
WOOD-PROCESSING		
Working of wood	Racloir (n= 1) Raclette (n= 1) Notch (n= 2) Piece with fine retouch (n= 2) Piece with abrupt retouch (n= 2)	8
BUTCHERING		
Butchering	Shape-defined sickle element (n= 2)	2
SKIN-PROCESSING		
Working of hard and dry skin	Blade (n= 1)	1
Working of skin	Racloir (n= 1) Tile Knife (n= 1) Denticulate (n= 1) Piece with fine retouch (n= 1) Piece with abrupt retouch (n= 1)	5
Cutting of skin	Shape-defined sickle element (n= 3) Tile Knife (n= 2) Notch (n= 2)	7
PLANT AND ANIMAL-PROCESSING		
Working of plants and skin	Pressure-flaked piece (n= 1) Tile Knife (n= 1)	2
Working of plants and used for butchering	Backed knife (n= 1)	1
Working of skin or hard silicious plants	Raclette (n= 1)	1

Table 5.20. Summary of the results of the microwear analysis.

Functional categories	Tool typology	N
BORING		
Boring of an unknown material, possibly wood	Borer (n= 1)	1
Boring of unknown material	Borer (n= 1)	1
CARVING		
Carving of hard material	Burin (n= 1)	1
PROCESSING OF UNKNOWN MATERIAL		
Cutting of unknown soft material	Racloir (n= 1)	1
Working of unknown soft material	Pressure-flaked piece (n= 2) Truncated piece (n= 2) Denticulate (n= 1)	5
NO POSITIVE RESULTS		
Probably used	Truncated piece (n= 2) End-scraper (n= 1) Steep-scraper (n= 1) Notch (n= 4) Denticulate (n= 3) Piece with fine retouch (n= 1) Piece with abrupt retouch (n= 4) Blade (n= 3) Flake (n= 3)	22
Not interpretable	Shape-defined sickle element (n= 3) Truncated piece (n= 1) Backed knife (n= 3) Tile Knife (n= 1) End-scraper (n= 1) Burin (n= 1) Borer (n= 1) Denticulate (n= 2) Piece with fine retouch (n= 1) Piece with abrupt retouch (n= 2) Blade (n= 1)	17
No traces	Pressure-flaked piece (n= 1) Shape-defined sickle element (n= 1) Truncated piece (n= 5) Backed knife (n= 2) Tile Knife (n= 1) Steep-scraper (n= 5) Burin (n= 9) Borer (n= 7) Notch (n= 2) Denticulate (n= 3) Blade (n= 1) Flake (n= 8)	45
Total		151

Table 5.20 (continued). Summary of the results of the microwear analysis.

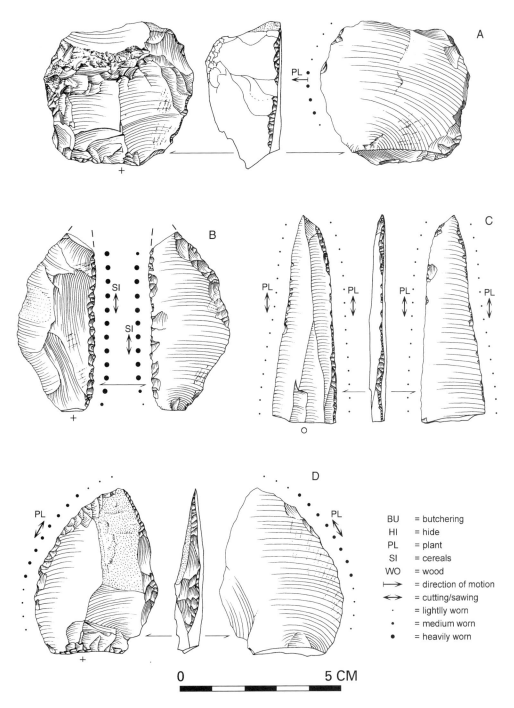

Figure 5.1. Implements displaying wear-traces inferred as being from processing plants. A: end-scraper with traces (streaks of plant-polish) of the working of hard silicious plants; B: lustred sickle element with traces of the cutting/harvesting of cereals ('sickle-gloss'); C: piece with fine retouch with traces of the cutting of plants; D: racloir with traces (streaks of plant-polish) of the cutting of hard silicious plants.

The movement of the plant-processing tools (table 5.21) varied, but of the attested motions (n= 29) the longitudinal movement is best-represented (n= 19). Following are the longitudinal and transverse (n= 4), the transverse (n= 3), the transverse to diagonal (n= 2) and the diagonal movements (n= 1). The lustred sickle elements, used for cutting grain, were all (n= 13) used in a longitudinal movement. For the working of plants the movements were most varied; apart from the transverse movement all distinguished kinds of movements were attested, perhaps indicating that various techniques of plant-processing were used. Most likely, different parts of plants were treated differently, or different types of processing required different actions. For the working of hard silicious plants (i.e. the Giant reed) with end-scrapers only the transverse movement was attested, which in fact is the most logical movement for scrapers. For the other plant-processing categories distinguished (cutting of plants, working of silicious plants, and cutting of hard silicious plants) the longitudinal movement was applied.

In table 5.22 the various analysed flint implements of level 6 and the activities they represent have been related to the provenance of these implements. It may be suggested that plant-processing tools were actively used as much as they were stored; half of the sample (n= 22) was recovered from activity areas, the other half from storage areas (cf. the analysis in chapter 6). Working of plants was carried out in open area 6 (i.e. east of building V), on the roof of building IV (where also grinding of food products seems to have taken place, cf. chapter 6), and most likely in room 2 of building IV. In this room also the working of silicious plants seems to have taken place. Room 1 of tholos VI was used for the cutting of plants. Perhaps the bin in the northeastern corner served as a receptacle for the processed plants? The activity areas where grain-cutting tools (five lustred sickle elements and one shape-defined sickle element) were found were all located in open areas; the sickle-elements inside buildings all came from storage areas. Most likely the sickle-elements were used in the fields (as a part of sickles) for harvesting grain, but their presence in open, unroofed activity areas might suggest that they were also used in other cutting-activities, which were apparently restricted to outdoor areas. Perhaps these activities consisted of cutting the straw off the ears of the harvested grain which was brought to the site.

Cereal remains make up the bulk of the Sabi Abyad plant record, and it can safely be assumed that cereals were the main crops of the Sabi Abyad farmers (Van Zeist and Waterbolk-Van Rooijen 1996:537). Therefore it can be suggested that a large part, if not the majority, of the plant-processing tools were used in relation to grain, especially wheat, since barley seems to have been of lesser importance. Emmer wheat was quantitatively the most important crop plant, the proportion of einkorn was rather modest, judging from the various plant samples (ibid.:537). The limited amount of pulse seeds (lentil, field pea, grass pea and bitter vetch) suggests that pulses only played a minor role in the food economy of Sabi Abyad.

The edible wild-plant record of Sabi Abyad is not particularly rich (ibid.:538), which most probably indicates that the collecting and subsequent processing was of minor importance only. For instance, wild fruits (e.g. pistachio) were brought to the site only occasionally. Apart from fruits, the wild plants have been divided into 'weed flora' and 'river-valley vegetation'. The weed flora, stemming from unprocessed crop supplies, was very scarce, suggesting that only few arable weeds were (unintentionally) harvested together with the crop. Among the retrieved field-weed taxa were *Lolium, Aegilops, Astragalus, Rumex pulcher* and *Prosopis* (for a complete list see Van Zeist and Waterbolk-Van Rooijen 1996:539 and tables 10.2 and 10.3). As to the river-valley vegetation it has already been mentioned that reeds may have been collected and processed for roofing. Furthermore some of the herbaceous ground flora may have been put to use: Van Zeist and Waterbolk-Van Rooijen suggest that sedge and spike-rush may have been used as litter for bedding (ibid.:540). Perhaps some of the plant-processing tools were used for collecting and processing these plants. Neither can it be ruled out that flint implements were used for fuel collecting, since the steppe surrounding Sabi Abyad contained relatively large numbers of shrubby perennials (e.g. *Helianthemum, Teucrium*) which could have been gathered as firewood. For these plants, too, flint implements may have been used.

Wood-processing

The working of wood accounts for ca. 5% of all the activities attested in the microwear analysis. Various implements were used for wood-processing: one racloir (fig. 5.3 A and 5.5 F), one raclette, and two of each of the following tools: notches (fig. 5.3 B), pieces with fine retouch and pieces with abrupt retouch. Most likely, these artefacts were applied to cut off branches, to scrape off bark and fibres, to produce even surfaces, to produce slits and grooves, to produce shapes, etc. Large and small wooden artefacts could have been produced with the help of these implements: rafters for the roof and other house furnishings such as doors, furniture, vessels, and implements such as pestles, handles for hammers and axes, sickle-hafts, sticks for digging, etc.[35]

The wood-processing tools were used in various directions (table 5.21): longitudinal (raclette, piece with abrupt retouch), transverse (2 notches, 1 piece with fine retouch), longitudinal and transverse (racloir, piece with fine retouch) and transverse to diagonally (piece with abrupt retouch). Presumably, the different motions corresponded to different actions upon the wood. One of the pieces with fine retouch (which was used on soft wood) had been used on three sides.

The analysed wood-processing tools from level 6 (n= 7) nearly all stem from activity areas; only one was found in a storage area (room 3 of building IV). Especially the open areas may have been used for the working of wood: three wood-processing tools stem from open area 6 and one from the central open area 3 (between buildings II and XII). The other two artefacts stem from room 3 of building III and room 2 of building IV. Perhaps, implements used for the working of wood were used and discarded rather than stored. This might be due to the circumstance that wood-working is a relatively heavy task in which the implement used may wear out or break. Storage for re-use is perhaps not as likely as for the other tools.

---

[35] Apart from burnt roof beams in level 6, no wooden artefacts have been recovered from Sabi Abyad. This is undoubtedly due to the fact that wooden objects are only preserved in exceptional circumstances (e.g. in wetland sites). In fires most small wooden objects disintegrate wholly, as seems to have been the case in the Burnt Village. Nevertheless, the amount of wooden objects in use at Sabi Abyad should not be overestimated: as nowadays, wood was presumably a relatively scarce product which would have been put to use in a very economical manner.

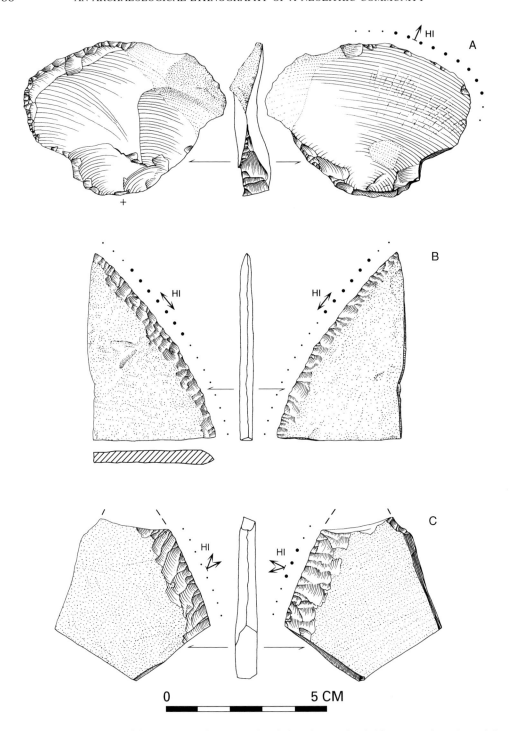

Figure 5.2. Implements with wear-traces interpreted as being the result of skin-processing. A: racloir with traces of skin-working (band of 'hide-polish'); B: 'Tile Knife' with traces of the cutting of skin; C: 'Tile Knife' with traces of skin-working (hide-polish) and traces of residue (most likely animal fat cells).

Van Zeist and Waterbolk-Van Rooijen (1996:539) suggest that along the Balikh a riverine forest was present during the sixth millennium B.C., with poplar, elm and ash as the main constituents. As they suggest, these woods must have provided the principal building timbers, and they would also have been used for the production of other artefacts, such as mentioned above. Other tree species, such as *Tamarix,* may also have been present but these have not been attested until now.

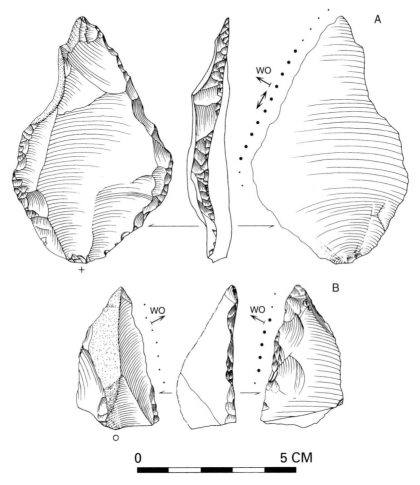

Figure 5.3. Implements displaying wear-traces inferred as being from processing wood. A: racloir with traces of wood-working; B: notched piece with traces of wood-working.

Butchering

Butchering (skinning, removing the flesh from bones, etc.) is attested by two 'shape-defined sickle elements' (table 5.24). Actually it was surprising that these relatively small implements were used for such a relatively heavy task; one would expect more robust artefacts such as heavy-duty types and backed knives. On one of these implements a longitudinal movement was noticed. On this specimen (fig. 5.4 B), stemming from room 12 of building II (a storage area), a clear skin-gloss (fig. 5.6 B) and spots with bone-gloss (fig. 5.6 C) were attested. On the other tool skin-gloss was also observed. Furthermore, both tools showed use-retouch.

Butchering, divided in cutmarks and chopmarks, has been noticed on sheep/goat bones, but also on the bones of cattle, pig, gazelle and (wild) horses (including onager).

As can be seen in table 5.23, butchering marks were most frequent on the bones of sheep/goat. The bones of the other species contained far fewer marks of this kind. Furthermore it appears that cutmarks are far more frequent than chopmarks. Actually this is not surprising, since chopping parts off a carcass is the initial and preparative task in the process of dismembering, which is mainly done by cutting.

Skin-processing

About 9% (n= 13) of the analysed tools were used for the processing of skins. This general activity has been divided in: (1) the working of hard and dry skin (dry hides are often scraped in order to soften them; Van Gijn 1990:27); (2) the working of soft skin (scraping) and (3) the cutting of skin (after disarticulation, during the subsequent skin-processing). For these tasks a variety of different artefacts was used, including fewer scrapers than expected (see fig. 5.2).

For the working of hard and dry skin a simple unretouched blade had been used. On one side of the implement a longitudinal movement was noted, whereas the other side gave evidence of a transverse movement. For the working of skin at least five different tools were used, and in the analysis each tool is represented by one specimen: a racloir, a 'Tile Knife', a denticulate, a piece with fine retouch and a piece with abrupt retouch (fig. 5.2). The wear-traces on the racloir (fig. 5.2 A and 5.5 D) indicated that to facilitate the removal of fat during scraping a substance (possibly sand, saw-dust or ochre) was used to absorb the fat during the process. On a 'Tile Knife' used for the working of skin residue was found, most likely representing animal fat (figs. 5.2 C and 5.5 E). The denticulate showed clear scraping traces. The working of skin was mainly done by moving the implement transversally (n= 3), but longitudinal and transverse (n= 1) and transverse to diagonal (n= 1) movements were also noted. The cutting of skin was apparently executed by a less-varied tool kit: three 'shape-defined sickle elements', two 'Tile Knives' and two notches have been ascribed to this activity. One of the notches was used on two sides. These implements were used in a longitudinal movement only (n= 6).

In the level 6 settlement, four of the six skin-processing tools which could be ascribed to a certain location stemmed from storage areas (table 5.22). This may suggest that, as opposed to the wood-processing tools, these implements were regularly re-used. Of the two specimens that stem from activity areas one came from the open area 15 in building II, and the other from the neighbouring room 18. Most likely, the working of skin (table 5.20) was carried out in these areas.

Plant and animal-processing

A few (n= 4) of the analysed tools were multi-functional: they were used for the processing of plants as well as of animal products. A pressure-flaked piece (used longitudinally) and a 'Tile Knife' (used transversally) were used for the working of plants and skin. One backed knife functioned as a plant-working and butchering implement. The working of plants proceeded in a longitudinal motion.

Functional categories	Movement				
	Long.	Transv.	Long.& transv.	Transv./ diag.	Diag.
PLANT-PROCESSING					
Working of plants	3		4	2	1
Cutting of plants	1				
Cutting of grain	13				
Working of silicious plants	1				
Working of hard silicious plants		3			
Cutting of hard silicious plants	1				
WOOD-PROCESSING					
Working of wood	2	2	2	1	
BUTCHERING					
Butchering	1				
SKIN-PROCESSING					
Working of hard and dry skin			1		
Working of skin		3	1	1	
Cutting of skin	6				
PLANT AND ANIMAL-PROCESSING					
Working of plants and skin	1	1			
Working of plants and used for butchering	1				
Working of skin or hard silicious plants	1				
CARVING					
Carving of hard material		1			

Table 5.21. Summary of the results of the microwear analysis: the relation of functional categories and movement of tools.

Functional categories	Movement				
	Long.	Transv.	Long.& transv.	Transv./ diag.	Diag.
PROCESSING OF UNKNOWN MATERIAL					
Cutting of unknown soft material	1				
Working of unknown soft material	3	1	1		
Total	35	11	9	4	1

Table 5.21 (continued). Summary of the results of the microwear analysis: the relation of functional categories and movement of tools.

One raclette was used for the working of skin, or the working of a hard silicious plant. This implement was moved longitudinally during use (fig. 5.4 A and 5.6 A). Both multi-functional tools were recovered from open areas in level 6: the plant/skin working tool from open area 15 in building II (the 'oven area'), and the plant/butchering tool from open area 6. These tools indicate the multi-functional character of the open activity areas in the level 6 settlement (see chapter 6). The raclette was found in storage area 5 in building I.

Boring

Only two of the analysed borers (n= 10) gave unequivocal evidence of boring. On seven 'borers' no traces at all were noted, and one was not interpretable due to damage by the level 6 fire (but it was noted that this implement was probably used as a reamer). Of the two identified borers one was possibly used on wood. This implement was recovered from room 8 of building V, an activity area. The material on which the other borer (from storage room 5 of building IV) was used could not be identified.

Carving

Apart from the borers, the burins, too, proved to be a problematic category; only of one burin (out of 11) burin could it be established that it was used for the carving of a hard material (bone?). The object, stemming from the open activity area 6, was moved trans-versally. Nine of the 'burins' showed no traces at all, and one was not interpretable due to too much gloss.

Actually the lack of microwear traces on the Sabi Abyad burins may not be that surprising. Vaughan (1985:322-324) notes for some late Paleolithic assemblages from Europe that often burin spalls seem to have been removed not to make a burin, but to refresh or remove damaged or broken working surfaces of artefacts. Furthermore, he suggests that the 'burin' spalls were removed to facilitate hafting (the side with the removed part would have been hafted). Here it should also be mentioned that the Sabi Abyad burins are not 'typical' burins which are known from many other sites, i.e. small implements with a clear spall removal facet. The 'burins' from Sabi Abyad appear in many different shapes (Copeland 1996:294). Therefore it may be suggested that many of the 'burins' are, as Vaughan suggested, artefacts that were re-sharpened, rather than burins.

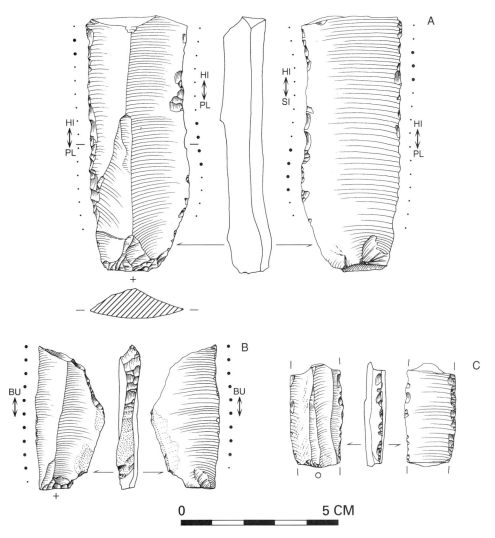

Figure 5.4. A: raclette with traces of skin-working or working of hard silicious plants; B: shape-defined sickle element with traces of butchering ('hide-polish' and spots with 'bone-polish'); C: piece with fine retouch, no traces (see fig. 5.6 D).

Processing of unknown material

Six flint artefacts (4%) were used for the processing of unknown soft material (possibly soft plants or meat). One racloir was used for the cutting of some soft material. This implement was used longitudinally. Five other implements were used for the working in general of an unknown soft material, i.e. two pressure-flaked pieces, two truncated pieces and one denticulate. Three of these artefacts were used longitudinally, one transversally and one longitudinally and transversally.

Of only two of the tools of this functional category the provenance could be established: one came from the storage room 8 of building I, the other from room 2 of building IV (an activity area).

Functional categories	Context	Function of unit	
		Activity area	Storage area
PLANT-PROCESSING			
Working of plants	Open area 6	1	
	Building II		1
	Building II, room 17		2
	Building IV, roof	1	
	Building IV, room 2	1	
	Building IV, room 6		1
Cutting of plants	Building V, room 7		1
	Building VI, room 1	1	
Cutting of grain	Open area 3	2	
	Building II, area 15	2	
	Open area 5	1	
	Open area 5	1	
	Building II, room 6		1
	Building II, room 7		1
	Building XII, room 2		1
Working of silicious plants	Building II, room 17		1
	Building IV, room 2	1	
	Building IV, room 3		1
Cutting of hard silicious plants	Building IV, room 3		1
WOOD-PROCESSING			
Working of wood	Open area 3	1	
	Open area 6	3	
	Building II, room 18	1	
	Building IV, room 2	1	
	Building IV, room 3		1
BUTCHERING			
Butchering	Building II, room 12		1
SKIN-PROCESSING			
Working of hard and dry skin	Building V	-	-
Working of skin	Building II, area 15	1	
	Building II, room 18	1	
	Building IV, room 4		1
	Building IV, room 9		1
Cutting of skin	Building II		2

Table 5.22. Summary of the results of the microwear analysis: the relation of functional categories and spatial distribution of tools, level 6.

Functional categories	Context	Function of unit	
		Activity area	Storage area
PLANT AND ANIMAL-PROCESSING			
Working of plants and skin	Open area 6	1	
Working of plants and used for butchering	Building II, area 15	1	
Working of skin or hard silicious plants	Building I, room 5		1
BORING			
Boring of an unknown material, possibly wood	Building V, room 8	1	
Boring of unknown material	Building IV, room 5		1
CARVING			
Carving of hard material	Open area 6	1	
PROCESSING OF UNKNOWN MATERIAL			
Cutting of unknown soft material	Building I, room 8		1
Working of unknown soft material	Building IV, room 2	1	
Total		24	20

Table 5.22 (continued). Summary of the results of the microwear analysis: the relation of functional categories and spatial distribution of tools, level 6.

No positive results

Over half (55.6%) of the analysed artefacts gave what has been called 'no positive results' (table 5.20); these implements were either 'probably used' (no usewear was observed, but visual inspection revealed that the implement was most probably used); 'not interpretable' (usewear could not be detected because the artefact was damaged); or they showed 'no traces' (no usewear was detected).

As appears from the above, an unequivocal form-function correlation can (as yet) not be established for most of the analysed Sabi Abyad flint artefacts. This may be due to (1) the limited sample size; perhaps the present sample is not representative for the flint artefact assemblage as a whole; or (2) the absence of a strong form-function correlation. Nevertheless, using table 5.24, some generalizations can be made. Clearly, the lustred sickle-elements were monofunctional tools used for the reaping of grain; the entire sample (n= 13) gave evidence of this activity. The 'Tile Knives' were most likely specialized skin-processing tools; the four tools of which the function was established all gave evidence of this activity. End-scrapers seem to have been especially used for the working of hard silicious plants. Apart from one end-scraper used for cutting grain, the specimens were most likely employed for cutting reeds or stems for making basketry. The blades were apparently mainly used for working plants. One blade, however, was used for the working of dry and hard skin. The majority (62.5%) of the pieces with fine retouch gave evidence of plant-processing, but also wood-working (25%) and skin-working (12.5%) were executed with these tools. Two pieces with abrupt retouch were used for the working of wood, one for the working of silicious plants, and one for the working of skin.

	Sheep/goat	Cattle	Pig	Gazelle	Horse
Cutmarks	107	4	13	5	9
Chopmarks	7	2	3	1	2
Total	114	6	16	6	11

Table 5.23. Butchering marks on the Sabi Abyad animal bones (after Cavallo 1997: table 6.5).

Tool	Functional category	Function	N
Pressure-flaked piece	PLANT AND ANIMAL-PROCESSING PROCESSING OF UNKNOWN MATERIAL	Working of plants and skin Working of unknown soft material	1 3
Lustred sickle-element	PLANT-PROCESSING	Cutting of grain	13
Shape-defined sickle element	PLANT-PROCESSING SKIN-PROCESSING BUTCHERING	Working of plants Cutting of grain Cutting of skin Butchering	2 1 3 2
Truncated piece	PROCESSING OF UNKNOWN MATERIAL	Working of unknown soft material	2
Backed knife	PLANT AND ANIMAL-PROCESSING	Working of plants and used for butchering	1
Racloir	PLANT-PROCESSING WOOD-PROCESSING SKIN-PROCESSING PROCESSING OF UNKNOWN MATERIAL	Cutting of hard silicious plants Working of wood Working of skin Cutting of unknown soft material	1 1 1 1
Raclette	PLANT AND ANIMAL-PROCESSING WOOD-PROCESSING	Working of skin or hard silicious plants Working of wood	1 1

Table 5.24. The function(s) of prehistoric flint tools at Sabi Abyad.

Tool	Functional category	Function	N
Tile Knife	SKIN-PROCESSING PLANT AND ANIMAL-PROCESSING	Working of skin Cutting of skin Working of plants and skin	1 2 1
End-scraper	PLANT-PROCESSING	Working of hard silicious plants Cutting of grain	3 1
Burin	CARVING	Carving of hard material	1
Borer	BORING	Boring of unknown material	2
Notch	WOOD-PROCESSING SKIN-PROCESSING	Working of wood Cutting of skin	2 2
Denticulate	SKIN-PROCESSING PROCESSING OF UNKNOWN MATERIAL	Working of skin Working of soft unknown material	1 1
Piece with fine retouch	PLANT-PROCESSING WOOD-PROCESSING SKIN-PROCESSING	Working of plants Cutting of plants Working of wood Working of skin	3 2 2 1
Piece with abrupt retouch	PLANT-PROCESSING WOOD-PROCESSING SKIN-PROCESSING	Working of silicious plants Working of wood Working of skin	1 2 1
Blade	PLANT-PROCESSING SKIN-PROCESSING	Working of plants Working of hard and dry skin	4 1

Table 5.24 (continued). The function(s) of prehistoric flint tools at Sabi Abyad.

The 'shape-defined sickle elements' and the racloirs seem to have been real multi-functional artefacts: both tool-classes were used for plant-processing as well as skin-processing. Furthermore butchering (shape-defined elements) and wood-working (racloirs) were among the uses of these implements. Of the notches (n= 4) half were used for working wood, and the other half for the cutting of skin.

Due to the limited number of attested functions of the remaining artefacts, generalizations are even more difficult to make than in the above instances. Pressure-flaked pieces were used for the working of an unknown soft material, as well as for the working of plants and skin. Both truncated pieces were used for the working of an unknown soft material. The backed knife was used for the working of plants and also for butchering. Of the flakes (n= 11) three have probably been used, but the remaining eight showed no traces, indicating that they were not used. The flakes of Sabi Abyad, therefore, should most likely be regarded as knapping refuse. Furthermore, relatively large numbers of the following artefacts showed no traces, and seem to have been unused: truncated pieces (n= 5), steep-scrapers (n= 5), burins (n= 9), borers (n= 7), denticulates (n= 3). It should, however, be taken into account that perhaps these objects were used after all, but that the traces are not visible because: (1) they were used too briefly for traces to develop; (2) they were used on materials which leave no (clear) traces; (3) traces that were once present have disappeared (due to various post-depositional processes); (4) (vague) traces that are present were perhaps not recognized.

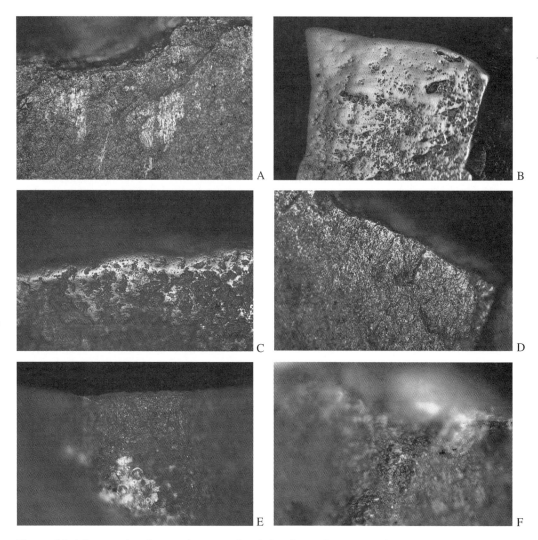

Figure 5.5. Micrographs of traces interpreted as being from plant-processing, wood-processing and skin-processing. A: fig. 5.1 A (100x); B: fig. 5.1 B (100x); C: fig. 5.1 D (200x); D: fig. 5.2 A (10x); E: fig. 5.2 C: (20x); F: fig. 5.3 A (20x).

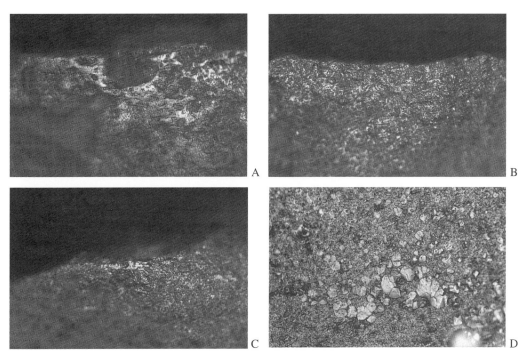

Figure 5.6. Micrographs of traces interpreted as being from skin-working or working of hard silicious plants, butchering and of potlids. A: fig. 5.4 A (200x); B: fig. 5.4 B: hide-polish (20x4); C: fig. 5.4 B: spots with bone-polish (20x4); D: fig. 5.4 C: potlids (200x).

4.3 Models of Functions of Flint and Obsidian Artefacts of the Burnt Village (tables 5.24-5.26)

As in section 2, the activity frequencies are dealt with only briefly here; in section 5 all the evidence (small finds, flint/obsidian tools and pottery) will be taken into account.

Of the subsistence activities in table 5.25 butchering, plant-processing and 'working the fields' are well-attested. Harvesting (mainly indicated by sickle elements) accounts for ca. 3% of the subsistence activities. The remaining activities (hunting, plant and animal processing and processing of unknown material) may not have been regularly executed, as indicated by the low number of artefacts.

The manufacturing/maintenance activities (table 5.26) are best-represented by the manufacture of bone and wooden artefacts, skin-processing, the production of leather items, and plant-processing. The other activities seem to have been carried out less regularly.

Activity	Artefacts	N total	%
Food procurement:			
Working the Fields	Heavy-duty type Unretouched piece: flake and blade/bladelet	730	26.5
Harvesting	Sickle elements (MWA) Shape-defined sickle element (MWA) End-scraper (MWA) Truncated piece?	88	3.2
Hunting	Arrowhead	3	0.1
Food preparation:			
Butchering	Shape-defined sickle element (MWA) Heavy-duty type Backed knife? Scraper Burin Retouch: fine and abrupt Unretouched piece: flake and blade/bladelet	944	34.3
Plant-processing	Shape-defined sickle element (MWA) Racloir (MWA) End-scraper (MWA) Backed knife? Scraper Retouch: fine and abrupt (MWA) Unretouched piece: flake and blade/bladelet	916	33.3
Grinding	-		
Food deposition	-		
Sun drying	-		
Fermenting	-		
Baking	-		
Cooking	-		
Plant and animal-processing	Pressure-flaked piece (MWA) Backed knife (MWA) Raclette (MWA) Tile Knife (MWA)	20	0.7

Table 5.25. Flint and obsidian artefacts: subsistence activities. MWA = according to the microwear analysis.

Activity	Artefacts	N total	%
Food serving	-		
Food storage:			
Dry	-		
Liquid	-		
Animal keeping	-		
Other:			
Processing of unknown material	Pressure-flaked piece (MWA) Truncated piece (MWA) Racloir (MWA) Denticulate (MWA)	53	1.9

Table 5.25 (continued). Flint and obsidian artefacts: subsistence activities. MWA = according to the microwear analysis.

Production of:	Artefacts	N total	%
Chipped stone artefacts	Heavy-duty type Splintered piece Polished or ground piece Core	83	1.5
Ground-stone artefacts	Heavy-duty type Backed knife? Borer Notch	96	1.8
Bone artefacts	Heavy-duty type Pressure-flaked piece? Truncated piece Backed knife? Scraper Burin Borer Notch Denticulate Retouch: fine and abrupt Unretouched piece: flake and Blade/bladelet	1035	19

Table 5.26. Flint and obsidian artefacts: manufacturing/maintenance activities.

Production of:	Artefacts	N total	%
Wooden Artefacts	Racloir (MWA) Raclette (MWA) Notch (MWA) Heavy-duty type Pressure-flaked piece? Truncated piece Backed knife? Scraper Burin Borer Denticulate Retouch: fine and abrupt (MWA) Unretouched piece: flake and blade/bladelet	1035	19
Shell artefacts	Backed knife? Scraper Borer	115	2.1
Unbaked clay artefacts	Borer	37	0.7
Baked clay artefacts	Borer	37	0.7
Pottery	Borer	37	0.7
Pigments	-		
Textiles	-		
Basketry	-		
Skin-processing	Pressure-flaked piece (MWA) Backed knife (MWA) Shape-defined sickle element (MWA) Racloir (MWA) Raclette (MWA) Tile Knife (MWA) Notch (MWA) Denticulate (MWA) Backed knife? Scraper Retouch: fine and abrupt (MWA) Unretouched piece: flake and blade/bladelet	988	18.2
Leather items	Backed knife? Scraper Borer Retouch: fine and abrupt Unretouched piece: flake and blade/bladelet	937	17.2

Table 5.26 (continued). Flint and obsidian artefacts: manufacturing/maintenance activities.

Production of:	Artefacts	N total	%
Plant-processing	Shape-defined sickle element (MWA) Racloir (MWA) End-scraper (MWA) Backed knife? Scraper Retouch: fine and abrupt (MWA) Unretouched piece: flake and blade/bladelet	916	16.9
Construction of buildings	Heavy-duty type	11	0.2
Other:			
Boring	Borer (MWA)	37	0.7
Carving	Burin (MWA)	17	0.3
Processing of unknown material	Pressure-flaked piece (MWA) Truncated piece (MWA) Racloir (MWA) Denticulate (MWA)	53	1

Table 5.26 (continued). Flint and obsidian artefacts: manufacturing/maintenance activities.

5 Activities in the Burnt Village

In this final section a short overview of the activities carried out in the Burnt Village will be given. It is recalled that only a rough picture can be sketched. The purpose is to give a general idea of the range and frequency of activities in the level 6 settlement. In chapter 6 a more detailed account will be presented.

In table 5.27 the numbers from the tables just mentioned have been combined in order to gain an insight into general activity frequencies. Pottery has been excluded, since the thousands of sherds would heavily skew the already rough calculations. However, it should be acknowledged that the absence of the pottery in the calculations results in the under-representation of the categories food preparation and storage (see section 3).

Omitting the pottery, it seems that manufacturing/maintenance activities account for 55% of all attested activities, for subsistence activities a figure of 34% was found, and administration, social life and 'other' represent 11% of the small finds and flint/obsidian artefacts. Undoubtedly, if pottery were included subsistence activities would be far better represented, but nevertheless manufacturing/maintenance activities seem to be well-represented. Of the subsistence activities plant-processing and butchering seem to have played a major role. The working of fields is relatively well-represented, too. Taking into account the absence of the pottery, indications of the storage of food are relatively strong.

Of the manufacturing/maintenance activities the manufacture of bone and wooden artefacts, skin-processing and the production of leather objects and plant-processing are well-attested (8 to 9.5%). Considering the much lower percentages related to the other activities, it may be suggested that these were executed less frequently.

Activity	N total	%
SUBSISTENCE ACTIVITIES		
Food procurement:		
Working the fields	750	6.5
Harvesting	88	0.8
Hunting	107	0.9
Food preparation:		
Butchering	962	8.4
Plant-processing/grinding	1120	9.7
Food deposition	35	0.3
Sun drying	35	0.3
Fermenting	35	0.3
Cooking	2	0.02
Plant and animal-processing	20	0.2
Food storage:		
Dry	358	3.1
Liquid	356	3.1
Other:		
Processing of unknown material	53	0.5
MANUFACTURING/MAINTENANCE ACTIVITIES		
Production of:		
Chipped stone artefacts	142	1.2
Ground-stone artefacts	159	1.4
Bone artefacts	1091	9.5
Wooden artefacts	1061	9.2
Shell artefacts	125	1.1
Unbaked clay artefacts	94	0.8
Baked clay artefacts	97	0.8
Pottery	97	0.8
Pigments	71	0.6
Textiles	176	1.5
Basketry	55	0.5
Skin-processing	1044	9.1
Leather items	1047	9.1
Plant-processing	916	8
Construction of buildings	73	0.6
Boring	37	0.3
Carving	17	0.1
Processing of unknown material	53	0.5
ADMINISTRATION, SOCIAL LIFE, OTHER		
Administration	560	4.9
Social life	324	2.8
Storage of non-food products	342	3

Table 5.27. Activities in the Burnt Village: the combined evidence from the small finds and flint/obsidian artefacts.

With 4.9% administration is well-represented. The relatively large number of sealings and tokens indicate that the storage of goods was a well-organized and regulated activity. The quantification related to 'social life' foremost refers to the categories personal adornment and ritual. The significance of ritual in the Burnt Village will be further explored in chapter 7. As was already indicated in section 2, the storage of non-food products is probably over-represented due to the inclusion of sealings.

The overall impression one gets is that in the Burnt Village the 'normal' (i.e. to be expected) range of domestic activities was carried out. The possible emphasis on butchering, skin-processing and production of leather objects perhaps indicates that animals and pastoralism played an important role in the economy. This, and the apparent importance of administrative practices, will be further elaborated upon in chapter 7.

CHAPTER 6

PUTTING THE PIECES TOGETHER: SPATIAL ANALYSIS

Part I

1 Introduction

This chapter presents the core of this book, i.e. the spatial and contextual analysis of the Burnt Village, which aims at a reconstruction of the function(s) of spatial units. The analysis has been divided into three parts. In part I the spatial analysis is introduced. Part II presents a detailed spatial analysis of the level 6 settlement. It starts with a visual inspection of distribution patterns (section 2). In sections 3 to 6 the functions of the various buildings, divided into rectangular (section 4) and round buildings (section 5), and open areas in the level 6 settlement (section 6) are discussed. In sections 7 and 8 a number of selected spatial units are compared. In part III (section 9) a general spatial analysis of the Halafian level 3A at Sabi Abyad is given. This analysis has been added in order to get a better grip on the processes of abandonment in the level 6 village (see chapter 4, section 4).

In appendix 2 the methodology of the fieldwork at Tell Sabi Abyad and the subsequent methods of analysis of the architecture and stratigraphy have been amplified. This was felt to be necessary because research methods are the first and most basic steps in archaeological interpretative frameworks like the present spatial analysis. In other words: the choices for particular research strategies influence all subsequent interpretations.

1.1 Functional Assessment

In chapter 2 a model (termed *from space to place*, cf. fig. 2.1) has been introduced by means of which the function of prehistoric spatial units can be designated. In this chapter the function of spatial units in the level 6 village is reconstructed according to the various steps of the model: (1) the visual inspection; (2) architectural analysis; (3) depositional analysis; (4) determination of 'object context'; (5) determination of general function; (6) 'functional analysis'. Of step 7, practice is mainly dealt with in this chapter, whereas structure and habitus are given attention in chapter 7. As an example, the functional assessment of room 1 of building I is presented in appendix 4.

It is stressed here that the functional reconstructions should be regarded as 'best-case scenarios', i.e. as the most likely options among a range of possibilities.

1.2 Editorial Remarks

Level 6 has been excavated over an extensive area, i.e. 800 m² (up to 1996). However, it should be taken into account that only buildings II, VI, VII, VIII and IX have been completely excavated (fig. 3.1). Nevertheless large areas of the other buildings have been unearthed, and it is felt that the functional assessments have validity.

Apart from the small finds, not all assemblages of all areas could be studied. The distribution of the small finds gives a truly representative picture of the areas excavated, since all of them have been analysed, and could therefore be included in the study. For the other assemblages the following remarks should be taken into account.

Of the pottery, the Mineral Coarse Ware has been studied from trench P15 only. Clearly this area (9 x 2 m) is too small for spatial analysis, and therefore, regrettably, this ware had to be omitted from the investigation. The other wares are included in the analysis. Analysis of the attributes of these wares, however, has been conducted for some areas only. Of the Standard Fine Ware, Fine Painted Ware and Orange Fine Ware in trench P15 and squares Q15, Q14 and Q13 and small selections from squares R12, R13 and S13, all attributes have been studied (e.g. fig. 6.13). For the other squares (P12, P13, Q12, R11 and S12), however, no information about attributes of these wares is available yet. Of the other wares (Dark-Faced Burnished Ware, Grey-Black Ware, Mineral Coarse Ware and Standard Ware) it is only of the pottery from trench P15 that all attributes have been analysed. In effect, the analysis of the distribution of the pottery will be largely limited to a study of the different wares; only in some instances has it been possible to look at it in more detail, e.g. to investigate the distribution of shape (i.e. jars, pots and bowls).

Of the flint and obsidian artefacts all but those from squares P12, P13, Q12 and R11 have been studied (e.g. fig. 6.15). This means that buildings IX, X, XI and XII had to be excluded from the analysis of the flint/obsidian distribution.

Of the level 6 animal remains only those from the narrow (9 x 2 m) trench P15 have been studied (Cavallo 1996). These remains, therefore, could not be included in the level 6 spatial analysis. In the spatial analysis of level 3A (section 9), however, animal remains have been included.

It should be borne in mind that not all areas in the level 6 settlement were affected by the fire. On fig. 3.2 in chapter 3 the area affected by the fire has been indicated. As can be seen, the fire swept past building XII, the northeastern part of building I, the northern chambers of building IV, tholoi VII, VIII and the western part of tholos IX, and ovens DA, AR, CR, CS, S and DG. Furthermore, the open areas in squares P12-P13, Q12, R11-R13, S12-S13 and Q15 were unaffected. It has already been mentioned that there are indications that building XII may already have been out of use and left to the elements when the other parts of the level 6 village were still in use. Most likely the deposits from this building do not represent in-situ contexts, and they have to be treated differently from the other supposedly in-situ deposits. It seems that the other unburnt structures, as well as the open areas, were, like the burnt buildings, all in use when the fire started. Even if the fire did not directly destroy all structures, there can be little doubt that the village was abandoned after the conflagration. In fact we have no evidence that structures were rebuilt or re-used. Therefore, it is argued that, apart from building XII, level 6 represents a vivid picture of the daily life of a prehistoric community.

Part II

2 Visual Inspection

Now, as the first step of the spatial analysis the results of the visual inspection of artefact distributions within the level 6 settlement are presented. The various assemblages will be discussed separately. It is recalled that the visual inspection should be regarded as a general presence/absence analysis of selected categories of data. Especially the various buildings will be compared. In order to keep the text readable, exact numbers and percentages have been omitted from the text. In appendix 3 the relevant figures per assemblage are presented.

Small finds

If a distribution map of all the small finds (n= 1422) is made it shows that these are generally more or less evenly distributed over the level 6 settlement (fig. 6.1). Closer inspection, however, tells us that they cluster in various areas, i.e. room 6 of building II, room 2 of building IV, rooms 3, 6 and 7 of building V and tholoi VI and IX.

Finds from floor contexts mainly occur in buildings I, II, IV, XI, and tholos VI. Objects recovered from fills cluster in building V. A plot of complete versus broken objects was made but no significant difference between both properties could be noticed. The objects of unbaked clay are largely present in building IV, in room 6 of building II, in rooms 1 and 2 of tholos VI and in tholos IX. In buildings X, XI and XII only a few objects of clay were present. If the distribution of clay objects is compared to that of the stone objects, it appears that buildings X, XI and XII have more stone than clay objects (n= 33 as compared to 10 clay objects in these buildings). Furthermore, stone objects are concentrated in room 6 of building II, room 2 of building IV, rooms 6 and 7 of building V, in tholos IX and in area 3 (an open area) in building I.

Bone tools appear to be largely absent from the open areas (fig. 6.2). The bone tools are concentrated in room 6 of building II, in building V and in tholos VI (fig. 6.2). They are largely absent from buildings IV, X, XI, XII.

Grinding slabs, grinders and pestles are more or less evenly distributed (fig. 6.3), but concentrations can be found in rooms 6, 17 and 18 of building II, room 2 of building IV, room 7 of building V and area 3 of building I. Pestles are found especially in the southern buildings IV and V (fig. 6.3). Mortar fragments were recovered mainly from building II, and it should be noted that, apart from a non-mobile mortar set in the floor of room 5 of building V, they are absent from buildings IV and V. Hammer stones are absent from building II, but in the other areas they seem evenly distributed.

Stone-bowl fragments are present throughout the village. Axes are absent from buildings X, XI and XII and are concentrated in buildings II and IV. Objects of unmodified stones (i.e. 'polishers', grinders, platforms, and so on) were mainly present in buildings II, IV and V; they were not found in building XII, nor in squares R13 and S12.

The sling missiles show an interesting distribution (fig. 6.4). They are wholly absent from buildings X, XI and XII, and only two were recovered from the upper fill of building IV. As can be seen on fig. 6.4, the missiles are clustered in room 6 of building II. In the other areas where the missiles occur (the eastern part of building V, open area 6 east of this building, building I, and tholoi VI and IX), they are more evenly distributed.

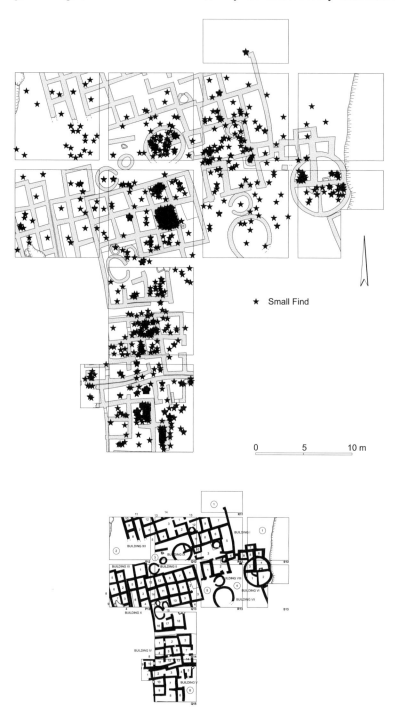

Figure 6.1. Level 6: distribution of small finds.

Spindle whorls are concentrated in room 6 of building II, they are absent from tholoi VII and VIII and buildings X, XI and XII (fig. 6.5). Pierced discs (which are also supposed to have functioned as spindle whorls, cf. appendix 1) are more evenly spread across the settlement (fig. 6.5), although they cluster in building IV. Two were present in building X, but from buildings XI and XII they are, like the spindle whorls, absent.

Objects which represent personal adornment (fig. 6.6) were largely absent from buildings X, XI and XII; one pendant was present in building XI, and one in building XII. The beads are more or less evenly distributed. The labrets cluster in room 6 of building II and in room 2 of building IV. Figurines (human and animal) are absent from buildings I, X and XII, tholoi VII and VIII, but they cluster in room 6 of building II, tholos IX and building V (fig. 6.7).

The distribution of the sealings (fig. 6.8) will be dealt with in detail in sections 4 and 7.1.

Ceramics

Distribution maps of the point locations of Standard Ware, Dark-Faced Burnished Ware, Grey-Black Ware and Standard Fine Ware show that these wares are more or less evenly distributed. However, when maps are made which indicate the quantity of the wares per spatial unit, another picture emerges. The largest numbers of Standard Ware (fig. 6.9) are found in buildings II, IV and V. The other buildings contain far fewer sherds of this ware. Dark-Faced Burnished Ware clusters in two areas, i.e. building IV and in the open area no. 2 in square P12 (fig. 6.10). Grey-Black Ware seems to be concentrated in building IV, the western part of building V, tholos IX and in the open area in the south of square P12 (fig. 6.11). The largest numbers of Standard Fine Ware come from room 6 of building II and from building IV (fig. 6.12). Smaller concentrations can be found in rooms 1 and 2 of tholos VI. From building X only a few Standard Fine Ware sherds have been recovered, whereas they are absent from buildings XI and XII. The distribution of Orange Fine Ware could only be analysed for trench P15 and squares Q13, Q14 and Q15. Nevertheless, an interesting picture emerges when the distribution of this ware is plotted (fig. 6.13): in building II in square Q13 only a few Orange Fine Ware fragments were found, however, in the southern buildings IV and V this ware is relatively well-represented. When the vessel shapes of Orange Fine Ware are analysed, it appears that building IV has jars only. Building V has bowls as well as jars.

The distribution of the (Standard Ware) husking trays indicates small concentrations in the southern wing of rooms of building II (rooms 16 to 18) and in the open area 3 of building I. Building II contained 10 fragments of husking trays. The analysed vessel shapes of Standard Ware are mainly represented by bowls. Furthermore, twelve jars and just one pot have been reported.[36] The largest number of bowls came from building I. Jars

---

[36] Research of the pottery is still in progress. So far, the vessel shape of the Standard Ware has been recorded for a small selection only; of all the excavated level 6 squares (see above) the vessel shape was noted for only 216 sherds (1.1%). Of this sample 123 sherds (0.6%) could be ascribed to buildings or open areas. The remaining sherds could not be used because they came from mixed units (see above).

Like the Orange Fine Ware (see above), the vessel shape of the Dark-Faced Burnished Ware, Grey-Black Ware, Fine Painted Ware and Standard Fine Ware assemblages was analysed only from ceramics from trench P15, and squares Q13 to Q15. Here, too, only a very small number of sherds were assigned to specific vessel shapes. It will be clear, therefore, that the vessel shape analysis presents only a very crude picture of the true situation!

were mainly recovered from buildings II, IV and V. Only one pot was reported from the open area 2. Of the Dark-Faced Burnished Ware only six vessel shapes were ascribed to buildings or open areas. Of the three jars two came from building IV and one from building V. One bowl and two pots also came from building V. Grey-Black Ware bowls (n= 6) are represented by four and two bowls from buildings IV and V respectively. Of the 19 Grey-Back Ware jars, ten were recovered from building IV. Pots were not represented. Of the Fine Painted Ware mainly jars were present (16 out of the 19 recorded vessel shapes). Half of these jars stem from buildings IV and V. Three jars were recovered from building I (two of which from room 6). Of the Standard Fine Ware, finally, 114 vessel shapes were reconstructed. The majority of the shapes is represented by jars. Again, most of the jars were located in buildings IV and V (24 specimens in each structure). From buildings II and III 14 and 12 jars were recorded respectively. Most of the bowls (n= 13) came from building IV. Both buildings II and V gave evidence of five bowls. The only pot was reported from building IV. In general it can be said that, of all wares, jars prevail in the southern buildings IV and V. It should be taken into account, however, that the numbers of bowls are much smaller than the numbers of jars. Pots are very rare. In the open areas only a small number of vessels could be assigned to a specific vessel shape. Open area 6 was best represented in this aspect, and here too jars prevailed (especially Standard Fine Ware jars; n= 8).

Flint/obsidian artefacts

On the whole, the flint/obsidian industry (n= 2555) is more or less randomly distributed over the level 6 village.[37] A distinction in distribution between the flint (85.9%) and obsidian artefacts (14.1%) was not observed. Likewise, plots of complete versus broken tools (blades included) and of tools versus debitage did not indicate any significant patterning.

Cores are mainly present in buildings IV and V, and they are scarce in the open areas (fig. 6.14). The same holds for by-products (e.g. decortification flakes).

Scrapers concentrate in building I, the northern part of building IV and open area 6 (fig. 6.15). Racloirs and raclettes are largely absent from building II (just one racloir in room 17). Pieces with fine retouch concentrate in buildings IV and V. The same holds for burins, borers, side-blow blade flakes, corner-thinned blades and sickles.

Interpretation

The visual inspection of the distribution of artefacts in the level 6 settlement has indicated a number of things. Looking at the level 6 plan, one is struck by the regularity and homogeneity of the architecture. A basic distinction can be made between round and rectangular buildings. The rectangular buildings all seem more or less the same. However, when studying the architecture in more detail, and looking at the distribution of artefacts, a different picture emerges. It seems that a distinction can be made between the various buildings, rectangular as well as round.

---

[37] It should be taken into account that in the flint/obsidian analysis squares P12, P13 and Q12 have not been analysed yet.

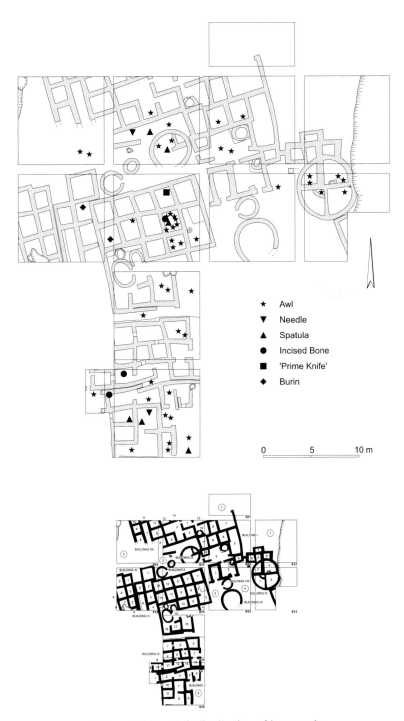

Figure 6.2. Level 6: distribution of bone tools.

Most notably, building II stands out because of the large concentration of small finds in room 6. The majority (ca. 70%) of these finds were broken. It seems as though room 6 functioned as a storage place for broken and damaged objects, most of which representing 'administrative' objects such as sealings and tokens (see also Akkermans and Verhoeven 1995:12-13). It is recalled that two-thirds of the sealings stem from building II (mainly room 6, but also nos. 1 and 7). Apart from rooms 1 and 7, the other rooms in this building contained few or very few artefacts. Hammer stones and racloirs and raclettes are absent from building II. Building II is also different from the other buildings in that it yielded the largest amount of sherds (21.7% of the assemblages from all buildings). Especially Standard Ware is well-represented in building II (22.7% of the assemblages from all buildings). In the western rooms 11, 12 and 14 deposits of charred grain (mainly emmer wheat and some einkorn wheat) were found. In addition some grain was found in room 7 (ibid.:12). In the other level 6 structures no grain was encountered. Room 18 is distinctive for its relatively large number of ground-stone tools (n= 5; grinding slab fragments, grinders, a pestle). Most likely this room 18 was unroofed (Verhoeven and Kranendonk 1996: fig. 2.15). Furthermore, small concentrations of husking-tray sherds were recovered from the southern rooms. Apart from its inventory, building II stands out as the most regular building, basically consisting of three rows of largely identical small square rooms. These three rows are separated from another row of five rooms in the south by an open area (no. 15) with ovens. The building was centrally located on the highest point of the prehistoric tell, and furthermore, it represents the largest excavated building.[38]

On a structural level, building I resembles building II because of the centrally located open area with an oven. In this area 3 a relatively large number of ground-stone tools were found. Furthermore, a small concentration of husking-tray fragments was recovered from area 3. Sling missiles were found in small quantities in the southern and central rooms of building I (fig. 6.4). It is recalled that apart from buildings II, V and VI sling missiles are largely absent from the Burnt Village. Furthermore, a variety of scrapers was recovered from building I, or from the direct vicinity of it (fig. 6.15). Architecturally speaking, building I is distinctive because of the relatively large ovens in rooms 2 and 12. Moreover, the small buttresses marking the entrances to rooms 1 and 2 are not present in any other level 6 structure.

The southern buildings IV and V are spatially segregated from the other buildings, i.e. they were founded about 2 m below the floor level of the other buildings. Like building II, the rooms in these two buildings were arranged in three rows. However, the shape and dimensions of the various rooms is less consistent; buildings IV and V are more irregular than building II (fig. 3.1). Apart from these architectural differences, there are differences between the inventories of these southern structures and those of the other level 6 buildings. In the first place, Orange Fine Ware was present in small numbers in the two buildings, whereas only few sherds appeared in building II.[39] Furthermore, all the other wares are numerically best represented in the two buildings (figs. 6.9-6.12).[40] Flint cores were present most often in buildings IV and V. Of the ground-stone tools, pestles are

[38] Note, however, that none of the other rectangular buildings have been completely excavated yet.
[39] It is recalled, however, that the distribution of Orange Fine Ware was only studied for squares Q13, Q14 and Q15, i.e. for buildings II to V.
[40] However, this may be largely due to the difference in preservation between buildings IV and V and the other level 6 buildings. Buildings IV and V were generally preserved to a height of ca. 1 m (and up to 1.40 m), whereas the other buildings were preserved to heights of ca. 50 cm or less.

concentrated in buildings IV and V. There are also differences between buildings IV and V. Of the (very few) reconstructed Orange Fine Ware vessel shapes, building IV gave evidence of jars only, building V of both jars and bowls. Sling missiles, for example, are largely absent from building IV, whereas they regularly appear in the eastern part of building V (and in the open area 6 east of it). The same holds for bone implements (fig. 6.2). Impressed sealings and tokens are clustered in rooms 6 and 7 of building V. In building IV just one sealing and a few tokens were present (fig. 6.8). Pierced discs and labrets, on the other hand, concentrate in building IV (figs. 6.5 and 6.6).

Buildings X, XI and XII, located in the northwestern part of the settlement (as excavated), are distinguished from the other level 6 buildings by the relative small amount of objects recovered from them (fig. 6.1).[41] In building X, however, a relatively large amount of pottery sherds (n= 1225, as opposed to 299 and 606 in buildings XI and XII) was present. With regard to building XII, however, it should be taken into account that this building was not affected by the fire, and that in-situ contexts are largely absent from this structure. In many instances, artefacts that are present in the other level 6 buildings, are absent or only represented by one or two specimens in buildings X to XII. This holds for e.g. sling missiles, items of personal adornment and bone tools (figs. 6.2, 6.4 and 6.6). It has already been indicated that in buildings X to XII stone objects are much more common than clay objects. A small concentration of ground-stone tools was found in room 6 of building XI (fig. 6.3).

Of the circular buildings, structures VII and VIII stand out for their small amount of objects (fig. 6.1). Tholos VI, however, gave evidence of a small number of each ceramic ware category and of various domestic implements (e.g. bone tools, ground-stone tools), most notably in the largest rooms 1 and 2. Tokens are relatively well-represented in building VI (fig. 6.8). Moreover, tholos VI is remarkable for at least two other reasons: (1) it is the largest tholos; (2) it was subdivided into four compartments, and a number of very small rectangular rooms had been added to the tholos. Tholos IX is also noteworthy, since it was not, like the other tholoi, located on the periphery of the rectangular buildings, but amidst them in an enclosed open area (fig. 3.1 in chapter 3). Moreover, relatively large numbers of finds were recovered from this tholos (fig. 6.1). Figurines are relatively well-represented (n= 6) in tholos IX. Figurines are a rare artefact class, and it is most likely that they served a special, probably ritual, purpose. Perhaps, therefore, tholos IX had some special function.

Summarizing the above we may conclude that this first and general analysis of artefact patterning indicates differences in assemblage composition between the various buildings, suggesting functional differences between them. The uniform architectural structure and layout of the buildings covers a variation in activities and social practices (see also Akkermans and Verhoeven 1995:29). In the following sections a more detailed view of these activities and practices is presented.

---

[41] It is recalled that the flint and obsidian industry from squares P12, P13 and Q12, i.e. from buildings X to XII, has not been analysed.

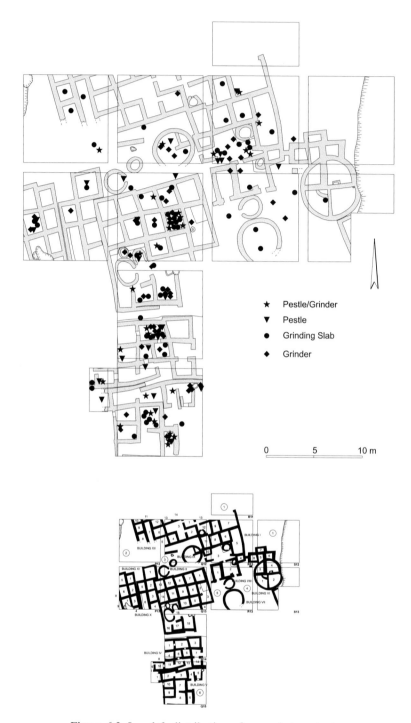

Figure 6.3. Level 6: distribution of ground-stone tools.

3 SPATIAL ANALYSIS

3.1 Introduction

As in previous publications about the Burnt Village (e.g. Akkermans and Verhoeven 1995), a basic distinction has been made between round and rectangular architecture. Within these two main classes the following types of buildings have been reconstructed:

1. Rectangular buildings:
    storage buildings with activity areas
    storage buildings

2. Circular buildings:
    activity areas with storage rooms
    activity area
    houses
    storage buildings

First the rectangular buildings (section 4), then the round buildings (section 5) will be discussed.

The discussion is a combined presentation of both scenarios 1 and 2 as presented in chapter 4, section 4.1. The tables summarizing the functional assessment (e.g. table 6.2) show that the designation according to the two scenarios is generally the same. As was to be expected, the main difference is that in scenario 2 evidence for subsistence activities which are related to the use of ceramics, and evidence for storage in pottery vessels, is largely absent, since the ceramic sherds (representing ca. 99.9% of the ceramic assemblage!) were omitted.

The various comparisons in the analysis below will be qualitative and quantitative.[42] In the quantitative, statistical, analysis similarity coefficients have been computed. Robinson's coefficient of agreement (or similarity) has been chosen because this test is suitable for groups of numerically few items (when divided, the various assemblages in spatial units at Sabi Abyad are rather small numerically). The test measures the degree of similarity between "... defined categories for pairs of archaeological assemblages" (Doran and Hodson 1975:139). In the test the maximal distance or difference between any two units is defined as 200%. "By subtracting any calculated distance or dissimilarity from 200 an equivalent measure of similarity or agreement is achieved" (ibid.:139). As will become clear, the coefficient is especially suitable to express the similarity between the composition of two assemblages, since differences are expressed as percentages. In table 6.1 an example of the calculation of the Degree of Similarity has been presented.

In the present study percentages of selected artefact categories are compared, the difference of presence is computed and this difference is subtracted from 200% to determine the degree of similarity. In formula the procedure is:

---

[42] A complete comparative analysis of all units is not feasible, since this would amount to hundreds of collations! Therefore selections were made, which have been based on the patterns observed in the visual inspection (section 2) and the actual functional assessments (section 3).

$$SR = 200 - \sum_{k=1}^{a} |Pik - Pjk|$$

where the presence of each functional class k in the assemblages i and j is expressed as a percentage P (see also Daviau 1993:66-68). The resulting percentage (from 1% to 100%) indicates the degree of similarity between two spatial units with regard to the compared inventories: the higher the percentage, the higher the degree of similarity. I have grouped the percentages into four classes which indicate the degree of similarity:

1-25%	dissimilar
25-50%	slightly similar
50-75%	quite similar
75-100%	similar

Area	Room 2	Area 3	Difference
Butchering	0.9	0	0.9
Plant-processing	2.4	2	0.4
Grinding	0.9	3	2.1
Baking/ cooking, preparation, deposition	87.2	91	3.8
Food serving	8.5	4	4.5
Total	100%	100%	11.7
Similarity: 200-11.7=188.3 or 94.2%			

Table 6.1. Example of the Degree of Similarity: subsistence activities in level 6 building I.

4 RECTANGULAR BUILDINGS

4.1 Storage Buildings with Activity Areas

BUILDING I (table 6.2)

Activity areas

Rooms 1-2, 7, 12 and open area 3 have been reconstructed as activity areas. From room 7 and its entrance only very few finds were recovered (n= 7); this possible activity area was either kept clean or looted. Another possibility is that in this area only very few artefacts were actually in use. In table 6.3 the various activities in the various rooms have been depicted. As can be seen the various activity areas in building I were multi-functional, giving evidence of subsistence as well as of manufacture/maintenance activities. Room 12, however, seems to have been used for the preparation of food only. Evidence for administration (i.e. record keeping and sealing: sealings and tokens) is lacking; the few tokens recovered from building I (fig. 6.8) were presumably not actively used, but lost.

Room	Function	Activity categories	Activities
1	Activity area (1, 2) A (1, 2), B (1, 2), D (1)	Subsistence (1, 2) Manufacture/maintenance (1, 2) Social life (1, 2)	Pottery: food preparation: deposition; serving? (1) Small finds: grinding (1); spinning (1, 2) Lithics: knapping?; working of plants; working of hard silicious plants; cutting of grain?; working of dry and hard skin (1, 2)
2	Activity area A (1, 2), B (1, 2), D (1)	Subsistence (1, 2) Manufacture/maintenance (1, 2) Social life (1, 2)	Pottery: food preparation: baking/cooking; deposition; serving? (1) Small finds: grinding (1, 2); working of ground-stone tools or flint artefacts (1, 2); spinning (1, 2) Lithics: knapping?; boring; grooving; butchering; working of plants; cutting of grain?; working of hard silicious plants; working of wood; working of skin; working of dry and hard skin (1, 2)
3	Activity area A (1, 2), B (1, 2), D (1)	Subsistence (1, 2) Manufacture/maintenance (1, 2) Social life (1, 2)	Pottery: food preparation: baking/cooking, sun drying, deposition; serving? (1) Small finds: grinding (1, 2); production of textiles and/or basketry and/or leather items (1); production of pigments (1, 2) Lithics: working of plants and cutting of plants; working of hard silicious plants; working of wood; working of skin; cutting of skin; working of dry and hard skin; working and cutting of soft material (1, 2)
4	Storage area G (1, 2), H (1): general storage area	Subsistence (1) Storage of non-food products (1, 2)	Storage of food and/or other products/objects (1, 2)
5	Primary: storage area G (1, 2), I (1): general storage area; Secondary: discard area K (1)	Subsistence (1) Storage of non-food products (1, 2)	Storage of food and/or other products/objects (1, 2)

Table 6.2. Assessment of the function of level 6 building I. The numeral one or two between brackets, (1) (2), indicates the activities according to the two scenarios; the capitals behind activity area, storage area and discard area, e.g. (A), indicate the specifiec type of area (cf. table 4.5); MWA = according to microwear analysis.

Room	Function	Activity categories	Activities
6	Primary: storage area G (1, 2), I (1): general storage area; Secondary: discard area K (1)	Subsistence (1) Storage of non-food products (1, 2)	Storage of food and/or other products/ objects (1, 2)
7	Primary: activity area A (1); Secondary: discard area K (1)	Subsistence (1) Manufacture/maintenance (1)	?
8	Storage area G (1, 2), I (1): general storage area	Subsistence (1, 2) Storage of non-food products (1, 2)	Storage of food and/or other products/ objects (1, 2)
9	Primary: storage area G (1, 2), I (1): general storage area; Secondary: discard area K (1)	Subsistence (1) Storage of non-food products (1, 2)	Storage of food and/or other products/ objects (1, 2)
10	Storage area G (1), I (1): general storage area	Subsistence (1) Storage of non-food products (1)	Storage of food and/or other products/ objects (1)
11	Storage area G (1, 2) I (1): general storage area	Subsistence (1, 2) Storage of non-food products (1)	Storage of food and/or other products/ objects (1, 2)
12	Primary: activity area A, B, D (1); Secondary: discard area K (1)	Subsistence (1)	Pottery: food preparation: baking/cooking; deposition? (1)
General function building I: STORAGE BUILDING WITH ACTIVITY AREAS			

Table 6.2 (continued). Assessment of the function of level 6 building I.

The activity category 'social life' refers to communal activities such as eating and drinking, meeting, the performance of rituals, etc. Sleeping has also been classified in this category, since it is felt that this basic human (non-) activity relates more to social life than to any of the other functional categories. Social life is associated with the amount of available space. Obviously, socializing cannot take place if an area is simply too small to contain a number of people. In rooms 1, 2, 7 and area 3 people may actually have met; room 12, however, is considered too small for socializing.

Food deposition (short term storage of food) and food serving in the various activity areas may be indicated by the fragments of pottery vessels of various wares. Most likely, during the preparation of food in the ovens ceramic vessels were used, and during the work light meals were probably taken in the activity areas.

Evidence for the preparation of food has been found in all the activity areas except for room 7 which was almost devoid of finds, and which seems to have been secondarily used as a discard area. Actually, rooms 2, 12 and area 3 contained ovens in which undoubtedly food was baked or boiled, since these ovens seem wholly unsuitable as pottery kilns.[43] The ovens of building I were domed clay ovens which resemble the *tabuns* such as can still be found in Jordan (Kana'an and Mc Quitty 1994:147; Lancaster and Lancaster 1997:40). *Tabuns* are multi-purpose: apart from bread all sorts of food is prepared in them.

Plant-processing, grinding and the temporary storage of food during processing were activities that were most likely carried out in rooms 1, 2 and area 3. In room 2, moreover, butchering may have taken place, as indicated by a backed knife and a shape-defined element. Perhaps food was served and consumed in rooms 1, 2 and area 3. Food serving seems rather unlikely for the very small room 12.

The oven, activities, serving of food, and presence of 'tableware' may indicate that room 2 served as a kitchen. The open area 3 contained the majority of the sherds in building I, i.e. 29.3%. Especially Standard Ware is well-represented (n= 298, 29.6%). One complete Standard Ware bowl has been recovered. Also Standard Fine Ware and Dark-Faced Burnished Ware are reasonably well-represented (30.6 and 27.6% respectively). It therefore seems that pottery was actively used in area 3, perhaps as containers for the foodstuffs that were apparently prepared in this area. Manufacture and maintenance activities seem to have been carried out in most activity areas. In room 12, however, no such activities were apparently carried out; this room was reserved for the preparation of food in oven AT. Not all the attested manufacturing/maintenance activities were evenly distributed. They were concentrated in area 3, and especially in room 2. Room 1 also gave evidence of such activities. Room 12 seems to have been used for food preparation only, considering the absence of manufacturing/maintenance tools. Evidence for spinning was found in rooms 1 and 2. Boring and grooving seem to have been restricted to room 2. Wooden artefacts were also produced in room 2, and also in area 3. Plant- processing and skin-processing are both assigned to rooms 1, 2 and area 3. The production of textiles and/ or basketry and/or leather items (as indicated by awls) was restricted to the open area 3 within building I. Likewise, evidence of production of pigments, as indicated by some lumps of red ochre, was only found in area 3. So-called social life 'activities', i.e. eating/ drinking and meeting, have been proposed for the relatively large open area 3.

As can be seen in table 6.3, area 3 and room 2, both containing ovens, show the greatest diversity in activities. In both areas, connected by a doorway, more or less the same range of activities was carried out. In order to assess the relation between both areas, the similarity coefficient has been computed.

---

[43] Pottery kilns are either pit kilns, updraft kilns or downdraft kilns (e.g. Rice 1987:158-163); clearly the ovens of building I do not belong to these types.

Area	1	2	3	7	12
Subsistence:					
Butchering	-	x (2)	-	?	-
Plant-processing	x (2)	x (5)	x (7)	?	-
Grinding	x (1)	x (2)	x (10)	?	-
Baking/cooking, preparation, deposition	x (218)	x (190)	x (311)	-	x (45)
Food serving	? (9)	? (12)	? (16)	?	-
Manufacture/maintenance:					
Stone and/or flint artefacts	-	x (1)	-	?	-
Knapping	-	-	-	?	-
Spinning	x (1)	x (1)	-	?	-
Boring	-	x (2)	-	?	-
Grooving	-	x (2)	-	?	-
Wooden artefacts	-	x (1)	x (3)	?	-
Plant-processing	x (2)	x (6)	x (7)	?	-
Skin-processing	x (1)	x (4)	x (9)	?	-
Textiles and/or basketry and/or leather	-	-	x (2)	?	-
Pigments	-	-	x (1)	?	-
Social life:					
Eating and drinking	?	?	x	?	-
Meeting	?	?	x	?	-
Sleeping	-	-	-	-	-

Table 6.3. Comparison of activities in level 6 building I. Key: - = absent, x = present, ? = possibly present. Between brackets the number of artefacts indicating an activity has been given.

Note that for the categories baking/cooking, preparation, deposition and food serving the relatively high numbers are due to the fact that here we are dealing with ceramic sherds. The categories baking/cooking, preparation and deposition could not be separated, since they are indicated by the same artefacts, i.e. sherds of Standard Ware and Dark-Faced Burnished Ware: both wares could have been used for these three types of activities. According to the models of chapter 4 serving is indicated by the other wares: Fine Painted Ware, Standard Fine Ware and Grey-Black Ware. Butchering, plant-processing, wood-processing, skin-processing, boring and grooving are activities indicated by the flint artefacts. The microwear analysis of selected flint artefacts (see chapter 5, section 4) has shown that the majority of flint artefacts were multi-functional. Scrapers, for instance, were used for the working of wood, skin and plants. Due to their multifunctional character the same flint artefacts have therefore often been counted twice or more. Moreover, all artefacts that were, according to the results of the microwear analysis, possibly used for the above activities have been counted. Consequently, the numbers between brackets represent maximum numbers.

It is interesting to note that while both areas are similar with respect to subsistence activities (the coefficient being 94.2%) they are not similar with respect to manufacturing/ maintenance activities (the coefficient being 61.1%). Apparently there was a clear difference between indoor and outdoor manufacturing/maintenance activities. With regard to the subsistence activities, however, it should be taken into account that the high degree of similarity is mainly caused by the ceramic sherds. Grinding was an important activity in area 3 as attested by the various ground-stone tools (fig. 6.3). In room 2, however, only two such tools were found. The oven, the possible serving of food and the presence of 'tableware' seem to indicate that room 2 served as a kitchen.

There seems to be a weak association between floorsurface and the number of different activities, as indicated by the following list:

room 1	(4.3 m²)	6 activities	(242 objects)
room 12	(5 m²)	1 activity	(48 objects)
room 2	(8.7 m²)	12 activities	(227 objects)
area 3	(17.9 m²)	10 activities	(374 objects)

If the number of objects indicates the frequency of activities, then it appears that room 12, most likely a mono-functional activity area, was used occasionally when compared with the other rooms in which many more objects were found. Rooms 1 and area 3 seem to have been used regularly, while room 2 was used most intensively. Room 1 contained a relatively large percentage (20.3%) of all ceramics of building I, perhaps indicating that pottery was regularly used in room 1. It seems that not only the number of activities, but also the frequency is proportionally related to floor surface.

Storage areas

In table 6.4 the seven reconstructed storage areas (rooms 4-6, 8-11) have been compared as to categories of finds: pottery, small finds and flint/obsidian artefacts. The storage areas are general storage rooms, i.e. multi-functional storage areas where various artefacts were stored. Pottery refers to subsistence activities, the small finds to manufac- turing/ maintenance activities and the flint/obsidian implements to both these functional classes. Clear differences between the various storerooms can be noted. First of all, 'rooms' 4 and 10 are marked by the few finds inside them. In the previous section it has been argued that such areas were used for the storage of bulk products, such as grain, since they are considered too small (1.25 x 1 m and 1.15 x 0.90 m respectively) for having contained other products. Only a few pots could have been stored and small objects would have been difficult to retrieve from these narrow spaces. Grain, however, (or perhaps fodder?) could easily have been scooped out. Perhaps these small areas were covered with a clay or wooden lid; the term bin would be more accurate than room. The (few) finds recovered from 'rooms' 4 and 10 were most likely intrusive.

Most of the finds were recovered from room 5. It has to be taken into account, however, that this room originally consisted of two rooms, which were not separated. On the basis of the very restricted size and on the basis of analogy with the other small rooms, the two units of which room 5 consisted were most likely storage units. The stone platform in the northern part (or room) may indicate that this area served a special purpose. From ethnographic accounts we know that platforms in storerooms serve to store items such as bedding, watersacks and cooking utensils. Platforms and benches are constructed in order

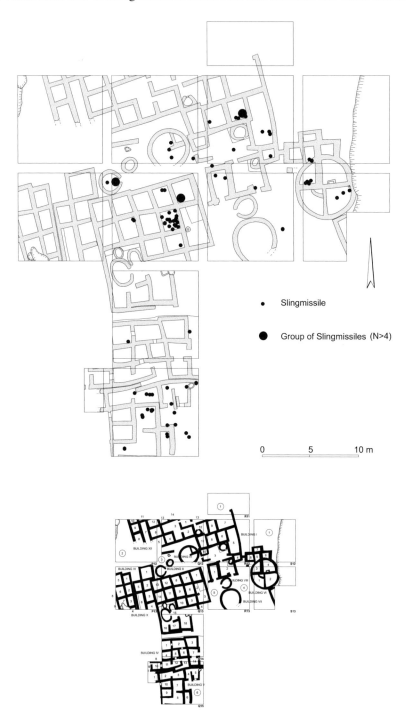

Figure 6.4. Level 6: distribution of sling missiles.

to protect objects from moisture (Krafeld-Daugherty 1994:132). Perhaps the platform in room 5 also served this purpose.

If the numbers given for room 5 are halved, with each half distributed over one unit, then both units seem to be comparable to rooms 6 and 8. In these rooms pottery accounts for the majority of the objects apparently stored. Small finds and flint/obsidian implements appear in limited numbers only. In rooms 9 and 11 no flint/obsidian was found at all, and in both rooms only few small finds were found.

Therefore it appears that in the storage rooms of building I mainly pottery was stored. Most likely these vessels contained food products, but this could not be assessed. It should be taken into account, however, that the pottery in the rooms is represented by sherds (only in room 8 a complete jar, a bowl and a small pot were found[44]); the actual number of vessels is unknown.[45] Small finds and flint/obsidian objects were present in limited numbers only. It appears that mainly subsistence products were stored.[46] Manufacturing/maintenance products were stored in smaller numbers.

Area	4		5		6		8		9		10		11	
Categories	N	%	N	%	N	%	N	%	N	%	N	%	N	%
Pottery	27	71.1	105	84.7	47	75.8	60	85.7	15	93.8	11	91.7	42	97.7
Small finds	1	2.6	7	5.6	5	8.1	4	5.7	1	6.2	1	8.3	1	2.3
Flint/ obsidian tools	1	2.6	5	4	3	4.8	1	1.4	0	0	0	0	0	0
Flint/ obsidian debitage	9	23.7	7	5.6	7	11.3	5	7.1	0	0	0	0	0	0
Total	38	100	124	100	62	100	70	100	16	100	12	100	43	100

Table 6.4. Comparison of storage rooms of level 6 building I.

The degree of similarity of rooms 6, 8, 9 and 11 has been compared. For rooms 6 and 8 a similarity coefficient of 90% has been calculated. As expected, rooms 8 and 11 are less similar, but still the areas are highly comparable: the coefficient is 82%. Rooms 6 and 11 are somewhat less comparable: now the coefficient is 78.1%. Rooms 9 and 11, however, are very similar: a coefficient of 96% has been computed for these areas.

By means of the Chi-squared (X^2) test it has been assessed whether there is an association between pottery wares and function of rooms.[47] First it was tested whether Standard

[44] Room 8 is one of the few rooms in level 6 where complete vessels were recovered. A large red-burnished and impressed bowl was found lying upside down, covering a red-burnished jar with a globular body and a flaring neck. Next to these vessels stood a coarsely finished pot with a low neck (Akkermans and Verhoeven 1995:12, fig. 10, nos. 14-16).

[45] In many instances rim sherds were not present, and moreover the minimum number of vessels could not be established since the radius has not been measured.

[46] Pottery is regarded as a subsistence product.

[47] In X^2 tests the 'null hypothesis' (H0) states that there is no association, while the 'one hypothesis' (H1) states that there is an association. Each time the 5% probability level has been chosen to judge the

Ware and Dark-Faced Burnished Ware, both presumably used for cooking as well as for storage, are associated with storage areas or to activity areas. The calculated X^2 is only 1.36 (df= 2), indicating that there is no significant association between Standard Ware and Dark-Faced Burnished Ware and area. Secondly it was tested whether the 'serving' and 'display' wares (Standard Fine Ware, Fine Painted Ware and Grey-Black Ware) are associated with storage or activity areas. The X^2 of 4.6 (df= 2) indicates that there is no significant association between 'serving' and 'display' ware and area.

The storage rooms were multi-purpose, containing all kinds of different artefacts (e.g. sling missiles, fragments of grinding slabs, etc.) and even flint/obsidian debitage. Of the small finds mainly sling missiles (n= 17) were stored. The other small finds consisted of just one to three artefacts or fragments of artefacts.

The various flint and obsidian tools, too, were each represented by one to two specimens only. The debitage, however, is represented by a few more items, especially flakes (n= 12). Actually, the storage of flint/obsidian debitage is surprising: one would not expect knapping debris to be stored. However, the term 'debitage' is misleading in this respect. As has been indicated in the microwear analysis of flint artefacts, flakes and especially blades seem to have been used for various activities, and can therefore not just be regarded as mere debitage or knapping debris. Most probably at least some, and perhaps a large percentage, of the flakes and blades within assemblages have been used as artefacts. 'Real' knapping debris, i.e. fragments that were not intentionally made as arte-facts, is represented by by-products, fragments, and debris (see Copeland 1996). Therefore the occurence of flakes and blades (and cores) within storage areas is not so surprising. What is peculiar, however, is the occurrence of the 'real' knapping debris indicated above. Why would one store this garbage? Most likely, these objects were not stored, but are present within storage areas due to various kinds of formation processes. These processes have been accounted for in the preceding analysis, and the finds ascribed to rooms were most likely indeed used or stored in these rooms. However, it is inevitable that some mixture of primary and secondary contexts is still present.

Discard areas

As has been indicated above, rooms 5, 6, 7, 9 and 12, of course originally functioning as activity or storage areas, may have been secondarily used as discard areas.

If regarded as discard areas, first of all the generally small amount of debris in the rooms is remarkable (n= 238, range: from 7 to 114 objects, sherds included). As has been indicated above, room 5 should be divided into two rooms and if the amount of finds from room 5 is halved, 57 is the highest number of discarded objects in these rooms. Room 7 in particular contained very few (n= 7) finds only. The large majority of the discarded items consist of sherds (n= 220, 92% of all objects from discard areas). Fragments of small finds and flint/obsidian debris is even absent from rooms 9 and 12, and room 7 only contained two pieces of lithic debris. It seems that the rooms were not used as refuse receptacles for a long time, and moreover that trash was not dumped in them in large amounts.

acceptability of the null hypothesis (Thomas 1986:266). The strength of the association has been expressed as a Cramer's V value (V takes a value between 0 and 1 with values close to 1 indicating a strong relationship; Fletcher and Lock 1994:120).

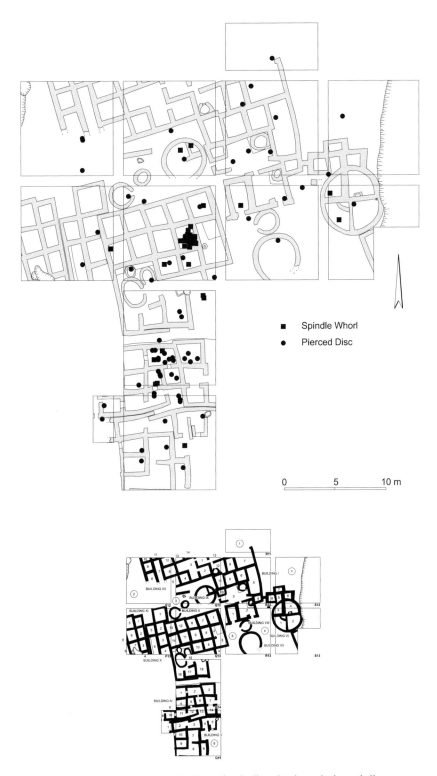

Figure 6.5. Level 6: distribution of spindle whorls and pierced discs.

BUILDING II (table 6.5)

Activity areas

The activity areas of building II are located south of the three rows of storage rooms; they are represented by open area 15 and room 18. Most likely, room 19 (fig. 3.3) also functioned as an activity area, since a mortar was sunk into the floor of this room. However, the inventory of this room (and of room 20) could not be analysed. Furthermore, the roof may have been used as an activity area.

Most likely, the use of open area 15 and room 18 (and 19?) was closely related to the use of ovens (CR, CS and S and DG) in the open area. Rooms 18, 19 and storage rooms 16 and 17 (see below) were all open to the north, i.e. towards open area 15; this may indicate a close functional connection between these rooms and the open area. The various ovens in open area 15 all seem to have been used for the preparation of food, since they seem wholly unsuitable for baking pottery (e.g. no separate combustion and heating chambers have been encountered), and moreover, no pottery wasters or the like have been found in or near the ovens. The four ovens each represent a different type. The large and domed oven S should perhaps be regarded, like the domed ovens of building I, as a *tabun*. Since bread was probably baked in the *tannur* CR, oven S may have been used for roasting meat, and perhaps for cooking food in large ceramic containers. Most likely the horseshoe-shaped 'oven' CS was not a closed construction, but an open hearthplace. In ethnographic context, hearths in the Near East may appear as more or less horseshoe-shaped constructions of clay, on which a vessel or baking plate can be set (e.g. Krafeld-Daugherty 1994:20-81; Van Der Kooij 1976: photo 13). On these 'horseshoe-hearths' bread is baked (on a *saj*), and soup, milk, *burghul*, etc. is cooked. Perhaps, then, food products like soup, milk and *burghul* were prepared on feature CS. Oven CR most likely was a bread oven or *tannur*, used for baking flat breads (Alford and Duguid 1995; Forbes 1958:63-64; Krafeld-Daugherty 1994; Voigt 1983:68, Wulff 1966). It should be mentioned, however, that *tannurs* generally have walls which are thinner than the wall of oven CR (which was ca. 25 cm thick). Oven CR was partly sunk into a clay bench against the façade of building II (fig. 3.1). Most likely this bench served for placing the dough to be baked, as is nowadays common for benches and platforms attached to *tannurs* (Hansen 1976: fig. 30, see also Van Der Kooij 1976: photo 11). Most likely, the hearthplace CS was constructed when oven CR was out of use; feature CS is situated directly in front of the *tannur*, making access to this structure very difficult. The small keyhole-shaped oven DG, finally, was perhaps used for special purposes only. The oven is peculiar in that it has two chambers, which were, however, not very articulated. In general two-chambered ovens are used for baking pottery, but clearly oven DG is much too small for such a procedure. So far, the precise function of oven DG remains enigmatic. Perhaps, though, this small oven was used for the melting of metals: in the northwest of room 19 (fig. 3.3), in the direct vicinity of oven DG, a small piece of copper ore was found, most likely indicating that copper (from Anatolia?) was being processed at Sabi Abyad, perhaps in oven DG! In table 6.6 the activities attested for open area 15, room 18 and the roof have been depicted.

Room	Function	Activity categories	Activities
1	Storage area G (1, 2), I (1): archive	Subsistence (1) Storage of non-food products (1, 2) Administration (1)	Storage of food and/or other products/objects (1, 2)
2	Storage area G (1, 2), I (1): general storage area	Subsistence (1) Storage of non-food products (1, 2)	Storage of food and/or other products/objects (1, 2)
3	Storage area G (1, 2), I (1): general storage area	Subsistence (1) Storage of non-food products (1, 2)	Storage of food and/or other products/objects (1, 2)
5	Storage area G (1, 2), I (1): general storage area	Subsistence (1) Storage of non-food products (1, 2)	Storage of food and/or other products/objects (1, 2)
6	Storage area H (1, 2), I (1): archive	Subsistence (1) Storage of non-food products (1, 2) Administration (1, 2)	Storage of food and/or other products/objects (1, 2)
7	Storage area G (1, 2), I (1): archive	Subsistence (1) Storage of non-food products (1, 2) Administration (1, 2)	Storage of food and/or other products/objects (1, 2)
8	Storage area G (1, 2), I (1): general storage area	Subsistence (1) Storage of non-food products (1, 2)	Storage of food and/or other products/objects (1, 2)
9	Storage area G (1, 2), I (1): general storage area	Subsistence (1) Storage of non-food products (2)	Storage of food and/or other products/objects (1, 2)
10	Storage area G (1, 2), I (1): general storage area	Subsistence (1) Storage of non-food products (1, 2)	Storage of food and/or other products/objects (1, 2)
11	Storage area H (1, 2): grain storage area	Subsistence (1, 2)	Storage of cereals in bulk (1, 2)
12	Storage area G (1, 2), H (1, 2), I (1): grain storage area	Subsistence (1, 2) Storage of non-food products (1, 2)	Storage of cereals in bulk, and storage of other products/objects (1, 2)

Table 6.5. Assessment of the function of level 6 building II.

Room	Function	Activity categories	Activities
13	Storage area G (1, 2), H (1, 2), I (1): grain storage area	Subsistence (1, 2) Storage of non-food products (1, 2)	Storage of cereals in bulk, and storage of other products/objects (1, 2)
14	Storage area H (1, 2): grain storage area	Subsistence (1, 2)	Storage of cereals in bulk (1, 2)
15	Activity area A (1, 2), B (1, 2), D (1), E (1)	Subsistence (1, 2) Social life (1, 2)	Pottery: food preparation: baking/ cooking; deposition; serving? (1) Small finds: grinding (1, 2) Lithics: working of plants (MWA); butchering (MWA); working of skin (MWA); grooving (1, 2)
16	Storage area H (1, 2), I (1): general storage area	Subsistence (1) Storage of non-food products (1, 2)	Storage of food and/or other products/ objects (1, 2) Storage of fuel?
17	Storage area G (1, 2), I (1): general storage area	Subsistence (1) Storage of non-food products (1, 2)	Storage of food and/or other products/ objects (1, 2)
18	Activity area A (1, 2), B (1, 2), D (1)	Subsistence (1, 2) Manufacture/maintenance (1, 2) Storage of food products (1) Social life (1, 2)	Pottery: food preparation: baking cooking; deposition; serving? (1) Small finds: grinding (1, 2); production of textiles and/or basketry and/or leather items (1); spinning (1, 2); knapping (1, 2) Lithics: knapping; butchering; working of plants; working of skin (MWA); working of wood (MWA); cutting of skin; working of dry and hard skin (1, 2)
General function of building II: STORAGE BUILDING WITH ACTIVITY AREAS			

Table 6.5 (continued). Assessment of the function of level 6 building II.

As can be seen in table 6.6, the same range of activities has been attested for open area 15 and room 18. Only spinning and the production of textiles and/or basketry and/or leather items were exclusive to room 18. The reconstructed subsistence activities are: butchering, plant-processing, baking/cooking, food preparation and deposition, grinding and possibly food serving. In open area 15 butchering and plant-processing have been attested on the basis of usewear traces of flint artefacts actually found in that area, and therefore the reconstruction of these activities is very certain!

The activities in open area 15 and room 18 have been compared, and a very high similarity coefficient, 96.1%, was obtained, indicating that the function of area 15 and room 18 was more or less similar. In open area 15, different kinds of food products were prepared in different structures. The artefacts found around the various ovens also point to food preparation. Standard Ware represents 92.3% of the recovered ceramics, most likely indicating that this 'cooking ware' (cf. chapter 5, section 3) was used for the heating (and holding) of food. Furthermore, ground-stone tools, among which two complete grinding slabs, were well-represented, cf. fig. 6.3.[48] These ground-stone implements seem to indicate that the grinding of food (e.g. grain, vegetables, herbs) was a major activity in open area 15. Also the two flint tools recovered, a backed knife and a denticulate, point to food processing (i.e. cutting).

Area	Area 15	Room 18
Subsistence:		
Butchering	MWA (1)	x (1)
Plant-processing	MWA (1)	x (7)
Baking/cooking, preparation, deposition	x (673)	x (276)
Food serving	? (27)	? (11)
Grinding	x (6)	x (6)
Manufacture/maintenance:		
Knapping	-	x (4)
Plant-processing	MWA (1)	x (7)
Skin-processing	MWA (2)	MWA (7)
Spinning	-	x (2)
Textiles and/or basketry and/or leather items	-	x (2)
Social life:		
Meeting	x	x

Table 6.6. Comparison of activities in level 6 building II. Key: - = absent, x = present, ? = possibly present, MWA = according to the microwear analysis of flint artefacts.

In table 6.6 it can be seen that room 18 was a multi-functional activity area where various subsistence and manufacturing/maintenance activities were carried out. As in the neighbouring open area 15, especially grinding activities were important in room 18; 6 out of the 13 small finds were grinding implements (fig. 6.3). Of the ceramic sherds (n= 287) ca. 95% were of Standard Ware. Seven pottery vessels were reconstructed, all of jars (2 of Grey-Black Ware, 1 of Fine Painted Ware and 4 of Standard Fine Ware). As in room 2, the presence of 'fine ware' jars may suggest that precious foodstuffs were kept in room 18.

[48] Grinding slab fragments were found in large numbers, but these are the only complete grinding slabs recovered from prehistoric Sabi Abyad.

Storage areas

The analysis of room function has indicated that the majority (85%) of the rooms of building II seem to have functioned as storage chambers. Three functionally related clusters of rooms have been defined:

1. Grain was stored in bulk in the western rooms 11, 12 and 14 (and in addition some grain was found in room 7); these rooms are designated as *grain storage rooms*;
2. Artefacts that are associated with administration (i.e. tokens and sealings) were stored in the northeastern rooms 1, 6 and 7; these rooms are called *archives*;
3. The other rooms (nos. 2, 3, 4, 5, 8, 9, 10, 13, 16 and 17) gave evidence of few finds only, mostly ground-stone tools; these rooms have been designated *general storage rooms*, i.e. multi-functional storage areas where various artefacts were stored.

In total 3528 objects were recovered from the various rooms. If the finds had been evenly distributed then 235 objects per room should have been recovered. As can be seen in table 6.7, however, the objects are not evenly distributed. If the three functional clusters of rooms defined above are compared, then it appears that they can be arranged according to numbers of objects found in them:

archives:
    n = 2509   (71%)
general storage rooms:
    n = 865   (24.5%)
grain storage rooms:
    n = 154   (4.5%)

The three defined classes of rooms are differentiated according to function and number of objects. Room 6 contained the majority of objects (n= 1652, 41% of the finds from building II (n= 4023)). In all three instances mainly pottery sherds account for the large number of objects.

As can be seen in table 6.8 Standard Ware makes up the largest amount of pottery. Only a few vessel shapes could be reconstructed, mostly from room 6. Of the different wares bowls (n= 17) and, as was to be expected, especially jars were stored (n= 25). Pots were not encountered, but this may be due to the limited sample size. It seems likely that most of the vessels were stored with contents. The nature of the contents, however, remains enigmatic. It is unlikely that grain was stored in them, since no traces of charred grain were found in room 6, and moreover, grain seems to have been stored in bulk in special grain storage rooms. Most likely, other foodstuffs, dry as well as liquid or semi-liquid (e.g. flour, herbs, dried vegetables, dried fruit, milk, yoghurt, etc.; cf. table A.3 in appendix 1) were stored in the vessels. Furthermore, it cannot be excluded that objects were stored in vessels, for example tokens (e.g. Akkermans and Duistermaat 1997:29). Products also seem to have been stored in containers made of perishable material, i.e. baskets and sacks. The abundant presence of the latter type of containers has been proved by the reverse of the numerous sealings found in building II (and in building V). Particularly baskets seem to have been used in large numbers (ibid.:20-21).

Category	Pottery		Small finds		Flint/ obsidian tools		Flint/ obsidian debitage		Total	
Room	N	%	N	%	N	%	N	%	N	%
1	505	18	10	2.1	2	6.7	26	12.5	543	15.4
2	37	1.3	1	0.2	1	3.3	6	2.9	45	1.3
3	75	2.7	2	0.4	3	10	19	9.1	99	2.8
5	20	0.7	1	0.2	1	3.3	0	0	22	0.6
6	1140	40.6	390	82.3	15	50	107	51.4	1652	46.8
7	251	8.9	45	9.5	1	3.3	17	8.2	314	8.9
8	29	1	2	0.4	0	0	3	1.4	34	1
9	52	1.9	3	0.6	2	6.7	5	2.4	62	1.8
10	175	6.2	7	1.5	2	6.7	15	7.2	199	5.6
11	0	0	0	0	0	0	0	0	0	0
12	116	4.1	4	0.8	0	0	10	4.8	130	3.7
13	142	5.1	6	1.3	0	0	0	0	148	4.2
14	24	0.9	0	0	0	0	0	0	24	0.7
16	15	0.5	1	0.2	0	0	0	0	16	0.5
17	227	8.1	2	0.4	3	10	8	3.8	240	6.8
Total	2808	100	474	100	30	100	208	100	3528	100

Table 6.7. Comparison of the storage rooms of level 6 building II.

As has already been said, the numbers of these ceramics differ widely, the archives having the majority of pottery (see table 6.8). In order to assess whether there is a difference in composition of the ceramic assemblages between the archives and the general storage rooms the degree of similarity between both room classes has been measured. The coefficient computed was 92.8%, indicating that both categories of rooms have more or less the same range of ceramic wares. If, however, the various small finds of the archives and the general storage rooms are compared then it appears that in both classes of rooms different artefacts were indeed stored; a similarity coefficient of only 35% was computed. Furthermore, it has been tested whether the range of the flint and obsidian tools differs between both classes of rooms. A coefficient of 58% was found, indicating that the range of lithic tools in archives and general storage rooms was largely, but not wholly, different.

	SW		SFW		FPW		GBW		DFBW		Total	
Room	N	%	N	%	N	%	N	%	N	%	N	%
1	481	18.4	9	10.7	2	10	3	16.7	10	13.7	505	18
2	37	1.4	0	0	0	0	0	0	0	0	37	1.3
3	50	1.9	3	3.6	0	0	1	5.6	21	28.8	75	2.7
4	0	0	0	0	0	0	0	0	0	0	0	0
5	17	0.7	0	0	0	0	0	0	3	4.1	20	0.7
6	1047	40.1	54	64.3	9	45	9	50	21	28.8	1140	40.6
7	244	9.3	3	3.6	0	0	1	5.6	3	4.1	251	8.9
8	27	1	0	0	2	10	0	0	0	0	29	1
9	46	1.8	2	2.4	2	10	0	0	2	2.7	52	1.9
10	165	6.3	3	3.6	0	0	0	0	7	9.6	175	6.2
11	0	0	0	0	0	0	0	0	0	0	0	0
12	113	4.3	1	1.2	0	0	1	5.6	1	1.4	116	4.1
13	135	5.2	4	4.8	1	5	2	11.1	0	0	142	5.1
14	22	0.8	1	1.2	0	0	0	0	1	1.4	24	0.9
16	11	0.4	3	3.6	0	0	0	0	1	1.4	15	0.5
17	218	8.3	1	1.2	4	20	1	5.6	3	4.1	227	8.1
Total	2613	100	84	100	20	100	18	100	73	100	2808	100

Table 6.8. Level 6, building II, storage rooms: ceramic wares.

In table 6.9 the stored objects have been listed. As can be seen, a large variety of different implements, used for subsistence-, manufacture/maintenance- and administrative activities was stored in the various rooms of building II. Large numbers of small finds were stored in building II, but it should be taken into account that the majority of the small finds (70.8 %) stems from room 6. Flint/obsidian tools were stored in smaller numbers. Relatively large numbers of debitage, however, especially flakes, were recovered from the various rooms. As has been argued for the storage rooms of building I, flakes and blades may indeed have been stored.

Now the various rooms within the three functional classes will be discussed in more detail.

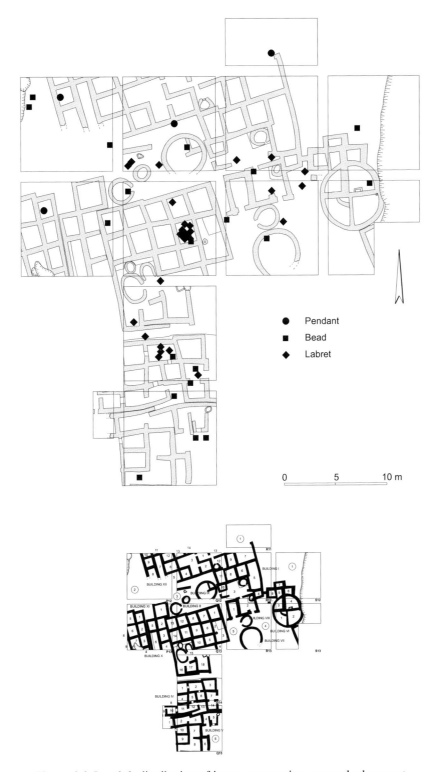

● Pendant

■ Bead

◆ Labret

0   5   10 m

Figure 6.6. Level 6: distribution of items representing personal adornment.

Category	Stored objects
Pottery	Of all wares: bowls, jars (and probably pots)
Small finds	Figurines (17b), awls (9b, 5c), axes (2b, 3c), beads (2c), bone 'burin' (1c), cloth (1b), stone disc (1b), grinders (4b, 10c), grinding slabs (25b), incised bone (1b), sealings (215b, 13c), labrets (5b, 6c), mortars (3b, 3c), pierced discs (3b, 9c), pestles (3c), stone bowls (5b, 1c), sling missiles (11b, 18c), spatula (1b), spindle whorls (8b, 17c), tokens (19b, 71c), polishing/rubbing stones (9b, 7c)
Flint and obsidian tools	Heavy-duty types (2), lustred sickle elements (5), shape-defined element (1), truncations (4), 'Tile Knives' (2), end-scrapers (2), steep-scrapers (2), burin (1), borer (1), notches (4), denticulates (4), pieces with fine and abrupt retouch (12), knife (1), side-blow blade flake (1), denticulate/borer (1), abrupt retouch/notch (1)
Flint and obsidian debitage	Cores (14), by-products (4), flakes (121), blades (24), bladelets (3), segments (21), fragments (79), debris (5)

Table 6.9. Objects stored in the storage rooms of level 6 building II (numbers between brackets, b = broken, c = complete).

(1) Grain storage rooms

Rooms 11, 12 and 14 have been reconstructed as grain storage rooms. In room 11 cereals were stored in bulk as proved by the quantities of charred grain (*Triticum* and *Hordeum*). Only few other finds were recovered from room 11. In room 14 the grain lay almost knee-high and was surrounded and partly covered by a layer of ashy white fibrous material of vegetable origin. The 22 sherds in this room may indicate that apart from grain in bulk other products were also stored in this room. Likewise in the neighbouring room 12, some small finds and a few pieces of flint/obsidian debitage were stored.

(2) Archives

On the basis of the presence of administrative objects, i.e. sealings, tokens and possibly figurines (cf. chapter 5) and the large quantity of small finds (table 6.10), rooms 1, 6 and 7 have been reconstructed as archives (fig. 6.8). Almost certainly the sealings were kept in the archives (see also building V) after they had been removed from the containers (Duistermaat 1994:56): (1) most of the sealings were broken; (2) room 6 of building II (the major archive room) seems too small to have contained the ca. 200 baskets and pottery vessels that were sealed by the sealings found there. Moreover, the one sealing which fitted a stone bowl (Duistermaat 1996: fig. 5.14, no. 1) was found in another room than the bowl.

Apart from the seven sealings recovered (all broken and impressed), a relatively large amount of sherds was recovered from room 1 (18% of the sherds in building II).

Room 6 is remarkable for its large number of small finds (n= 390; fig. 6.1). The best represented items within this assemblage are sealings (n= 201) and tokens (n= 73), i.e. artefacts that are related to administration. Furthermore, spindle whorls (n= 19), sling missiles (n= 18) and human and animal figurines are well-represented (n= 17). Room 6 also stands out for its large number (n= 1140) of sherds (40.6% of all sherds recovered from building II!; table 6.8). Especially the finer ('luxurious') wares, i.e. Standard Fine Ware and Dark-Faced Burnished Ware are well-represented (n being 54 and 21 respectively). The reconstructed vessel shapes indicate bowls as well as jars: two Fine Painted Ware jars, three bowls and six jars of Standard Fine Ware, and four bowls and two jars of Standard Ware. These (and the other non-reconstructed vessels) may have contained food or products other than food, or perhaps they were simply stored by themselves. The flint/ obsidian industry (table 6.10) is mainly represented by debitage (51.4%), especially flakes and fragments (32 and 27 pieces respectively). Three cores were present. Among the tools lustred sickle elements (n= 3), end-scrapers (n= 2) and pieces with fine retouch (n= 5) were most numerous.

Of the sealings in room 6 all but one of the reverse types have been recognized, but the majority of the sealings sealed basketry (90 pieces, 52.9%). Ca. 33% of the sealings were used on pottery. Furthermore plaited mats (n= 3), stone bowls (n= 4), bags (n= 2) and especially ceramic vessels (n= 56, ca. 33%) were sealed. Seventeen sealings were used on unidentified objects. Twenty-two of the 27 different seal designs in use at Sabi Abyad have been attested in room 6.

The second largest number of sealings was found in room 7 (n= 19). The sealings were used to close baskets (n= 7), plaited mats (n= 3) and pottery (n= 5). Seven seal designs have been distinguished in room 7. As in room 6, the sealings are associated with tokens (n= 16). Some charred grain was found in room 7, indicating that grain was also stored here. A relatively large number of Standard Ware sherds were recovered from room 7, and, unlike in room 6, the finer wares are represented by a few sherds only (table 6.8).

Category	Stored objects
Pottery	Standard Ware (1047), Standard Fine Ware (54), Fine Painted ware (9), Grey-Black Ware (9), Dark-Faced Burnished Ware (21)
Small finds	Animal figurines (3b), awls (4b, 5c), axes (1b, 1c), bead (1c), cloth (1b), figurines (human or animal) (3b), grinders (1b, 2c), grinding slabs (9b), human figurines (11b), incised (or notched) bone (1b), labrets (5b, 5c), mortars (2c), pestle/grinders (1b, 3c), pierced discs (1b, 7c), pestle (1c), pierced stone (1c), sealings (201b), stone bowls (2b), sling missiles (5b, 13c), spatula (1b), spindle whorls (7b, 12c), small stand? (1c), tokens (11b, 62c), polishing stones? (4b, 2c)
Flint and obsidian tools	Heavy-duty type (1), lustred sickle elements (3), end-scrapers (2), steep-scraper (1), notch (1), denticulate (1), pieces with fine retouch (5), end-scraper/denticulate (1)
Flint and obsidian debitage	Cores (3), by-products (2), flakes (51), blades (10), segments (11), fragments (27), debris (3)

Table 6.10. Objects stored in room 6 of level 6 building II (numbers between brackets, b = broken, c = complete).

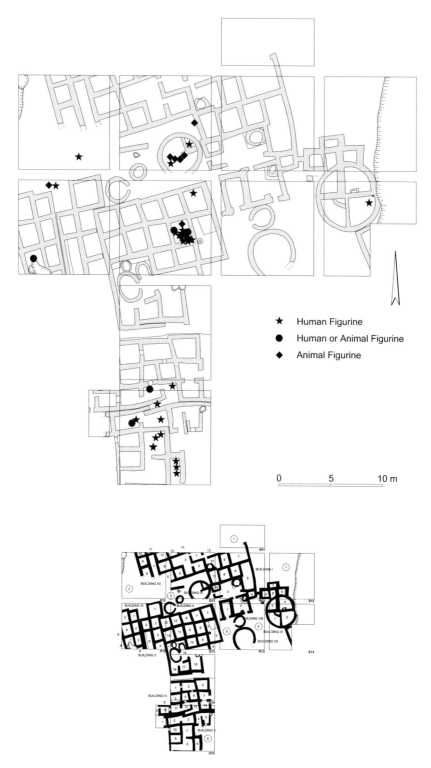

Figure 6.7. Level 6: distribution of figurines.

Summarizing, it seems that in the archives, apart from administrative objects (sealings, tokens, and possibly figurines) large numbers of other artefacts were kept. Presumably the objects were stored on shelves along the walls, since this would be the most convenient method for storing so many different items. It is interesting to note that so many broken objects were present (66% of the small finds, 99% of the pottery), and it could be argued that the archives were in fact dumps. This would indicate that the rooms were out of use when the fire broke out, since it is highly unlikely that the rooms were designed as dumps. Moreover, the clustered appearance of sealings in only three rooms of building II indicates that these objects, broken off containers, were deliberately deposited in certain rooms which were not just trash receptacles. Also the fact that 44% of the small finds in the archives were complete does not substantiate the view that they were dumps. Like the sealings, these various complete objects must still have had a certain value. The broken objects in the rooms were perhaps kept as provisional refuse (cf. chapter 4), awaiting final discard in a dump outside the rooms.

Broken artefacts may indeed have been stored. If so, an economical use of material culture can be proposed: when broken, artefacts were apparently not immediately discarded, but kept among complete tools in storage areas. It is furthermore not unlikely that broken tools were kept to be re-shaped into new or other artefacts. Actually, the temporary storage of discard (as provisional refuse) is not as strange as it may seem at first sight. In our own kitchens, for example, the waste of food preparation, etc. is deposited in a bin (as provisional refuse), which is regularly emptied, the debris becoming final discard when it is dumped on a garbage heap (e.g. Rathje and Murphy 1992). Perhaps broken objects were kept in storage rooms in anticipation of future re-use, and became final discard after a long cycle of use and re-use only. How about the large numbers of sherds? A number of possible explanations can be offered:

First, the large numbers of ceramic fragments (cf. table 6.8) might indicate the provisional storage of sherds, to be used as temper for new vessels. However, the temper of the Transitional pottery is characterized by plant inclusions and mineral inclusions, and sherds do not seem to have been regularly used as temper (Le Mière and Nieuwenhuyse 1996:149-150).

Second, it may be suggested that the sherds in the archives are intrusive or were dumped there, and have no relation to the function of the archives. However, the association between the occurrence of the large numbers of sherds, sealings and tokens and other small finds in only three rooms of building II is conspicuous. Why were the largest numbers of sherds recovered from building II spatially related to sealings and tokens? As was argued above, it is not likely that the archives were dumps for secondary refuse, considering the complete objects and the improbability of constructing rooms especially for dumping debris.

Third (although as yet the number of vessels cannot be reconstructed; cf. chapter 4), it seems reasonable to assume that the number of sherds gives an indication of the number of vessels originally present in the rooms. It could be argued, therefore, that the many sherds (1140 sherds in room 6, 505 in room 1, and 251 in room 7) demonstrate that many vessels were present in the archives. However, the rooms are very small (3.40 m^2 on average), and would have had no place for large numbers of vessels. For example, in the case of room 6, if a vessel would be represented by, say, 25 sherds ca. 46 vessels must have been present, implying ca. 14 vessels per square metre! Moreover, if so many complete ceramic vessels were present a high degree of reconstructability would be expected, and this was certainly not the case.

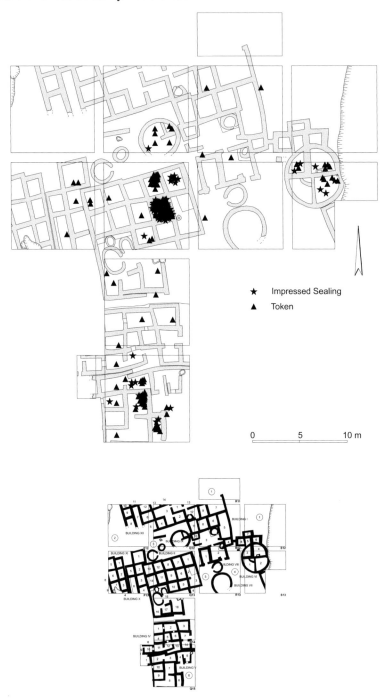

Figure 6.8. Level 6: distribution of sealings and tokens.

Considering all this, I suggest that the large number of sherds cannot be explained by just one factor, but are the result of various processes. Probably the sherds partly represent vessels stored in the areas, dumped refuse, etc. (see chapter 4, section 4 and section 9.6. in this chapter).

(3) General storage rooms

The majority of rooms (nos. 2-5, 8-10, 13, 16 and 17) have been reconstructed as general storage rooms. In room 2 the pivot-stone indicates that this room could be closed by a (wooden) door. None of the other small and rectangular compartments have shown such pivot-stones. This door seems to indicate that room 2 was a special area. In fact this is reinforced by the presence of a low platform. As has been suggested for room 5 of building I, perhaps the platform served to protect stored objects from moisture. If so, then it follows that food (being susceptible to decay) was stored on the platform presumably in baskets or sacks. Furthermore room 2 is the only room of building II where, leaving one Dark-Faced Burnished Ware fragment aside, only (a few) Standard Ware sherds were found.

Room 5 gave evidence of a relatively large number of Dark-Faced Burnished Ware sherds (n= 21, 30.4%). Perhaps the user of this room could dispose of this special 'luxury ware'. One Standard Fine Ware Jar could be reconstructed, maybe indicating food storage.

Rooms 1, 2, 8, 9 and 13 each contained few finds only. The relatively large amount of Standard Ware sherds (n= 135) in room 13 perhaps indicates storage of food in vessels.

The southern rooms 16 and 17 were most likely used for the storage of various goods. Room 16 has been reconstructed as a small storage area where perhaps fuel for the ovens in open area 15 was stored, since artefacts were very scarce in this area. The small room 17 contained 227 sherds, 96% of which are Standard Ware fragments. The reconstructed vessels are represented by three jars (of the 'fine wares': Fine Painted Ware, Standard Fine Ware and Orange Fine Ware), and one Orange Fine Ware bowl. Furthermore three flint tools and three ground-stone tools seem to have been stored in room 17.

BUILDING IV (table 6.11)

Activity areas

As can be seen in table 6.12, by far the largest number of finds (n= 1573) have been ascribed to the roof. This was done on the basis of their position high (e.g. 1 m) in the fill, and moreover, a number of objects were found amidst and above the charred remains of roof beams (Verhoeven and Kranendonk 1996:44). Of the other areas of building IV room 2 contained the second largest number (after room 1) of objects (n= 380, 9.3% of all finds from building IV). Especially small finds and flint/obsidian implements are well represented (n= 115). In table 6.13 the various activities apparently carried out in room 2 and on the roof, and the number of objects indicating these activities, have been listed.

Already during excavation it became clear that ground-stone tools were a major characteristic of room 2. Sixteen such implements were found (fig. 6.3 and see fig. 6 in Akkermans and Verhoeven 1996). Most likely these implements were used for the grinding of foodstuffs (e.g. grain, herbs), but perhaps also for the pulverizing of other materials such as ochre.[49] Furthermore, a relatively large number of other stone objects was recovered: a hammer, a fragment of a possible axe, a macehead, a stone disc, a fragment of a platform and three complete platforms. Probably these objects were used for the production and maintenance of other artefacts. Most likely, other stone objects were also produced in room 2: the single bead and the four labrets found (in a finished state) may have been produced with the help of a grooved stone, which was also found in the room. Flint knapping is indicated by four flint cores, one of which is a large flint nodule. Therefore the use and production of stone objects seems to have been a major activity in room 2. However, it cannot be excluded that all these stone objects were stored in room 2; perhaps it functioned as a sort of communal storage chamber, especially for stone tools.

Room	Function	Activity categories	Activities
1	Storage area G (1, 2), I (1): general storage area	Subsistence (1) Storage of non-food products (1, 2)	Storage of food and/or other products/objects (1, 2)
2	Activity area A (1, 2), B (1, 2), D (1)	Subsistence (1, 2) Manufacture/mainte- nance (1, 2) Social life (1, 2)	Pottery: food preparation: preparation; deposition; grinding; serving? (1, 2) Small finds: production of stone objects/knapping (1, 2); production of wooden objects (1, 2); spinning (1, 2); pottery manufacture (1, 2) Lithics: knapping; working of plants; cutting of plants; working of wood (MWA); working of hard silicious plants (MWA); working of soft material; working of skin (1, 2)
3	Storage area G (1, 2), I (1): general storage area	Subsistence (1) Storage of non-food products (1, 2)	Storage of food and/or other products/objects (1, 2)
4	Storage area G (1, 2), I (1): general storage area	Subsistence (1) Storage of non-food products (1, 2)	Storage of food and/or other products/objects (1, 2)
5	Storage area G (1, 2), I (1): general storage area	Subsistence (1) Storage of non-food products (1, 2)	Storage of food and/or other products/objects (1, 2)
6	Storage area G (1, 2), H (2), I (1): general storage area	Subsistence (1, 2) Storage of non-food products (1, 2)	Storage of food and other products/objects (1, 2) Eight potstands

Table 6.11. Assessment of the function of level 6 building IV.

[49] On a mortar and a few pestles found in Sabi Abyad (not in building IV) traces of red ochre have been found.

Room	Function	Activity categories	Activities
7	Storage area G (1, 2), I (1): general storage area	Subsistence (1) Storage of non-food products (1, 2)	Storage of food and/or other products/objects (1, 2)
10	Storage area? G (1, 2), I (1): general storage area	Subsistence (1) Storage of non-food products (2)	Storage of food and/or other products/objects (1, 2)
11	Storage area G (1, 2), I (1): general storage area	Subsistence (1) Storage of non-food products (1, 2)	Storage of food and/or other products/objects (1, 2)
12	Storage area G (1, 2), I (1): general storage area	Subsistence (1) storage of non-food products (1, 2)	Storage of food and/or other products/objects (1, 2)
13	Storage area G (1, 2), I (1): general storage area	Subsistence (1) Storage of non-food products (1, 2)	Storage of food and/or other products/objects (1, 2)
14	Storage area G (1, 2), I (1): general storage area	Subsistence (1) Storage of non-food products (1, 2)	Storage of food and/or other products/objects (1, 2)
Roof	Activity area? A (1, 2), B (1, 2), D (1)	Subsistence (1, 2) Sanufacture/maintenance (1, 2) Administration (1) Social life (1, 2)	Pottery: food preparation: deposition (1) Small finds: grinding (1); production of textiles and/or basketry and/or leather items (1, 2); working of wood and/or bone and/or stone and/or hide (1, 2); spinning (1, 2) Lithics: knapping; butchering; working and cutting of plants; working of silicious plants; working and cutting of skin; working of dry and hard skin; working of soft material; wood-working (1, 2)
General function of building IV: STORAGE BUILDING WITH AN ACTIVITY AREA			

Table 6.11 (continued). Assessment of the function of level 6 building IV.

Apart from grinding, the subsistence activities in room 2 seem to have consisted of plant-processing (i.e cutting of plants other than grain), food preparation and possibly food serving. Food preparation is indicated by the find of a complete strainer, indicating the use of semi-liquid products in room 2. After room 1, room 2 contained the largest number of sherds (n= 265); vessels seem to have been present in reasonable numbers in room 2. Perhaps they were used for (temporarily) storing the prepared foodstuffs. The vessels that could be reconstructed were: a Standard Fine Ware bowl, a Standard Fine Ware jar, a Fine Painted Ware bowl, a Grey-Black Ware bowl and two Grey-Black Ware jars. Especially Dark-Faced Burnished Ware is well-represented (n= 33, 35% of this ware in building IV).

Besides the production of stone tools, manufacturing/maintenance activities consisted of the processing of wood, plants and skins, spinning and pottery manufacture. This last activity is indicated by eight more or less circular loamers made of large sherds. They were found clustered in the southeast corner of room 2. These loamers were made of bodysherds of Coarse Ware vessels. They were ca. 10 cm in diameter and their sides were smoothed. Presumably these loamers were used to smoothen the outer surfaces of 'leather-hard' ceramic vessels.

Like the other activity areas of level 6, room 2 was multi-functional, but in two respects the room stands out among the other activity areas of the level 6 village: the room is the largest interior area (measuring 2.65 x 1.50 m), and it is marked by a relatively large number and variety of stone objects. Another difference is that room 2 was accessible through two portholes. In other rooms with such openings (in buildings IV and V) only one porthole was present.

Room	Pottery N	Pottery %	Small finds N	Small finds %	Flint/obsidian tools N	Flint/obsidian tools %	Flint/obsidian debitage N	Flint/obsidian debitage %	Total N	Total %
1	398	13.7	0	0	16	14	41	8.4	455	11.2
2	265	9.1	46	38.3	16	14	53	10.9	380	9.3
3	88	3	1	0.8	5	4.4	16	3.3	110	2.7
4	53	1.8	2	1.7	2	1.8	5	1	59	1.4
5	137	4.7	4	3.3	5	4.4	18	3.7	164	4
6	181	6.2	15	12.5	4	3.5	22	4.5	222	5.5
7	100	3.5	8	6.7	4	3.5	14	2.9	126	3.1
9	182	6.3	0	0	1	0.9	4	0.8	187	4.6
10	141	4.9	1	0.8	2	1.8	8	1.6	152	3.7
11	90	3.1	3	2.5	3	2.6	10	2.1	106	2.6
12	18	0.6	3	2.5	0	0	0	0	21	0.5
13	27	0.9	7	5.8	1	0.9	5	1	40	1
14	18	0.6	1	0.8	1	0.9	3	0.6	23	0.6
Roof	1199	41.4	32	26.7	54	47.4	288	59.1	1573	38.6
Total	2897	100	120	100	114	100	487	100	4073	100

Table 6.12. Number of objects in the rooms of level 6 building IV.

The largest number of artefacts connected with building IV (n= 1573) has been assigned to the roof. It has already been mentioned that post-depositional factors may account for the occurrence of so many items on the roof, but most likely the roof was indeed a locus of many activities. The reconstructed subsistence activities on the roof consisted of butchering, plant-processing, baking/cooking, food preparation, food deposition, food serving and grinding (table 6.13 and see Verhoeven and Kranendonk 1996 fig. 2.15). The large number of sherds (n= 1199), especially Standard Ware (80.4%), may indicate that the preparation of food was one of the most important activities on the roof. As in the other activity areas without ovens and hearths baking/cooking of food is unlikely, but one is reminded that portable braziers may have been used for these purposes (Hansen 1976:23; Jacobs 1979:188; Vossen and Ebert 1986).

Apart from subsistence activities various manufacturing/maintenance activities have been attested. First of all flint knapping may have been a regular activity on the roof, since 12 flint cores were recovered. The other flint implements, six borers and one burin, indicate boring and grooving, most probably of wood, but perhaps also of bone, stone and shell. Wood-working seems to be demonstrated by a relatively large number of flint tools. Spinning and the manufacture of textiles and/or basketry and/or leather items are most likely indicated by, respectively, a spindle whorl and eight pierced discs, and two bone awls.

When the two activity areas of building IV are compared (table 6.13) it becomes clear that a more or less similar range of activities was carried out in both areas. Some activities, however, were exclusive to one of the areas: butchering, boring, grooving and the manufacture of textiles and/or basketry and/or leather items only occurred on the roof, whereas the production/maintenance of (stone) tools and the smoothening of pottery was done in room 2 only. A similarity coefficient of 86.5% was obtained for both areas, pointing out that the range (and intensity) of activities was indeed comparable, the main difference being the absence of stone tool production (other than flint) on the roof.

Storage areas

Rooms 1 and 3 to 14 of building IV have been reconstructed as storage rooms. In building II three types of such rooms could be distinguished, i.e. grain storage rooms, archives and general storage rooms. Grain storage rooms and archives seem absent from building IV; the rooms seem to have functioned as general storage rooms, used for storing pottery, small finds, and flint/obsidian artefacts. In order to detect patterning within this class of rooms a simple nearest neighbour test has been carried out. In the test the composition of the pottery assemblages in the various storage rooms has been compared. The percentages of Standard Ware, Standard Fine Ware, Fine Painted Ware, Grey-Black Ware and Dark-Faced Burnished Ware were compared by means of similarity coefficients, and then rooms were grouped into classes. The following results were obtained.

In general the composition of the pottery assemblages in the rooms is highly com–parable. In all instances Standard Ware accounts for the largest number of sherds; taken together a mean of 89.3 % for all storage areas was obtained (range: 78-99.3%). The other wares make up 11% of the assemblages in the storage areas at the most. This pattern holds for the other level 6 buildings as well. Apparently the composition of the pottery assemblages in the rooms was more or less the same. This points to a uniformity in room use, a topic that will be elaborated upon later (chapter 7). However, - slight - differences do exist in the pottery assemblages from the storage rooms. The percentages expressing the differences can be divided into 'nearest neighbour classes', i.e. into categories which express the degree of similarity (the lower the percentage, the higher the degree of similarity). However, when the rooms are divided into equal classes, and when the number of occurrences of each room within these classes is counted, no clear pattern emerges.[50] Apart from rooms 3 and 14 all rooms closely resemble each other with regard to the composition of their pottery assemblages. Rooms 3 and 14 show rather high percentages (up to 43%), indicating that their pottery inventories differ from those of the other rooms.

The few reconstructed vessel shapes suggest that of all wares mainly jars were stored (n= 32). Bowls, however, were also stored regularly, as indicated by 21 reconstructed vessels (39% of the reconstructed vessel shapes).

A most interesting discovery in room 6 were eight sawn-off jar necks (Standard Fine Ware: Le Mière and Nieuwenhuyse 1996:163) which were standing upside-down (along the walls in the southeast corner). Most likely they were used as potstands. This seems to indicate that vessels with round bodies, most likely jars with fluid or semi-fluid contents, were stored here.

The number of small finds in the storage rooms is very limited: the largest number is 15, found in room 6. Obviously, the storage of small finds was not a regular or major activity in building IV. Actually this also holds for the storage rooms in the other buildings: apart from the archives (in buildings II and V) only small numbers of small finds were recovered from individual rooms. Apparently small finds were stored in limited numbers only. Pierced discs and polishing/rubbing stones, represented by five objects each, were the most common items. The other artefacts are represented by one to three items at the most. A relatively large variety of flint/obsidian artefacts (19 different implements) was stored, each in small numbers only. Pieces with fine retouch and abrupt retouch (n= 10) and borers (n= 6) were the best-represented. Of the debitage, flakes and fragments were most numerous.

If the total number of finds from the various storage rooms are compared (see table 6.12), three classes of rooms appear. The first class consists of rooms with few finds only: rooms 4, 12, 13 and 14. Rooms 12 to 14 are located in a row next to each other. The second class, representing the majority of rooms (nos. 3 and 5 to 11), is represented by rooms which contain 106 (room 11) to 222 finds (room 6). The third class is represented by one room with a relatively large amount of objects: room 1 with 455 finds. No small finds were recovered from room 1; this area seems to have been used for storing relatively large numbers of pottery (398 sherds, 23.4% of all sherds in building IV) and flint/obsidian artefacts only. Especially Dark-Faced Burnished Ware is well-represented (20 fragments; 21% of all Dark-Faced Burnished Ware sherds in building IV), perhaps indicating storage of imported goods. Small finds were also absent from room 9, but perhaps this is due to

---

[50] This 'nearest neighbour analysis' expresses the distance between rooms with regard to pottery assemblages. Consequently, when divided into classes (of percentages expressing similarity), rooms can occur in different classes. Room 1, for instance, has its nearest neighbours in class I, but also some in class II. However, by counting and comparing the number of occurrences of rooms within the classes, rooms can be assigned to a particular class. Room 1, for instance, occurs eight times in class I, only two times in class II, and it is absent from class III; room I, therefore, is more closely related to class I than to the other classes.

the incomplete exposure of this area. From room 12 no flint/obsidian objects at all were recovered; obviously these items were not stored here.

Another way of classifying the rooms of building IV is to assess which rooms are linked by openings. Then the following pattern appears:

1. Rooms linked by portholes:

a) 1+2, 2+6
b) 5+12
c) 10+11

2. Rooms linked by openings at floorlevel:

a) 11+12, 12+13
b) 7+14
c) 9+10

3. Rooms which are not connected to other rooms:

   3, 4

The rooms which were connected were obviously related, perhaps functionally, or maybe they were used by specific persons or groups. The rooms connected by the portholes and the rooms with no openings in their walls were probably accessible via the roof.[51] Two 'chains' of rooms can be detected: (1) rooms 11, 12 and 13 were linked directly by openings at floorlevel (and to rooms 5 and 10 by a porthole); (2) rooms 1, 2 and 6 were connected by portholes. On the basis of the numbers of small finds, it has already been suggested above that rooms 12 and 13 may have been related.

Discard area

As has been indicated above, it is likely that the material from room 10 represents secondary refuse. Apart from a single Fine Painted Ware sherd, all 'fine' wares are absent from room 10; apart from this sherd the pottery is represented by 140 Standard Ware sherds. Besides sherds, two tools and eight pieces of flint/obsidian debitage were recovered from room 10.

---

[51] The walls of building IV (and V) were among the best-preserved in the Burnt Village, and were standing up to heights of 1.40 m; surely entrances in these walls would have been detected. Absence of openings in these walls therefore indicates a true absence of such openings, and consequently suggests openings in the roof to gain entrance.

Area	Area 15	Room 18
Subsistence:		
Butchering	-	x (5)
Plant-processing	MWA (11)	x (55)
Baking/cooking, preparation, deposition	x (251)	x (1058)
Food serving	? (14)	? (141)
Grinding	x (16)	x (3)
Manufacture/maintenance:		
Stone and/or flint artefacts	x (15)	-
Knapping	x (4)	x (12)
Boring	-	x (6)
Grooving	-	x (1)
Wood-processing	MWA (8)	x (20)
Plant-processing	MWA (11)	x (55)
Skin-processing	x (14)	x (51)
Spinning	x (8)	x (9)
Textiles and/or basketry and/or leather items	-	x (2)
Pottery	x (8)	-
Social life:		
Meeting	x	x

Table 6.13. Comparison of activities in level 6 building IV. Key: - = absent, x = present, ? = possibly present, MWA = according to the microwear analysis of flint artefacts.

BUILDING V (table 6.14)

Activity areas

In rooms 5 and 8, both reconstructed as activity areas, respectively 108 and 152 finds were present, accounting for 3.6 and 5.1% of all finds ascribed to building V (table 6.15). The majority of the objects from the rooms are sherds; as can be seen in table 6.15, only few small finds and flint/obsidian objects were recovered from these rooms. Objects high up in the fill have been assigned to the roof (38% of all finds from building V came from this area). In a number of instances objects, most notably ground-stone tools, have been found above the burnt roof remains (i.e. baulks, and pieces of clay with impressions of poles and reeds). In table 6.16 the reconstructed activities in these three activity areas are listed, viz. plant-processing, food preparation, and possibly food serving and grinding. Knapping, plant, wood and skin-processing, boring, spinning and the production of textiles and/or basketry and/or leather items (with awls) are the reconstructed manufacturing/maintenance activities.

In room 5 a *tannur* was present, most likely indicating the baking of flat loaves. In the floor of this room a large boulder mortar had been sunk, presumably for grinding food (Verhoeven and Kranendonk 1996: fig. 2.12). Apart from the large mortar three grinders were recovered from room 5. Clearly these fixed features and the grinders indicate that room 5 served as a food preparation area or kitchen. Compared with the two other activity areas, manufacturing/ maintenance activities (i.e. plant-processing and skin-processing) are not well-attested in room 5. This seems to indicate that room 5 was a specialized activity area.

Considering its size, the relatively large room 8 (4.50 x 3 m) was most likely used as an activity area. However, only seven small finds and 14 Standard Fine Ware sherds were recovered from this area. Perhaps the small amount of finds indicates that room 8 was not intensively used, or that it was kept clean. Perhaps liquids to be served were contained in the five reconstructable Standard Fine Ware jars.

When compared to room 5, room 8 was more multi-functional, giving evidence of plant-processing, possibly food serving, wood and skin-processing, boring and the production of textiles and/or basketry and/or leather items. Evidence for grinding activities was absent from room 8.

The artefacts assigned to the roof suggest that this area was used for the regular performance of different activities. Flint knapping and spinning were activities that were only carried out on the roof: in rooms 5 and 8 these activities were not attested. Given the larger number of artefacts, the other activities were carried out more regularly than in rooms 5 and 8. Grinding in particular must have been an important task on the roof: 25 objects (grinders, pestles and fragments of grinding slabs) indicated this activity. Furthermore, the processing of plants, wood and skins seem to have been carried out on the roof. Of the ceramics especially Standard Ware sherds were well-represented on the roof (n= 690, 84.1%); as has been argued (chapter 5, section 3), this ware may have served principally for the preparation and storage of food. Most likely the Standard Ware and the other wares represented were used in association with the food preparation activities. The relatively large number of jars may indicate that the processed foods were deposited in jars, which, most likely, were eventually to be stored in the small rooms.

In addition to this possible domestic use, it seems that some sort of ritual activity was practised on the roof of building V, as indicated by eleven large and rather curiously-shaped clay objects surrounding two skeletons. In chapter 7 (section 5) this possible ritual activity on the roof of building V is discussed in detail.

Room	Function	Activity categories	Activities
1	Storage area G (1, 2), I (1): general storage area	Subsistence (1) Storage of non-food products (1, 2)	Storage of food and/or other products/objects (1, 2)
2	Storage area G (1, 2), I (1): general storage area activity area? A, B, D	Subsistence (1) Storage of non-food products (1, 2)	Pottery: food storage/deposition; storage of non-food products/objects; food preparation: deposition; serving? (1) Lithics: knapping; boring; working of silicious plants; working of wood; working of skin (1, 2)
3	Storage area G (1, 2), I (1): archive activity area? A, B, D	Subsistence (1) Storage of non-food products (1, 2) Administration (1, 2)	Pottery: food storage/deposition; food preparation: deposition; serving? (1) Small finds: grinding (1); production of textiles and/or basketry and/or leather items (1); spinning (1, 2); administration? (1, 2) Lithics: knapping; grooving; butchering; working of plants; working of silicious plants; working of hard silicious plants; cutting of grain?; working of wood; working of skin; working of dry and hard skin (1, 2)
4	Storage area G (1, 2), I (1): general storage area	Subsistence (1) Storage of non-food products (1, 2)	Storage of food and/or other products/objects (1, 2)
5	Activity area A (1, 2), B (1, 2), C (1, 2), D (1)	Subsistence (1, 2) Storage of non-food products (1, 2) Administration ? (1)	Pottery: food preparation: baking; deposition; serving (1) Small finds: grinding (1, 2); working of wood and/or bone and/or stone and/or hide (1); spinning (1, 2) Lithics: knapping?; working of plants; working of dry and hard skin (1, 2) mortar sunk into floor
6	Storage area G (1, 2), H (1, 2), I (1): archive	Subsistence (1) Storage of non-food products (1, 2) Administration (1, 2)	Storage of food and/or other products/objects (1, 2)
7	Storage area F (1, 2), G (1, 2), H (1, 2), I (1): archive activity area? A, B, D	Subsistence (1) Storage of non-food products (1, 2) Administration (1, 2)	Pottery: food storage/deposition; storage of non-food products (1) Small finds: food preparation: grinding (1, 2); production of textiles and/or basketry and/or leather items (1, 2); spinning (1, 2); knapping (1, 2); working of wood and/or bone and/or stone and/or hide (1, 2) Lithics: butchering; working and cutting of plants; working of wood; working of skin; working of dry and hard skin; working of soft material (1, 2)

Table 6.14. Assessment of the function of level 6 building V.

Room	Function	Activity categories	Activities
8	Activity area A (1, 2), B (1, 2), D (1)	Subsistence (, 21) Manufacture/maintenance (1, 2) Storage of non-food products (1, 2) Social life (1, 2)	Pottery: storage of non-food products?; food preparation: deposition; serving? (1) Small finds: production of textiles and/or basketry and/or leather items (1) Lithics: boring (MWA); knapping?; working of silicious plants; working of wood; working of skin; working of dry and hard skin (1, 2)
9	Storage area G (1, 2), I (1): general storage area	Subsistence (1) Storage of non-food products (1, 2)	Storage of food and/or other products/objects (1, 2)
10	Storage area G (1, 2), I (1): general storage area	Subsistence (1) Storage of non-food products (1, 2)	Storage of food and/or other products/objects (1, 2)
Roof	Activity area A (1, 2), B (1, 2), D (1)	Subsistence (1) Manufacture/maintenance (1, 2) Social life (1, 2)	Pottery: food preparation: food deposition (1) Small finds: grinding (1, 2); production of textiles and/or basketry and/or leather items (1, 2); spinning (1, 2); working of wood and/or bone and/or stone and/or hide (1, 2) Lithics: knapping; boring; working and cutting of plants; working of silicious plants; working of hard silicious plants; working of wood; working and cutting of skin; working of dry and hard skin; working of soft material (1, 2)
General function of building V: STORAGE BUILDING WITH ACTIVITY AREAS			

Table 6.14 (continued). Assessment of the function of level 6 building V.

Storage areas

Rooms 1-4, 6-7 and 9-10 have been reconstructed as storage areas. As in building IV three clusters of room can be distinguished with regard to amount of artefacts (pottery, small finds and flint/obsidian artefacts). Rooms 2, 4, 9 and 10 contained few artefacts only (40-64). Rooms 1 and 6 had 154 and 166 finds respectively. Rooms 3 and 7 gave evidence of the largest assemblages: 599 and 468 artefacts respectively. As in the other buildings, pottery makes up the majority of the assemblages (ca. 70%). Standard Ware accounts for the largest number of sherds (up to 95%), followed by (in decreasing order) Standard Fine Ware, Dark-Faced Burnished Ware, Grey-Black Ware and Fine Painted Ware.

As in building IV the few reconstructed vessel shapes suggest that of all wares mainly jars were stored (n= 42). Bowls, however, were also stored regularly, as indicated by 15 reconstructed vessels (25% of the reconstructed vessel shapes). Only two pots (both from room 1) were reconstructed.

Small finds and especially flint/obsidian tools were present in small numbers only. Flint/obsidian debitage, on the other hand, was found in somewhat larger quantities (see the above - building I - discussion of debitage). A large variety of small finds and flint/obsidian tools was recovered from the various storage chambers. Most categories are represented by few objects only.

A distinction can be made between (1) archives and (2) general storage rooms.

Room	Pottery		Small finds		Flint/ obsidian tools		Flint/ obsidian debitage		Total	
	N	%	N	%	N	%	N	%	N	%
1	130	5.7	3	1.3	1	1.5	20	4.9	154	5.1
2	40	1.7	3	1.3	2	3	19	4.6	64	2.1
3	537	23.5	17	7.4	10	14.9	35	8.6	599	20
4	39	1.7	3	1.3	5	7.5	16	3.9	63	2.1
5	90	0.2	7	3.1	1	1.5	10	2.4	108	3.6
6	109	4.8	34	14.8	4	6	19	4.6	166	5.5
7	333	14.6	81	35.4	8	11.9	46	11.2	468	15.6
8	124	5.4	3	1.3	2	3	23	5.6	152	5.1
9	33	1.4	0	0	1	1.5	6	1.5	40	1.3
10	33	1.4	4	1.7	0	0	5	1.2	42	1.4
Roof	820	35.8	74	32.3	33	49.3	210	51.3	1137	38
Total	2288	100	299	100	67	100	409	100	2993	100

Table 6.15. Number of objects in the rooms of level 6 building V.

(1) Archives

Of the small finds sealings, figurines and tokens are well-represented. These objects were not evenly distributed, and cluster in room 6 (21 sealings, 6 tokens) and room 7 (36 sealings, 26 tokens, 3 human figurines), and to a lesser degree in room 3 (3 sealings, 5 tokens and 1 human figurine). The same pattern as in building II appears: three small adjoining rooms with varying amounts of sealings which are associated with tokens and/ or figurines (figs. 6.7 and 6.8). Furthermore, as in building II, these rooms contain the majority of artefacts of building V (roof excluded: 66%, roof included: 41%). On the basis of their inventories, therefore, rooms 3, 6 and 7 have been reconstructed as archives. Especially the inventory of room 6 closely parallels that of room 6 of building II; in both rooms a relatively large number of broken small finds were stored, and sealings, tokens and figurines are associated (cf. figs. 6.7 and 6.8). As has been argued, these objects were kept for administrative purposes. Many of the small finds were found high up in the fill, and it is suggested that they were stored on shelves along the walls. A position on the roof is unlikely, since the small finds were generally found below the remains of roof beams. The small find assemblages of rooms 6 and 7 have been compared. A similarity coefficient of 69% has been computed, indicating that rooms 6 and 7 are rather comparable with regard to their small find assemblages.

The sealings of building V (and the other level 6 buildings) have been discussed in detail by Duistermaat (1994, and see Akkermans and Duistermaat 1997). Three sealings came from room 3 (stored containers unidentifiable, 2 different seal designs). In room 10 one sealing without impression was found, which had been used on pottery. The majority of the sealings, however, was found in rooms 6 and 7. Of the sealings in room 6 (n= 21), concentrated in the northern part of the room, nine were used to seal basketry, five on pottery, the others sealed unidentified objects. Eighteen of the twenty-one sealings do not carry any seal impression. In contrast to room 6 of building II, room 6 gave evidence of seven 'fine ware' sherds only; the majority (n= 101) of the sherds consisted of Standard Ware.

The sealings in room 7 (n= 36) were mainly (n= 29) found in the northern part and northeastern corner of the room. The remaining seven sealings were mainly situated in the southern part of the chamber. The sealings in room 7 closed baskets (n= 2), bags (n= 1), and especially pottery (n= 21). The remaining eleven sealings sealed unidentified objects. Three quarters (n= 27) of the sealings did not show seal impressions. It appears that these sealings were mainly (n= 19) used to seal pottery or objects of indeterminable type. Only one was used to seal basketry. Duistermaat (1994:57) suggests that these unimpressed pottery sealings may have been used as lids for closing vessels, rather than as sealings participating in the sealing system. Only eight sealings carry a recognizable impression. A relatively large amount of sherds (n= 333, 22.5% of all sherds of building V) was recovered from room 7. As in room 6 of building II the 'finer wares' are relatively well-represented, especially Standard Fine Ware (n= 13). Room 7 is a relatively large area, and it was well-accessible through a dooropening which connected room 7 with room 8. It is therefore suggested that room 7 was a storage area where activities may have been carried out (i.e. a type F storage area; cf. table 4.5 in chapter 4). If the (small) objects such as sealings were stored on shelves along the walls there would have been enough space to conduct activities.

When rooms 6 and 7 are compared according to sealing reverses, a striking difference can be observed: "The sealings found in room 7 sealed mostly pottery (83.3%...), while the sealings from room 6 originated mostly from other container types (68.8%), mainly basketry (56.3%), while pottery has a share of 31.2%..." (Duistermaat 1994:58). In both rooms different kinds of sealings were stored, indicating a functional difference.

Of the archives, and of the storage chambers in general, room 3 contained the largest number of finds (n= 599). In contrast to rooms 6 and 7, however, small finds make up only a small part of the assemblage from this room (2.9%, as opposed to 20.5% and 17.3% in rooms 6 and 7 respectively). The objects pointing to administration, for instance, were present in small numbers only (i.e. three sealings, five tokens and one human figurine). In this respect room 3 resembles rooms 1 and 7 of building II, both reconstructed as archives with relatively few sealings (when compared with room 6 of building II). The other small finds in room 3 are: a fragment of an awl, a pierced disc, a fragment of a pestle and of a pestle/grinder, a piece of a perforated stone, a fragment of a stone bowl, and one complete and two broken sling missiles. Pottery sherds (n= 537, 89.6%) account for the majority of objects from room 3, most likely indicating that mainly ceramic vessels were stored in this room. As usual, Standard Ware is most frequent (n= 509), followed by Standard Fine Ware, Dark-Faced Burnished Ware, Grey-Black Ware and Fine Painted Ware. The (very few) reconstructed vessel shapes are: a Standard Ware bowl, a Standard Ware jar, an Orange Fine Ware jar and a Standard Ware jar. Flint/obsidian tools and debitage were found in limited numbers.

Room 3 was a relatively large area (3 x 1.35 m) which at first sight might account for the large number of finds. However, room 2, to which room 3 was connected by an opening at floorlevel, was also relatively large (2.35 x 1.30 m), but here only 64 objects were found. Surface area alone, therefore, does not seem to account for the difference. As in building II, archives simply contained far more (and other) objects than other rooms: they are differentiated not only by kind of artefact (sealings, figurines and tokens), but also by the larger number of accompanying objects.

Area	Room 5	Room 8	Roof
Subsistence:			
Plant-processing	x (2)	x (4)	x (34)
Baking/cooking, preparation, deposition	x (85)	x (108)	x (736)
Food serving	? (5)	x (16)	x (84)
Grinding	x (3)	-	x (25)
Manufacture/maintenance:			
Knapping	-	-	x (11)
Plant-processing	x (2)	x (4)	x (34)
Wood-processing	-	x (2)	x (12)
Skin-processing	x (2)	x (6)	x (40)
Boring	-	MWA (1)	x (5)
Spinning	-	-	x (2)
Textiles and/or basketry and/or leather items	-	x (1)	x (1)
Social life:			
Meeting	-	x	x
Sleeping	-	?	x
Performance of ritual	-	-	x (11)

Table 6.16. Comparison of activities in level 6 building V. Key: - = absent, x = present, ? = possibly present, MWA = according to the microwear analysis of flint artefacts.

(2) General storage rooms

Rooms 1, 2, 4, 9 and 10, i.e. the majority of storage rooms, have been reconstructed as general storage rooms, which are rooms that did not specifically function as archives or grain storage rooms (like in building II), but for the storage or keeping of various items. These rooms have already been dealt with in the general discussion, but some extra observations can be made. First of all, rooms 9 and 10 stand out as two rooms with very few finds (respectively 40 and 42 objects, including sherds). In room 9 only three small finds were present, and in room 10 no flint/obsidian tools were present (table 6.15). Furthermore, they were not, like the other excavated rooms of building V (apart from room 1), connected by doorways at floorlevel; most probably they were accessible via the roof. In the western wall of room 10 a porthole was present.

Room 4 was closely connected to activity area room no. 5. Both rooms could be entered from open area 6, but they were also connected by a doorway in the north. Perhaps goods to be processed in room 5 were stored in room 4. Remember that a link between an activity area and a storage area was also observed for rooms 2 and 6 of building IV.

A similar link is attested in the case of rooms 7 and 8 of building V. In fact, as in building IV, the rooms of building V can be arranged in classes according to accessibility:

1. Rooms linked by openings at floorlevel:

a) 2+3
b) 4+5
c) 7+8

2. Rooms which are not connected to other rooms:

a) 1?, 9
b) 6 (connected to open area 6)

3. Rooms connected by a porthole:

    10+? (room to the west, not excavated)

From the above the following pattern emerges. Room 2 differs from room 3, since room 2 is a general purpose storeroom with few artefacts, and room 3 is an archive with many artefacts. Room 4 differs from room 5 in that room 4 is a storage area, room 5 an activity area. For rooms 7 and 8, finally, the same holds. Perhaps these pairs of connected but different rooms were functionally related, or maybe, as was also suggested for rooms of building IV, clusters of rooms belonged to different persons or groups.

Figure 6.9. Level 6: distribution of Standard Ware.

4.2 Storage Buildings

BUILDING X (table 6.17)

Storage rooms

Only very few small finds were recovered from the various rooms of building X (n= 14): a complete bone burin, a broken grinder, a fragment of a grinding slab, a fragment of a pestle, a possible loamer, a broken and a complete pierced disc, a stone bowl fragment, four tokens, a fragment of a polishing/rubbing stone and a piece of 'white ware'. Apparently small finds were stored (in general storage rooms) in limited quantities in building X; they were absent from room 4.

Room	Function	Activity categories	Activities
1	Storage area? G (1, 2), I (1): general storage area	Subsistence (1) Storage of non-food products (1, 2)	Storage of food and/or other products/objects (1, 2)
2	Storage area G (1, 2), I (1): general storage area	Subsistence (1) Storage of non-food products (1, 2)	Storage of food and/or other products/objects (1, 2)
3	Storage area G (1), I (1): general storage area	Subsistence (1) Storage of non-food products (1)	Storage of food and/or other products/objects (1)
4	Storage area G (1, 2), I (1): grain storage area	Subsistence (1, 2) Storage of non-food products (1)	Storage of food and/or other products/objects (1, 2)
5	Storage area G (1), I (1): general storage area	Subsistence (1) Storage of non-food products (1)	Storage of food and/or other products/objects (1)
6	Storage area H (1, 2): grain storage area	Subsistence (1, 2) Storage of non-food products (1, 2)	Storage of food and/or other products/objects (1, 2)
General function of building X: STORAGE BUILDING			

Table 6.17. Assessment of the function of level 6 building X.

Pottery sherds, however, were found in relatively large numbers: a total of 1225 sherds is reported. Standard Ware (n= 1131) accounts for 92.3% of all ceramics. Next come Dark-Faced Burnished Ware (n= 57, 4.6%), Standard Fine Ware (n= 31, 2.5%), and Grey Black Ware (n= 6, 0.5%). Fine Painted Ware is wholly lacking. A thin layer of burnt cereals was found on the floor of room 4, and small amounts of burnt grain were also found on the floor of room 6. Only very few other finds were recovered from these rooms (10 sherds in room 4; 18 sherds and a piece of 'white ware' in room 6). It may be suggested that the sherds were accidental and that both rooms served for the storage of grain in bulk, as in the nearby western rooms of building II. If so, then two classes of rooms may be distinguished within building X: (1) general storage rooms, and (2) grain storage rooms. Unlike the other rooms of building X, room 6 has a long-drawn shape and a bench against its western wall. As yet the function of this bench remains enigmatic.

Room	Function	Activity categories	Activities
1	Storage area G (1), I (1): general storage area	Subsistence (1) Storage of non-food products (1)	Storage of food and/or other products/objects (1)
2	Storage area G (1), I (1): general storage area	Subsistence (1) Storage of non-food products (1)	Storage of food and/or other products/objects (1)
3	Storage area G (1), I (1): general storage area	Subsistence (1) Storage of non-food products (1)	Storage of food and/or other products/objects (1)
5	Storage area G (1), I (1): general storage area	Subsistence (1) Storage of non-food products (1)	Storage of food and/or other products/objects (1)
6	Storage area G (1, 2), I (1): general storage area	Subsistence (1) Storage of non-food products (1, 2)	Storage of food and/or other products/objects (1, 2)
7	Storage area G (1), I (1): general storage area	Subsistence (1) Storage of non-food products (1)	Storage of food and/or other products/objects (1)
General function of building XI: STORAGE BUILDING			

Table 6.18. Assessment of the function of level 6 building XI.

Discard area

Room 1 contained the large majority (81.1%) of all sherds of building X. A relatively large percentage of these sherds consists of Dark Faced Burnished Ware (n= 23, 60.5% of the sherds of room 1). It is true that room 1 was a relatively large room (2.25 x 2.25 m), but the difference in size between room 1 and the other areas of building X is not large enough to account for the difference in sherd quantities. It is suggested that the difference in sherd distribution within building X is due to depositional contexts; it seems to be significant that the only room unaffected by the fire contains so many sherds. Perhaps room 1 was out of use and used as a dump when the fire started. With respect to the large amount of sherds from an unburnt context, room 1 resembles the unburnt tholos VII. The two areas resemble each other in at least two other respects: in both units Dark-Faced Burnished Ware is well-represented and in both areas almost all small finds are broken (from room 1 only one complete small find was recovered). It is here suggested that both areas were used as dumps of secondary refuse.

BUILDING XI (table 6.18)

Like in building X, only limited amounts of artefacts were recovered from building XI. Rooms 1 and 5 contained very few finds only. These northern areas were unburnt, and we do not seem to be dealing with in-situ contexts here. In building X the unburnt room 1 was most likely used as a dump. The limited amount of finds from rooms 1 and 5 of building XI, however, seem to indicate that these areas were not (regularly) used as dumps.

The small finds (n= 14) recovered from building XI are: a figurine fragment, two complete grinders, a complete grinding slab, and a fragment of a slab, a broken hammer, two pestles, a pendant, two fragments of stone bowls, and one fagment of a stone platform.

Building XI yielded far fewer sherds than building X (n= 299). As in building X, Fine Painted Ware was wholly absent. Standard Ware accounts for the majority of sherds (n= 270, 90.3%), the other wares are present in small numbers only. Room 6 contained the majority of finds (n= 40), the other chambers contained even fewer sherds (e.g. 8 in room 5).

BUILDING XII (table 6.19)

Building XII was not affected by the fire, and seems to have been out of use while the other level 6 buildings were still in use (Verhoeven and Kranendonk 1996:58 and fig. 2.9). Only few finds were recovered from the building; especially the small finds appeared in limited numbers. Only two complete objects were recovered from building XII. Most likely the objects from building 12 should be regarded as secondary refuse, perhaps some represent abandonment stage refuse. The building seems to have consisted of small store-rooms only, therefore building XII is reconstructed as a storage building. When it was abandoned, most likely before the fire started, this storage building was secondarily used for dumping refuse. The finds should in all probability be regarded as secondary refuse.

The few small finds (n= 10) consisted of: a fragment of plaster coating (perhaps of a vessel or a pit), a complete grinder, three fragments of grinding slabs, a fragment of a mortar, a piece of a pendant, four fragments of stone bowls, and a piece of 'white ware'. The northern rooms 10, 11, 13 and 14 have been excavated over very limited areas only (see fig. 3.1), and no finds (small finds nor ceramics) were recovered from these areas. The completely excavated rooms 1, 3, and 5, however, also lacked small finds; the other rooms each had one or two small finds only.

Room	Function	Activity categories	Activities
1	Primary: storage area: general storage area; Secondary: discard area K (1)	Storage of non-food products	Storage of non-food products
2	Primary: storage area: general storage area; Secondary: discard area K (1)	Subsistence Storage of non-food products	Storage of food and/or other products/objects
3	Storage area: general storage area	Subsistence Storage of non-food products	Storage of food and/or other products/objects
4	Storage area: general storage area	Subsistence storage of non-food products	Storage of food and/or other products/objects
5	Storage area: general storage area	Subsistence Storage of non-food products	storage of food and/or other products/objects
6	Primary: storage area: general storage area; Secondary: discard area K (1)	Subsistence Storage of non-food products	Storage of food and/or other products/objects
7	Primary: storage area: general storage area; Secondary: discard area K (1)	Subsistence Storage of non-food products	Storage of food and/or other products/objects
8	Primary: storage area: general storage area; Secondary: discard area K (1)	Subsistence Storage of non-food products	Storage of food and/or other products/objects

Table 6.19 Assessment of the function of level 6 building XII.

Room	Function	Activity categories	Activities
9	Storage area: general storage area	Subsistence Storage of non-food products	Storage of food and/or other products/objects
12	Primary: storage area: general storage area; Secondary: discard area K (1)	Subsistence Storage of non-food products	Storage of food and/or other products/objects
General function of building XII: STORAGE BUILDING			

Table 6.19 (continued). Assessment of the function of level 6 building XII.

A total of 606 ceramic fragments was recovered from building XII, largely represented by Standard Ware (n= 538, 88.8%). Dark-Faced Burnished Ware, Standard Fine Ware and Grey-Black Ware were present in small quantities only (respectively 41, 21 and 6 sherds). Remarkably, as in buildings X and XI, Fine Painted Ware was absent. Only 269 sherds could be assigned to specific rooms. To rooms 1, 3, 4, 10-11, 13 and 14 no sherds could be assigned. Areas 2, 5, 6, 8 and 12 contained few sherds only (3 to 49); from room 7, however, 48.3% (n= 130) of the sherds was recovered. Apart from room 5, the rooms that showed small finds also gave evidence of pottery.

Room 7 is remarkable in a number of aspects: unlike the other rooms, room 7 has a long-drawn shape; its walls were plastered; the large majority of finds from building XII were recovered from it. The architectural information may indicate that room 7 functioned as a special storage area. Perhaps the thick (ca. 0.5 cm) interior plaster acted as a vermin-resistant coating, indicating that in room 7 agricultural products were stored.

Room 8 showed a carefully made floor of sherds which was covered by a thick white plaster. This may indicate that this area had some special purpose. Since the room is regarded as too small for an activity area, such a purpose was most likely related to storage. Perhaps the floor indicates that liquids were stored and handled in room 8; in case of spilling the floor could be easily cleaned, thus preventing the spread of vermin. Compared to the other rooms of building XII a reasonable number of sherds (n= 46) were recovered from room 8.

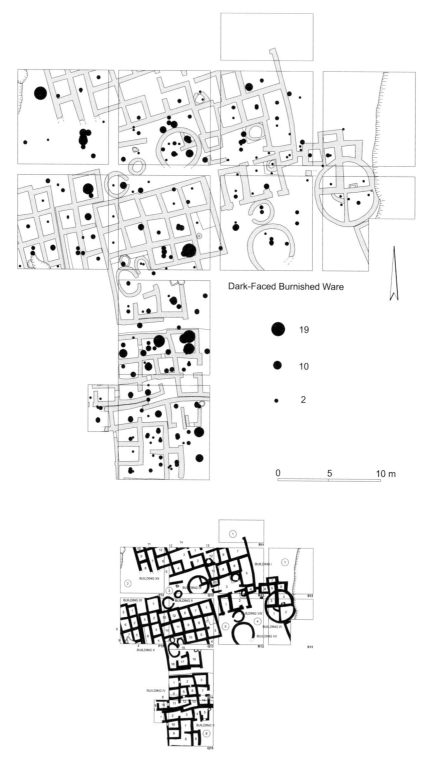

Figure 6.10. Level 6: distribution of Dark-Faced Burnished Ware.

5 ROUND BUILDINGS

5.1 Activity Areas with Storage Rooms

BUILDING VI (table 6.20)

Activity areas

In table 6.21 the activities in rooms 1, 2 and 4 have been listed and compared. In general it seems that the activity areas were multifunctional, giving evidence of both subsistence and manufacturing/maintenance activities. In room 1 plant-processing is suggested by one piece with fine retouch which, according to the microwear analysis, was used for the cutting of plants. Furthermore two blades and a racloir may have been used for plant-working. Preparation and/or deposition of food is indicated by the presence of a relatively large amount of Standard Ware sherds. Serving of food may be indicated by seven Standard Fine Ware sherds. The other activities, i.e. grinding, production of stone and/or flint artefacts, wood processing, skin-processing, spinning and the manufacture of textiles and/or basketry and/or leather items are marked by just one or two artefacts. The clay bin in the northeastern corner of room 1 must have served as a receptacle, most likely for goods processed or to be processed in room 1.

In room 2 a somewhat different range of activities was carried out: no evidence of the manufacture of stone and/or flint objects, wood and skin-processing and spinning, all attested for room 1, was found. The preparation and deposition of food may have been of less importance than in room 1, since fewer Standard Ware sherds (n= 85) were found. Food serving, however, is relatively well-represented by 21 sherds (Standard Fine Ware, Fine Painted Ware and Grey-Black Ware). As in room 1 the other activities (grinding, knapping, plant-processing, boring, manufacture of textiles and/or basketry and/or leather items) were indicated by few objects only.

Compared with the other rooms, room 4 gave evidence of few activities only: plant processing, food preparation/deposition, perhaps skin-processing and the making of textiles and/or basketry and/or leather items (see table 6.21). Like room 1, this activity area was marked by an architectural feature, i.e. a small gypsum-plastered basin sunk before the southern wall of this room. Undoubtedly this basin was closely related to the use of room 4. Its function, however, remains enigmatic.

Noticeable is the occurence of sealings and tokens in rooms 2 and 4 (fig. 6.8). In room 2 two sealings and five tokens were found. One of these sealings was used to seal pottery. The other sealing (with a smooth reverse) was impressed with a design that does not appear in the other buildings. In room 2 a fragment of a human figurine accompanied the sealings and tokens: yet another indication that sealings, tokens and figurines are associated (cf. figs. 6.7 and 6.8). From room 4 one unimpressed sealing used on pottery and four tokens were recovered. Rooms 2 and 4 of tholos VI are the only two activity areas where these 'administrative' objects were found: in the other instances they only occurred in storage rooms or archives. It is most tempting to suggest that activities in which tokens, sealings and perhaps also figurines were actually used took place in rooms 2 and 4. Perhaps, just before the fire that destroyed the village started, administrative transactions were taking place in these rooms: (pottery) containers were perhaps unsealed, their contents and/or the value of their contents expressed by the tokens? And maybe food was being served and consumed during this transaction, as indicated by the relatively large amount of 'serving ware' in room 2.

By means of similarity coefficients the relation between the three activity areas within tholos VI has been investigated. For rooms 1 and 2 a coefficient of 78.9% was calculated, for rooms 2 and 4 64.7% and for rooms 1 and 4 86.3%: rooms 1 and 4 are the most alike, and rooms 1 and 2 are rather comparable too. Rooms 2 and 4, despite their supposed administrative activities, are the least similar (ca. 65%). In fact a number of activities carried out in the other two rooms were apparently absent from room 4: food serving, grinding, manufacturing/maintenance of stone and/or flint artefacts, knapping, wood processing, boring and spinning. In room 4 a less varied range of activities was carried out: five activities as opposed to ten activities in room 1 and nine in room 2.

Room	Function	Activity categories	Activities
1	Activity area A (1, 2), B (1, 2), D (1)	Subsistence (1) Manufacture/maintenance (1, 2) Storage of non-food products (1, 2)	Pottery: food preparation: deposition; serving? (1) Small finds: grinding (1); production of textiles and/or basketry and/or leather items (1); working of wood and/or bone and/or stone and/or hide (1, 2); production of ground-stone tools and/or flint artefacts (1, 2) Lithics: knapping?; working and cutting of plants; working of hard silicious plants; wood working; working of skin; working of dry and hard skin; cutting of soft material (1, 2)
2	Activity area A (1, 2), B (1, 2), D (1)	Subsistence (1) Manufacture/maintenance (1, 2) Storage of non-food products (1, 2) Administration (1, 2)	Pottery: food preparation: deposition; serving (1) Small finds: grinding (1, 2); production of textiles and/or basketry and/or leather items (1); working of wood and/or bone and/or stone and/or hide (1, 2) Lithics: knapping; working and cutting of plants; wood-processing; working of skin; working of dry and hard skin; cutting of soft material (1, 2)
3	Storage area G (1, 2), I (1): general storage area	Subsistence (1) Storage of non-food products (1, 2) Administration? (1, 2)	Food storage; storage of non-food products (1, 2)
4	Activity area A (1, 2), B (1, 2), D (1)	Subsistence (1) Manufacture/maintenance (1, 2) Storage of non-food products (1, 2) Administration (1, 2)	Pottery: food deposition?; deposition of non-food products?; food preparation: deposition; serving? (1) Small finds: production of textiles and/or basketry and/or leather items (1) Lithics: knapping?; working of plants; working of dry and hard skin (1, 2) administration?

Table 6.20. Assessment of the function of level 6 building VI.

Room	Function	Activity categories	Activities
5	Storage area G (1, 2), H (1, 2), I (1): general storage area	Subsistence (1) Storage of non-food products (1, 2)	Storage of food and/or other products/objects (1, 2)
6	Storage area G (1, 2), H (1, 2), I (1): general storage area	Subsistence (1) Storage of non-food products (1, 2	Storage of food and/or other products/objects (1, 2)
7	Storage area G (1, 2), H (1, 2), I (1): general storage area	Subsistence (1) Storage of non-food products (1, 2)	Storage of food and/or other products/objects (1, 2)
8	Storage area G (1, 2), H (1, 2), I (1): general storage area	Subsistence (1) Storage of non-food products (1, 2)	Storage of food and/or other products/objects (1, 2)
General function of building VI: ACTIVITY AREAS WITH STORAGE ROOMS			

Table 6.20 (continued). Assessment of the function of level 6 building VI.

Storage areas

The very small rooms 3, 5, 6, 7 and 8 have been reconstructed as general storage. Room 3, the largest of these areas and situated within the building proper, contained 59 objects. The very small 'room' 8 (probably a re-used oven) gave evidence of 63 artefacts. From the other rooms, probably added to the round building after its original construction, very few artefacts were recovered; cf. fig. 6.1. As in the other rooms in the level 6 village, pottery accounts for the majority of objects. Room 8 is very small, and one wonders whether the finds from this area were not accidental to its use. In fact rooms 5, 6, 7 and 8 are so small that they could not be entered; most likely they should be regarded as a series of bins with openings in the top (like 'room' 4 of building I). Perhaps these bins were used for the storage of bulk products, such as grain. The finds within these bins may indicate secondary use, or maybe they are due to post-depositional processes.

Area	Room 5	Room 8	Roof
Subsistence:			
Plant-processing	MWA (4)	x (2)	x (1)
Baking/cooking, preparation, deposition	x (116)	x (85)	x (87)
Food serving	x (7)	x (21)	-
Grinding	x (1)	x (2)	-
Manufacture/maintenance:			
Stone and/or flint artefacts	x (1)	-	-
Knapping	-	x (1)	-
Plant-processing	MWA (4)	x (2)	x (1)
Wood-processing	x (1)	-	-
Skin-processing	x (2)	-	X (1)
Boring	-	x (2)	-
Spinning	x (1)	-	-
Textiles and/or basketry and/or leather items	x (1)	x (1)	x (2)
Administration	-	x (7)	? (5)
Social life:			
Meeting	-	?	-

Table 6.21. Comparison of activities in level 6 building VI. Key: - = absent, x = present, ? = possibly present, MWA = according to the microwear analysis of flint artefacts.

5.2 Activity Area

BUILDING VII

The relatively large size of building (or tholos) VII suggests that it was used as an activity area where subsistence and manufacturing/maintenance activities were carried out. In this respect it is recalled that almost all of the larger areas in the level 6 village were activity areas, whereas the smaller areas are regarded to have functioned as storage chambers. Due to the absence of an in-situ context (tholos VII was not affected by the fire), however, the exact function of tholos VII cannot be ascertained on the basis of the objects. Nevertheless the suggested function of tholos VII according to scenarios 1 and 2 is similar. Due to the lack of the characteristic burnt deposit, it was not possible to make a clear distinction between possible provisional refuse and subsequent secondary refuse, or dump. It is suggested, however, that we are mainly dealing with secondary refuse, since four out of the five small finds are broken (and the single complete small find is a bead which may have been lost). In general, the other areas with in-situ contexts gave evidence of a larger number of complete artefacts. From tholos VII a relatively large amount of sherds were recovered (n= 404). As in the majority of spatial units, the majority of these sherds (n= 378) are Standard Ware. The large number of sherds may indicate that tholos VII was used as a dump after the fire. Or, perhaps, the sherds are - partly - to be regarded as residual provisional refuse, which would indicate extensive use of Standard Fine Ware in tholos VII (probably for storage and preparation of food).

The possible activities carried out inside tholos VII are: food preparation (Standard Ware and Dark-Faced Burnished Ware sherds were found), serving (Standard Fine Ware, Grey-Black Ware), grinding (two fragments of grinding slabs, a piece of a mortar) and spinning (a broken pierced disc). The flint/obsidian artefacts (a denticulate, a piece with fine retouch, a racloir, a notch, a borer, a side-blow blade flake, six flakes and a blade), moreover, may indicate plant, skin and wood processing, and boring. Like the other activity areas, building VII was multifunctional.

5.3 Houses[52]

During the 1996 campaign new evidence was found with regard to the function of tholoi at Sabi Abyad. As was indicated in chapter 3 (section 2), in square R13 two superimposed tholoi with hearths and in square R14 another tholos with a hearth (cf. the colour plate; the tholos east of building IV) were recovered. Regrettably, the artefact inventories of these new tholoi could not be taken into account in the present study, since the stratigraphy is not available as yet and the finds have not been analysed in detail. Nevertheless, I suggest that these structures should be viewed as dwelling units.

First, a number of interpretations of the function of Halafian tholoi should be mentioned. One of the most comprehensive studies about tholoi is Breniquet's (1996) analysis of Halafian architecture (pp. 80-96). Breniquet distinguishes three types of tholoi: (1) undivided round tholoi; (2) multi-roomed tholoi; (3) tholoi with annexes. The tholoi with annexes have been divided into buildings with: (a) series of small walls which may have served as platforms for grain storage silos; (b) small rectangular 'cells'; (c) one large rectangular antechamber (e.g. tholos TT8 of Arpachiyah). On the basis of size, architectural features (e.g. presence or absence of hearths), finds, and ethnographic parallels from agricultural African villages consisting of circular 'huts', Breniquet has reconstructed different functions for Halafian tholoi (ibid.:93-94). Small tholoi (i.e. 2-3m²) are considered to have functioned as storerooms, or perhaps as stables for small, or small numbers of, domestic animals. The larger tholoi may have served as houses or for domestic purposes. These tholoi can be subdivided into the tholoi with antechambers (having a floor surface of minimally 28 m²), and tholoi without such antechambers, but occasionally with annexes. In an ethnographic account Van Der Kooij (1976:128) reports that in the village Mir-hussein, ca. 20 km east of Aleppo, the large beehive-shaped tholoi (ibid.: photo 101) are used for living (including cooking), the small ones are used for storage and as chicken coops.

---

[52] *Building* is a general term, whereas *house* specifically refers to a dwelling.

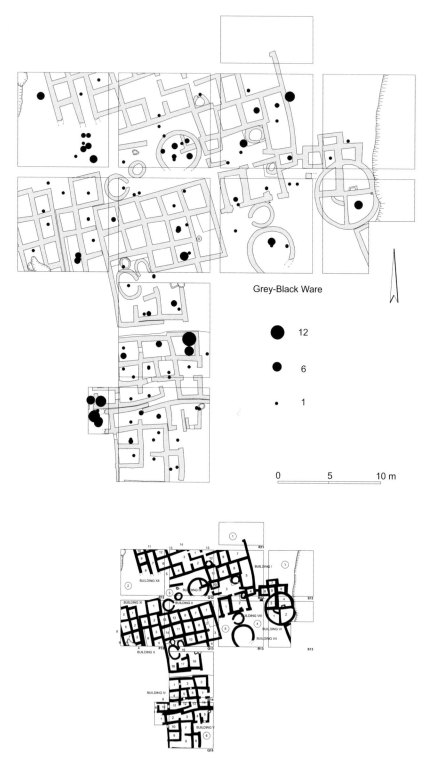

Figure 6.11. Level 6: distribution of Grey-Black Ware.

Other investigators, too, interpret most large tholoi (i.e. larger than 3 m in diameter) as dwellings (e.g. Seeden 1982; Watson and LeBlanc 1990), and the other rectangular buildings are commonly interpreted as auxiliary structures used for storage, etc. (Akkermans 1989a:68). David and Joan Oates (1976:107) have pointed out that the common appearance of tholoi at Halaf sites, and the distribution of artefacts in and around them suggests a domestic function for these buildings. Most of the numerous tholoi of Halafian Yarim Tepe II (and III), the largest with a diameter of over 5 meters, have been interpreted as basic living units, i.e. as houses (Merpert and Munchaev 1993b:131, 1993c:186-192). Some of the smaller tholoi may have served as granaries or storage rooms (Merpert et al. 1977:91; Munchaev and Merpert 1971:19; Munchaev and Merpert 1973:12).

Second, I argue that the newly discovered tholoi in level 6 at Sabi Abyad represent houses because they represent the largest interior spaces in level 6; they are (unlike all other spaces) suited for habitation and such domestic activities as eating, sleeping, entertaining, etc. As has been indicated in the present spatial analysis (and as suggested by Breniquet), the smaller tholoi without hearths (i.e. tholoi VIII and IX) seem to have served for storage.

Third, the presence of hearths serves as a key indicator for dwellings; they are heating facilities, as well as areas where food is prepared and shared. In the earliest level 9 at Yarim Tepe II four of the six excavated tholoi had ovens and/or hearths in their interior, clearly indicating a domestic use (Merpert and Munchaev 1993b:139 and fig. 8.8). At Girikihaçyan a large keyhole-shaped tholos (interior diameter of 4.5 m) showed a hearth in the circular part. The other large keyhole-shaped tholos 4 showed traces of burning which also suggests a hearth. Furthermore, in tholos 4 a considerable array of pottery and several grinding stones were found in situ (Watson and LeBlanc 1990:33 and figs. 2.7 and 2.15). In Shams ed-Din Tannira, finally, one tholos, ca. 4.50 m in diameter, had a fireplace near the wall opposite the entrance.

Khirbet esh-Shenef, a small later Halaf site (dated at ca. 4800 B.C., 5600 cal BC) at a distance of ca. 5 km south of Tell Sabi Abyad, should also be mentioned here. In level 3 a large number of tholoi were recovered, small ones with diameters of ca. 2 to 3 m, as well as large ones with diameters of up to 4.5 m. Besides the tholoi, a multi-roomed rectangular building with exterior buttresses and an area with circular ovens was recovered (Akkermans and Wittmann 1993:149-156). In three of the five larger tholoi (nos. XII, XIII and XIV) circular ovens or hearths were found. The remaining two tholoi have not been completely excavated, and it cannot be excluded that here hearths were present as well. In tholos XIV the hearth was lined with mudbricks. The interior buttresses in this large tholos (ca. 4.5 m) may have served as benches and/or as decorative elements. Clearly these large tholoi at Shenef would have been suitable as dwellings, as indicated by their relatively large size (the other buildings and rooms are very small) and presence of hearths. In one of the smaller tholoi (no. XX) a mud-brick lined oven or hearth perhaps indicates that this building (with a diameter of ca. 3 m) was used as a 'cooking hut', since it seems rather small for a true dwelling. In the remaining small tholoi no ovens or hearths were present. It should be noted that tholos XX is the largest of the smaller tholoi, which most likely served for storage, since they are far too small (diameters of ca. 2 m) to have functioned as activity areas or dwellings.

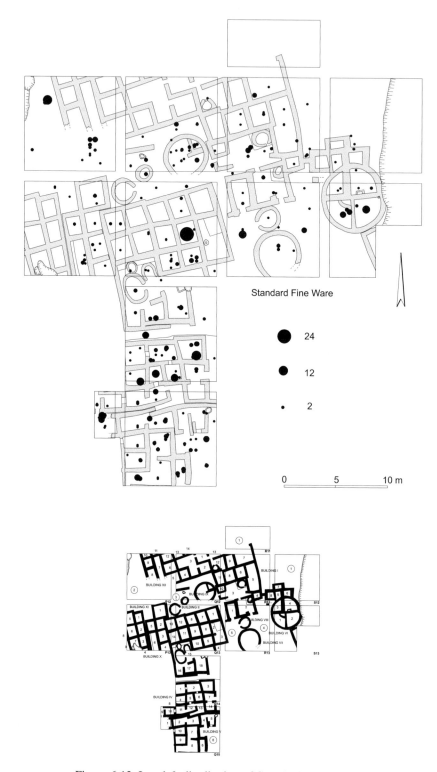

Figure 6.12. Level 6: distribution of Standard Fine Ware.

Fourth, the other rectangular spaces, but also the other tholoi (i.e. tholoi VIII and IX) in level 6, seem to be much too small for dwellings (and, as has been argued in chapter 3, section 2, there is no evidence for upper storeys in which living areas may have been located). Sleeping, for instance, is nearly impossible in most of the level 6 rooms (at least in an extended position). Exterior sleeping platforms may, of course, have been used, but other interior sleeping arrangements must have been present to secure a place when it rained, or when it was too cold to sleep outside. The large tholoi, with hearths for warmth, may have provided such space. The large tholoi with hearths at Khirbet esh-Shenef have already been mentioned. As yet they are the only candidates for dwellings at this site. In this respect the evidence from Yarim Tepe II is also significant: if one looks at the plan of the earliest level 9 (i.e. Merpert and Munchaev 1993b: fig. 8.8), a clear distinction can be made between the relatively large tholoi, which can be used as dwellings and/or domestic units, and a series of clustered and vary narrow rectangular areas, which only seem to be suited for storage.

Finally, it should be mentioned that a first preliminary analysis of the artefact inventories of the tholoi mentioned here suggests that they were not used for storage. Large storage jars, for instance, seem to be largely absent, and a clustered appearance of objects has not been noticed either. In fact, storage of food products would not be expected in these tholoi, considering the negative effect of the heat of the hearths on the storage life of products like grain, milk, etc.

Both tholoi VI and VII, reconstructed as a building consisting of activity areas and storerooms (VI) and an activity area (VII), may have been dwellings as well. Tholos VI was relatively large (6.75 m in diameter), and it may orginally have been designed as a house. Later, then, it was subdivided into a number of small compartments which seem rather small for living units. A hearth has not been found in tholos VI, but perhaps this feature was removed when the building was subdivided.

Likewise tholos VII, without a fixed hearth, was perhaps a dwelling (see fig. 7.4A), certainly being large enough for it (i.e. it has a diameter of 4.50 m).

Summarizing, it is most likely that the newly discovered tholoi in squares R13 and R14 were used as dwellings. Tholoi VI and VII seem to be large enough to have functioned as living units, but in the absence of hearths such a designation is more problematic. These tholoi, then, have been reconstructed as, respectively, activity areas with storage rooms and an activity area. The smaller tholoi VIII and IX were designated storage buildings. Like the rectangular buildings, which were largely used for storage but also for conducting domestic activities, tholoi at Sabi Abyad therefore seem to have had different functions. The largest ones most likely served as houses, whereas the smaller ones were used for storage, which was presumably related to the households in the houses. Indeed, as was indicated above, other researchers have also suggested different functions for different kinds (i.e. large and small) of tholoi.

At first sight, the incorporated small open areas or 'courtyards' with ovens in the rectangular buildings I, II and the one between buildings I and XII reflect traditional ancient Near Eastern courtyard houses (see e.g. the various plans in Aurenche 1981, Eichmann 1991, and Wright 1985). However, such houses are characteristic of the Bronze Age (see e.g. examples in Meijer 1989 and Veenhof, ed., 1996) rather than of the Neolithic (among the few examples from the Neolithic the houses of Çatal Hüyük, Haçilar, Bouqras and perhaps Hassuna may be mentioned; see Eichmann 1991). However, in these instances the rooms surrounding the courtyard (and the courtyards themselves) are much larger (and therefore suited for living) than the 'cells' of Sabi Abyad. The rectangular buildings with courtyards at Sabi Abyad, then, are not to be regarded as houses (and moreover, as already indicated, no evidence for the presence of living spaces in upper storeys has been found).

In chapter 7 the relationships between the circular houses, households, and the rectangular storage buildings will be discussed in detail.

5.4 Storage Buildings

BUILDING VIII

Tholos VIII, like nearby tholos VII, was not affected by the fire. Due to the lack of the characteristic burnt deposit, it was not possible to make a clear distinction between possible provisional refuse and subsequent secondary refuse, or dump. The virtual absence of (complete) small finds, however (just one stone bowl fragment was found), may suggest that we are mainly dealing with secondary refuse. This lack can, however, also be explained by the idea that goods (e.g. grain or animal fodder) were perhaps stored in bulk in tholos VIII. Both scenarios indicate the same function. The small size of building (or tholos) VIII suggests that it was used as a storage area, perhaps for bulk goods. Due to the absence of an in-situ context, however, the exact function of tholos VII cannot be ascertained on the basis of the objects. The objects recovered from tholos VIII are: pottery sherds (102 Standard Ware sherds, a Standard Fine Ware fragment, four Dark-Faced Burnished Ware sherds), flint/obsidian: an end-scraper, a corner-thinned blade, four flakes.

Category	Stored objects
Pottery	Of all wares: bowls, jars (and probably pots)
Small finds	Animal figurines (4b), awls (2b), axe (1c), bead (1c), grinders (2c, 1b), grinding slab (1b), hammer (1b), human figurines (2b), impressed pieces of clay (2b), pierced disc (1b), pendant (1c), pestle (1b), stone platforms (2b, 2c), sling missiles (3c), spatula (1c), spindle whorl (2c), stamp (1b), tokens (6b), polishing/rubbing stones (2c), impressed sealing (1b)

Table 6.22. Objects stored in level 6 building IX (numbers between brackets, b = broken, c = complete).

BUILDING IX

On the basis of its size and the presence of the small bin, tholos IX could be regarded as an activity area. However, the kind and large variety of small finds closely parallel the situation in some of the archives, i.e. rooms 1, 6 and 7 in building II, and rooms 3, 6 and 7 of building V. Therefore it is most likely that tholos IX was a storage area. Both scenarios 1 and 2 indicate this function. Perhaps the small bin, or compartment, served as a container of particular items.

Area	Function	Activity categories	Activities
1	Activity area A (1, 2), B (1, 2), D (1), E (1)	Subsistence (1, 2) Manufacture/mainte-nance (1, 2) Social life (1, 2)	Pottery: food preparation: baking/cooking; deposition; serving? (1) Small finds: grinding (1); spinning (1, 2); working of wood and/or bone and/or stone and/or hide (1) Lithics: working of plants; working of dry and hard skin (1, 2)
2	Activity area A (1, 2), B (1, 2), D (1), E (1) discard area K (1)	Subsistence (1, 2) Manufacture/mainte-nance (1, 2) Administration? (1) Social life (1, 2)	Pottery: food preparation: baking/cooking; deposition; serving? (1) Small finds: grinding (1); spinning (1, 2); working of wood and/or bone and/or stone and/or hide (1)
3	Activity area A (1, 2), B (1, 2), D (1), E (1)	Subsistence (1) Manufacture/mainte-nance (1, 2) Administration? (1) Social life (1, 2)	Pottery: food preparation: baking/cooking; deposition; serving? (1) Small finds: production of textiles and/or basketry and/or leather items (1); working of wood and/or bone and/or stone and/or hide (1, 2) Lithics: wood working (MWA); processing of grain? (MWA) (1, 2)
4	Activity area A (1, 2), B (1, 2), D (1), E (1)	Subsistence (1) Manufacture/mainte-nance (1, 2) Social life (1, 2)	Pottery: food preparation: baking/cooking; deposition; serving? (1) Small finds: grinding (1, 2); production of textiles and/or basketry and/or leather items; spinning (1, 2) Lithics: knapping; grooving; working of hard silicious plants; working of wood; working and cutting of skin; working of dry and hard skin; cutting of soft material (1, 2)
5	Activity area A (1, 2), B (1, 2), D (1), E (1)	Subsistence (1) Manufacture/mainte-nance (1, 2) Administration? (1, 2) Social life (1, 2)	Pottery: food preparation: baking/cooking; deposition; serving? (1) Small finds: grinding (1); spinning (1, 2); Lithics: knapping?; cutting of grain? (MWA); working of sili-cious plants; working of hard silicious plants; working of wood; working of skin; working of dry and hard skin; working of unknown material (1, 2)
6	Activity area A (1, 2), B (1, 2), D (1), E (1)	Subsistence (1) Manufacture/mainte-nance (1, 2) Administration? (1) Social life (1, 2)	Pottery: food preparation: baking/cooking; deposition; serving? (1) Small finds: grinding (1); production of textiles and/or basketry and/or leather items (1); working of wood and/or bone and/or stone and/or hide (1, 2); spinning (1, 2) Lithics: knapping; boring; grooving; butchering; work-ing and cutting of plants; working of silicious plants; working of hard silicious plants; working of wood (MWA); working of skin; working of dry and hard skin; working of soft material (1, 2)

Table 6.23. Assessment of the function of open areas in the level 6 settlement.

In table 6.22 the small finds stored in tholos IX (n= 57) have been listed (the flint/obsidian industry from square Q12, in which tholos IX is situated, has not been analysed).

As can be seen, a rather large variety of small finds was present within tholos IX. Interesting to note is the relatively large number of human and animal figurines (n= 6, all broken); apart from room 6 of building II (with 17 figurines) no other area of the level 6 village gave evidence of so many figurines. Again, the figurines are associated with tokens (cf. figs. 6.7 and 6.8). However, in this instance, only one sealing was found (in the other areas just mentioned sealings were invariably part of the association). The tokens and the figurines are considered to have played a role in administrative transactions. The sealing found in the tholos was used on basketry and it showed a seal impression.

Remarkable is the large amount of sherds recovered from the relatively limited area of tholos IX (n= 980, 5.5% of the sherds of all buildings). As expected, Standard Ware is best represented (n= 871, 88.9% of all sherds of tholos IX), but the finer wares are remarkably well-attested, too: 39 Standard Fine Ware sherds, 15 Grey-Black Ware sherds and 53 Dark-Faced Burnished Ware fragments were recovered (but only two Fine Painted Ware sherds). Together with the figurines, this seems to underscore the special character and perhaps significance of tholos IX. In this respect the situation of the tholos should be mentioned as well; whereas the other tholoi (including those excavated in 1996) were located outside and peripheral to the rectangular buildings, tholos IX is located centrally in a small courtyard, surrounded by rectangular buildings.

6 OPEN AREAS (table 6.23)

All six open areas distinguished (fig. 3.1) have, according to both scenarios 1 and 2, been reconstructed as activity areas (table 6.23). Apart from area 3, none of these areas has been completely excavated. The open areas were largely unaffected by the level 6 fire, and most likely the majority of the objects recovered from the open areas have to be regarded as primary and secondary refuse. Nevertheless a functional assessment of these areas has been attempted. While some objects may have been stored in open areas, it is not expected that these areas were regularly used for storage; (1) storage in the open air is unlikely, since the stored products would be very susceptible to human and natural disturbance; (2) storage facilities abound in the various buildings; (3) according to ethnographic studies (cf. appendix 1) storage in the open air is not very likely (but sometimes things such as dung, firewood, agricultural equipment are stored in the open).

The open areas as excavated from 1991 to 1995 do not appear to have been intensively used for dumping debris. Only in two instances we found what may be termed a trash heap, or a dumping place (marked by concentrations of broken objects). In the small more or less triangular area between tholos IX, open area 3 in building I, and room 1 of building II a relatively large amount of animal bones, sherds, and remains of possible organic refuse was found. From the rest of open area 1 such concentrations of debris were absent. From open area 2 a relatively large amount of sherds, more or less clustered in one place, was recovered: probably the sherds were deliberately dumped here. In the other areas, however, such concentrations were not encountered. Apparently, debris was not regularly and/or sytematically dumped in the investigated open areas.

Figure 6.13. Level 6: distribution of Orange Fine Ware.

However, during the 1997 excavations in the level 6 settlement (which could not be taken into account in the present study) a large dumping area (ca. 80 m^2!) was found in square T12 (ca. 12 m east of building I). This open area was characterized by thick accumulations of ashes which included many animal bones and large numbers of sherds. This area, most likely located on the periphery of the settlement, indicates that refuse was dumped somewhat away from the main buildings.

Storage and dumping, however, do not seem to have been major processes within the open areas analysed in the present study. Activities, on the other hand, were undoubtedly carried out in open areas: (1) from ethnographic accounts we know that open areas (e.g.

courtyards) are used for many activities, manufacturing/maintenance as well as susbsistence and social activities (entertaining, recreation, meeting, assembling); (2) the open areas present the largest spaces in the excavated level 6 village, enabling activities which require relatively much space, and/or 'messy' activities; (3) at least three of the open areas (no. 3, and the open areas nos. 3 and 15 within buildings I and II) are marked by ovens, indicating domestic activities. In the following, therefore, the open areas are regarded as activity areas, while not excluding that they occasionally witnessed storage or the dumping of debris.

In all areas the subsistence activities baking/cooking, deposition and serving have been reconstructed on the basis of the suggested functions of the various ceramic wares (baking/cooking: Standard Ware, Dark-Faced Burnished Ware; deposition: Standard Ware; serving: Standard Fine Ware, Fine Painted Ware, Grey-Black Ware). However, as has been indicated above, it should in mind that at least some of the fragments of these wares were not used but dumped in open areas.

In most areas objects that could refer to 'social life activities' (e.g. meeting, recreation, display: Krafeld-Daugherty 1994; Kramer 1982; Van Der Kooij 1976; Watson 1979) have been found in small numbers. Perhaps, though, these small objects were simply lost here (beads, once belonging to necklaces, for instance, represent isolated objects). In some areas just one or two tokens or sling missiles were found (i.e. areas 3, 4, 5 and 6).

7 Buildings: General Comparison

In this general comparison of buildings the pottery is dealt with first. As has already been indicated, the composition of the level 6 ceramic assemblage as a whole, as well as in buildings, rooms and open areas, shows the following pattern: Standard Ware is most numerous (ca. 90%), followed by Dark-Faced Burnished Ware (ca. 5%), Standard Fine Ware (ca. 3%), Fine Painted Ware (ca. 1%) and Grey-Black Ware (ca. 1%). Henceforth these wares other than Standard Ware will be called the 'finer wares'.

Building II contained the majority (26.8%) of all the wares within buildings (it has to be taken into account, however, that this is the largest building, which was moreover completely excavated). Buildings IV and V yielded large numbers of sherds as well: 22.1 and 17.3% respectively (table 6.24). In building I 9.4% of the pottery was present. In the other buildings, including all tholoi, much smaller numbers of pottery fragments were present. This pattern for the ceramic assemblage as a whole also holds for the Standard Ware component. As to the other wares, however, some deviations from this general pattern can be noticed. Buildings IV and V stand out for their relatively large numbers of the finer wares (cf. table 6.24). Especially Grey-Black Ware is well-represented in building IV (n= 101). As has already been noticed in the visual inspection, Orange Fine Ware was well-represented in buildings IV and V, whereas only two sherds of this ware were recovered from building II. Apparently, the people using the two southern buildings, especially building IV, used and stored vessels of these imported (Dark-Faced Burnished Ware) and/or generally well-made and more luxurious wares more intensively than the people related to the other level 6 buildings.

Tholoi VI, VII and VIII each contained small amounts of sherds only. The finer wares are present in small numbers, especially Fine Painted Ware, which is even absent from tholoi VII and VIII. Tholos IX, however, stands out for its relatively large number of sherds (n= 980), Standard Ware as well as the finer wares. The possible special character of this building will be further discussed below.

Considering its relatively limited size, building X gave evidence of quite a large number of sherds (n= 1225); pottery seems to have been stored in large quantities here. In the neighbouring buildings XI and XII far fewer ceramic fragments were found. Fine Painted Ware was wholly absent from buildings X, XI and XII. Apparently, this 'serving ware' was not deposited in these storage buildings.

The location and orientation of the various buildings suggest a number of 'building clusters':

1. I-XII   (storage building with activity areas-storage building)
2. II   (storage building with activity areas)
3. IV-V   (storage building with activity areas)
4. VI-VII-VIII   (storage building with activity areas-activity area-storage building)
5. IX   (storage building)
6. X-XI   (storage buildings)

Moreover, an X^2 test showed that with regard to their complete artefact inventories the clusters differ in a statistically significant manner: there is an association between the various buildings (X^2= 291.2, df= 25, V= 0.06). This suggests that the clusters are indeed meaningful: the buildings in the various clusters were most likely functionally and economically related.

Within the clusters consisting of more than one building there are differences. In each of these clusters one of the buildings stands out for having a larger number of finds than the other buildings in that cluster: building IV of cluster 3, building VI of cluster 4, and building X of cluster 6. Within functionally related building groups, then, some specialization may have been present (the clusters are discussed in more detail in chapter 7).

7.1 Comparison of Selected Buildings

Now that we have an idea of the association of various buildings we can proceed with a statistical comparison of specific buildings. As has already been said not all buildings can be compared: on the basis of the above analysis the 'most interesting' cases have been selected, i.e. building I vs. building II, building IV vs. V, building II vs. V, tholos IX vs. tholos VI, and building X vs. building XI. The architectural structure, the number of activity areas and storage areas, the activities carried out and the products stored in these various buildings will be compared.

Building I vs. building II

Buildings I and II, both reconstructed as storage buildings with activity areas, differ in some respects, while at the same time they are alike in other respects (it has to be taken into account, however, that building I has not been completely excavated). Their main resemblance is that they are both mainly made up of small storage areas, and both have a central open area with ovens. In both buildings the activity areas were multi-functional, giving evidence of subsistence as well as manufacture/maintenance activities. Comparison of subsistence activities (i.e. baking/cooking, food deposition, serving, grinding, working of plant and butchering) gave a similarity coefficient of 98.7%. Comparison of the manufacturing/maintenance activities (i.e. spinning, working of wood, working of plants, working of skin, knapping, production of stone tools, boring, grooving, and working of textiles/basketry/leather) gave a coefficient of 81.4%. This indicates that the activities in buildings I and II are highly comparable, especially the subsistence activities. In fact all activities attested in building I are also present in building II.

	SW		SFW		FPW		GBW		DFBW		Total	
Building	N	%	N	%	N	%	N	%	N	%	N	%
I	1510	9.5	54	8.4	13	6	27	11.4	78	9.2	1682	9.4
II	4437	28	122	18.9	30	13.8	39	16.6	154	18.2	4773	26.8
IV	3293	20.8	179	27.8	83	38.2	101	42.8	273	32.3	3929	22.1
V	2690	17	142	22	86	39.6	19	8.1	138	16.3	3075	17.3
VI	642	4	47	7.3	3	1.4	10	4.2	15	1.8	717	4
VII	378	2.4	7	1.1	0	0	9	3.8	10	1.2	404	2.3
VIII	102	0.6	1	0.2	0	0	0	0	4	0.5	107	0.6
IX	871	5.5	39	6	2	0.9	15	6.4	53	6.3	980	5.5
X	1131	7.1	31	4.9	0	0	6	2.5	57	6.7	1225	6.9
XI	270	1.7	2	0.3	0	0	4	1.7	23	2.7	299	1.7
XII	538	3.4	21	3.3	0	0	6	2.5	41	4.8	606	3.4
Total	15862	100	645	100	217	100	236	100	846	100	17806	100

Table 6.24. The ceramic wares in the level 6 buildings.

With regard to storage one important difference immediately presents itself: building I lacks archives, such as room 6 of building II which was crammed with all kinds of objects. Furthermore, no rooms where grain was stored in bulk (as found in building II) were found in building I. The other storage rooms of both buildings were multi-functional. In buildings I and II a large variety of objects was stored in the general storage rooms: different pottery wares, small finds and flint/obsidian tools as well as debitage. If taken together these storage rooms of both buildings are rather alike as indicated by a similarity coefficient 87.2%.

Consequently, buildings I and II are similar in some respects, but different in others. The similarities consist of (1) the general structure of the buildings, consisting of small rectangular rooms, and both with a central open area with ovens; (2) the range of activities carried out in the activity areas; (3) the range of objects stored in the general storage rooms. The differences consist of (1) the fact that building II lacks buttresses and it is more regular, (2) the fact that building I lacks archives.

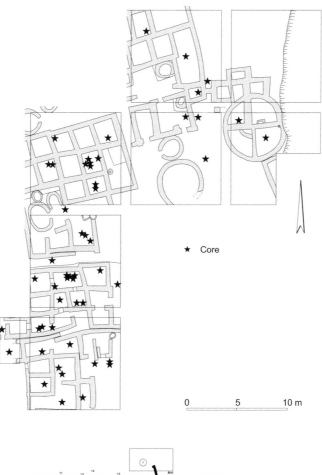

★ Core

0   5   10 m

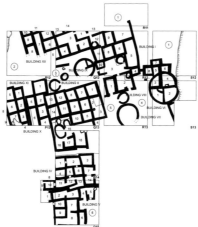

Figure 6.14. Level 6: distribution of flint cores.

Building IV vs. building V

At first sight, buildings IV and V (both not completely excavated) seem rather similar. In order to get a more detailed picture first the activities (carried out in and on the roofs of the buildings) will be compared. When comparing the subsistence activities (i.e. baking/ cooking, food deposition, serving, grinding, working of plants and butchering) a similarity coefficient of 97.7% is obtained, indicating that subsistence activities were largely identical in both buildings. The manufacturing/ maintenance activities in both buildings, however, were less similar: now a coefficient of 79% was calculated. A number of activities attested for building IV were absent from building V: the production of stone objects (as in room 2 of building IV), the production of wooden objects, and grooving of stone, wood or bone. Moreover, spinning was represented by 17 whorls and pierced discs in building IV, but only two such tools were found in building V. Furthermore, wood-working, skin-working, plant-working and knapping were probably carried out less regularly in/on building V, as suggested by the smaller numbers when compared to building IV.

The storage facilities of buildings IV and V contrast due to the absence of archives in building IV; this seems to be an important difference between both buildings. The other 'general storage rooms' of both buildings have been compared, giving a similarity coefficient of 73.6%. With regard to the other coefficients calculated in this section, this coefficient expresses the largest difference noted as yet. But still the inventories of the general storage rooms of buildings IV and V resemble each other in a significant way (ca. 75%). However, when comparing the rooms it should be taken into account that 13 rooms of building IV are compared with 5 rooms of building V. In general objects were stored in much smaller numbers in building V. Objects that were stored in building IV, but were absent from building V are: labrets, spindle whorls, polishing/rubbing stones, pottery stands, backed knives, cores for side-blow blade flakes and lustred sickle elements.

Therefore, as buildings I and II, buildings IV and V resemble each other but also differ in some important respects. The similarities consist of (1) the general structure of the buildings, consisting of small rectangular rooms, various portholes in building IV, and at least one in building V; (2) the range of subsistence activities carried out in the activity areas. The differences consist of: (1) the fact that building V seems to be somewhat less regular than building IV; (2) the absence of ovens or other features in building IV; (3) a somewhat different range of manufacturing/maintenance activities; (4) the lack of archives in building IV.

Building II vs. building V

Buildings II and V have been compared because they are the two structures where sealings, tokens and figurines were stored in large numbers in the so-called archives (fig. 6.8).

Let us start with the range of activities carried out in both buildings. What should be noted first in this respect is the absence of a cluster of ovens in a specialized activity area in building V (such as area 15 in building II). In building V just one very small and isolated *tannur* was found (however, it should be taken into account that Building V has not been completely excavated). When the subsistence activities (i.e. food deposition, serving, grinding, working of plants and butchering) are compared a very high similarity coefficient 98.2% is obtained. Apart from the oven-related activities, subsistence activities in both buildings are therfore more or less similar. The manufacturing/maintenance activities (i.e. spinning, wood-working, plant-working, skin-processing, knapping, boring, grooving, working of textiles/baskets/leather, working of wood/bone/stone), however, differ more, as expressed by a coefficient of 78.3%. In general fewer artefacts related to these activities were found in building II, and some activities are rather well-represented in building V, i.e. the processing of wood, skin and plants.

An important difference between buildings II and V is that grain storage rooms were absent from building V. General storage rooms, however, were present in both buildings. When comparing these chambers a similarity coefficient of 84.2% is obtained, indicating that in the general storage rooms of buildings II and V more or less the same range of artefacts was stored. A number of flint/obsidian artefacts that were stored in building V, however, were lacking from building II: borers, pieces with fine and with abrupt retouch, notches, truncations, denticulates, corner-thinned blades and shape-defined elements. Moreover, axes were not present in the general storage rooms of building II.

As has been argued above, with respect to administrative objects and subsequently administrative practices, buildings II and V seem associated. The archives of building II contained far more finds than those of building V (n= 2480 vs. n= 1256). The majority of objects (n= 1652) stem from room 6 of building II. In both buildings complete, but especially broken items were stored. Between the inventories of the archives of the two buildings a number of differences can be observed, the most important of which are:

- animal figurines are absent from building V, and human figurines are not as well-represented in building V;
- awls occur in smaller numbers in building V;
- grinding tools are present in smaller numbers in building V;
- labrets are absent from building V;
- spindle whorls are absent from building V, whereas pierced discs (most likely used as spindle whorls) occur in small numbers only in this building;
- sealings and tokens occur in smaller numbers in building V;
- sling missiles are found occasionally only in building V;
- backed knives, side-blow blade-flakes and pieces with abrupt retouch are absent from building II.

Apparently, the use of the archives of both buildings was not identical: different products were stored in different quantities. Relevant in this respect is Duistermaat's analysis of the distribution of the sealings (Duistermaat 1994:56-62; Akkermans and Duistermaat 1997:19). Two major differences are observed.

First, most sealings of building II carried stamp-seal impressions, while the other features contained much larger amounts of sealings without impressions.

Second, the sealings stored in building II show different impressions from the ones found in building V (although they occasionally show a similar - but not identical - general type of design), suggesting that buildings II and V were used by different sealing agencies. Moreover, most sealings in building V are unimpressed (Duistermaat 1994:60).

Thus, although buildings II and V are related on the basis of the presence of archives, the range of activities and the range of objects stored in general storage rooms, the above analysis has clearly indicated that major differences are present. This suggests a contrast in use and function of both buildings.

Tholos VI vs. tholos IX

Tholoi VI and IX are rather different, as pointed out by their location, architectural structure and function. If the stored objects in both buildings are compared, first of all the large difference in number (139 objects in tholos VI and 1022 in tholos IX, sherds included) must be noted.[53] The relatively high similarity coefficient (88.7%) is mainly due to the composition of the ceramic assemblage which is rather alike in both units. A relatively large number of other finds was peculiar to one of either buildings, indicating differences between the use of both buildings as storage units. As a result of their small quantity (n= 1 to 3), however, this difference is not clearly expressed in the coefficient. Objects that were stored in tholos IX but that were lacking from tholos VI are: an axe, an impressed piece of clay, a fragment of a pestle, a hammer, six fragments of figurines, three fragments of stone platforms, and a bone spatula. This indicates that a far larger variety of objects was stored in tholos IX than in tholos VI. As has already been argued, in view of the presence (and association) of sealings, tokens and figurines, tholos IX might be regarded as an archive. It was also suggested that the tokens, sealings and the single figurine in the activity areas of tholos VI were not stored, but that they may have been actually used here in administrative transactions.

Building X vs. building XI

Buildings X and XI, situated directly east of the central building II, both not completely excavated, have been reconstructed as storage buildings. On the floors of rooms 4 and 6 small amounts of grain were found, most likely indicating that these areas functioned as grain storage rooms. This represents the main difference between buildings X and XI: in building XI no traces of charred grain were recovered. Another difference is the presence of a long-drawn chamber (no. 6) in building X. In order to verify the difference suggested above between buildings X and XI the assemblages from the general storage rooms of both structures have been compared.[54] A high similarity coefficient of 93.4% was calculated, suggesting a significant resemblance. However, as was the case for tholoi VI and IX, this high number is mainly caused by the composition of the ceramic assemblage which is rather similar in both units. The fact that in building X 1225 and in building XI only 299 ceramic sherds were found already seems to indicate an important difference (in both buildings not a single fragment of Fine Painted Ware was found). Objects that were present in building XI, but not in building X are: figurines, hammers, pendants and stone platforms. Building XI lacked loamers, pierced discs, tokens, polishing stones and 'white ware'.

---

[53] Flint/obsidian excluded: from tholos IX these artefacts were not analysed.
[54] Flint/obsidian artefacts excluded, since the flint/obsidian assemblages of buildings X and XI were not analysed.

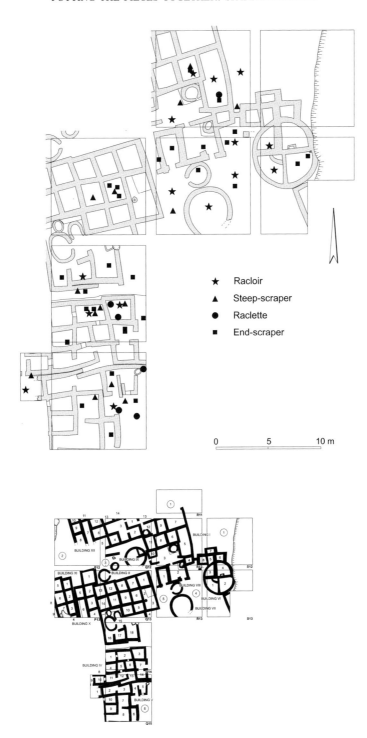

Figure 6.15. Level 6: distribution of flint scrapers.

8 Buildings and Open Areas compared

In this section objects from all buildings are compared to objects from all open areas. First pottery, then the small finds and finally flint/obsidian artefacts will be taken into account.

In a X^2 test where the assemblages (ceramics, small finds, flint/obsidian tools and flint/obsidian debitage) of both buildings and open areas are compared a high X^2 of 120 was obtained, indicating that at the 5% level (and even at the 10% level) there is evidence for a significant association between the assemblages in both kinds of spatial units. A low Cramer's V of 0.069, however, indicates that the association is very weak.

In table 6.25 the distribution of the various wares over buildings and open areas has been depicted. The majority of the pottery (78%) is immediately seen to stem from the various buildings. This is an important difference, most likely indicating (not surprisingly) that pottery was used (stored) in much larger numbers in buildings than in open areas. However, the composition of the ceramic assemblages from the buildings and open areas is more or less the same. Similarly, if only the activity areas in buildings are compared with the open areas (which are regarded as activity areas; see above), the same pattern emerges. This suggests that the distribution of the various wares within the level 6 settlement was more or less uniform. The functions of buildings and open areas, however, differ strongly: as has been argued, most buildings were storage buildings, whereas the open areas seem to have been used as activity areas and in certain instances for the dumping of debris. How, then, to explain the correspondence in ceramic assemblage composition between these two different sets of activity areas? Apparently the ceramic assemblage, consisting of various ceramic wares, is in this instance too coarse a category to express differences: as will be shown below, finer distinctions within the small finds and flint/obsidian assemblages do indicate differences! Most likely, if pottery shapes of the various wares were to be analysed differences between the ceramic assemblages of buildings and open areas would appear (e.g. more storage vessels - jars - are expected to occur in the buildings than in the open areas). As has already been indicated, however, a representative shape analysis of the pottery has not yet been carried out, and in the present study therefore vessel shape could not be incorporated in the analysis. It should be mentioned, though, that the limited number of reconstructed vessel shapes of all wares indeed seems to indicate differences between buildings and open areas: bowls are more evenly distributed than jars, which concentrate in buildings. However, the number of reconstructed vessel shapes is too small to allow any firm conclusions yet.

The number of small finds in buildings and open areas differs largely. In the buildings a total of 1203 small finds was found, whereas in the open areas only 71 such objects appeared. Statistical comparison between buildings and open areas with regard to the small finds assemblage gives a similarity coefficient of 62%, indicating that indeed there is a significant, but not very large, difference in assemblage composition between buildings and open areas. With regard to the composition the main difference between the two spatial categories is due to differences in the number of items of personal adornment, stone vessels, sealings and tokens. Items of personal adornment and stone vessel fragments were relatively well-represented in the open areas. Most likely these small objects had been lost in these areas. Sealings and tokens are peculiar to (some) buildings and hardly occur in open areas, and therefore these two (administrative) categories particularly account for the difference between open areas and buildings. However, a number of the objects ascribed to the open areas may have been dumped rather than used there. If, for instance, the objects from area 2, which probably represent dump, are left out then a lower coefficient of 60.2% is obtained.

If only the activity areas in the buildings are compared with the open areas, a higher coefficient of 73.9% is obtained, indicating that the activities are comparable. The main differences are represented by: items of personal adornment, which are relatively well-represented in the open areas (again: probably lost); grinding implements, which seem to have been used more intensively in the indoor activity areas; stone vessel fragments, which were found more often in open areas (lost, discarded); and the 11 large oval clay objects which were unique to the roof of building V.

	Buildings			Open areas		Total	
Ware	N	%	N	%	N	%
SW	15862	89.1	4539	91.1	20401	89.5
SFW	645	3.6	140	2.8	785	3.4
FPW	217	1.2	24	0.5	241	1.1
GBW	236	1.3	33	0.7	269	1.2
DFBW	846	4.8	246	4.9	1092	4.8
Total	17806	100	4982	100	22788	100

Table 6.25. Buildings and open areas: ceramic wares.

The general composition of the flint/obsidian assemblage in both open areas and buildings is more or less the same: ca. three quarters of the assemblage consist of tools, whereas another quarter is made up of debitage.[55] Looking at it in more detail, however, we see that in the buildings 17% consist of tools and 83% of debitage, whereas in the open areas 26% are tools and 74% is debitage: within the assemblages relatively more tools are present in the open areas than in the buildings. Perhaps this indicates that in the open areas flint/obsidian tools were used more frequently than in the buildings. When the flint/obsidian assemblage is broken up into the most important categories, it appears that with regard to the composition of the assemblages the buildings and open areas are rather comparable: a similarity coefficient of 89% was computed.

If only the activity areas in the buildings are compared with the open areas, a coefficient of 90.7% is obtained, indicating that the range of activities in activity areas within and outside buildings in which flint/obsidian tools were used was more or less comparable.

In conclusion, it can be said that the composition of the pottery, small finds and flint/obsidian assemblage of buildings and open areas is in all cases more or less similar. As

[55] The reader is reminded that open areas 2 and 3 have not been analysed and are therefore not taken into account.

has been argued, this most probably indicates that the functions carried out in both classes of spatial units were comparable as well. The major difference between buildings and open areas is that the buildings mainly functioned as storage units.

In the above analysis the architectural features, i.e. the ovens, have not been included. In the open areas only two ovens (DA and AR) were encountered (in area 3). However, in the two open areas within buildings (areas 3 and 15 in buildings I and II) more ovens are present (fig. 3.1). In this respect these areas resemble open area 3. Another resemblance is that open area 3 is a secluded and central space. In function and character open area 3 is therefore closer to the open areas within buildings than to the other open areas. Contrary to area 3, the other open areas are located around the buildings, they are not secluded (but open), and they acted as passageways.

Part III

9 SPATIAL ANALYSIS OF THE LEVEL 3 SETTLEMENT

9.1 Introduction

In the following, a restricted analysis of artefact distributions within the Halafian level 3 settlement will be presented. This settlement represents an unburnt village which seems to have been gradually and 'normally' abandoned. The study is restricted in the sense that it is specifically aimed at obtaining insight in the general character of artefact distributions in the level 3 village, which is then compared to that of the level 6 Burnt Village in order to get a grip on the processes of abandonment of the level 6 settlement (cf. chapter 4, section 4). First the level 3 village will be briefly introduced.

9.2 The Level 3 Village

Level 3 is part of the Early Halaf or Balikh IIIB period, and is dated at around 5100-5050 B.C. (5910-5840 cal BC). Level 3 has been divided into three closely related sublevels, i.e. 3C to 3A (Verhoeven and Kranendonk 1996:85-106). The earliest level 3C is represented by an impressive stone wall. In level 3B times this wall was incorporated into the large, multi-roomed building I. This building is associated with a series of circular structures (tholoi) and other features. Level 3A starts with the construction of the small rectangular building IV south of the main building I (the latter structure remaining in use). So far, the generally well-preserved level 3 remains, which occasionally stood to a height of about 1.50 m, have been traced over an extensive area, i.e. around 875 m².

In the present analysis I will largely restrict myself to level 3A (fig. 6.16), since this phase represents the period during which the level 3 settlement was being abandoned: in level 3A formation processes and processes of abandonment can be studied and compared with those of the level 6 settlement.

Level 3C is represented by parts of a large stone wall, which could be traced over a distance of at least 18 m. The wall was oriented more or less east-west. The wall was constructed of several rows of roughly hewn, gypsum boulders. The width of the wall varied between 0.75 and 1.75 m. It partly stood to a height of 1.20-1.50 m. The southern, exterior façade was plastered. Akkermans (1993:57) suggested that this wall served as a kind of retaining wall supporting a terrace on the top and along the northern slope of the southeastern mound. The north-south oriented parts then may have served as 'grips', clamping the support wall into the terrace and strengthening it to withstand the terrace's lateral thrust.

In level 3B times, the level 3C stone terrace wall was incorporated into the large, rectangular and multi-roomed building I (fig. 6.16). Building I was oriented NWW-SEE. The extensive structure consisted of a large western wing measuring ca. 15.90 x 9.50 m, which was separated from an elongated eastern wing measuring 16.50 x 3.20 m by a small court measuring ca. 6 x 2 m. The building comprised 22 small rooms, varying in size between about 2.60 m² (room 14) and 5 m² (room 4). The area west of rooms 1 and 2 seems to have been a courtyard.

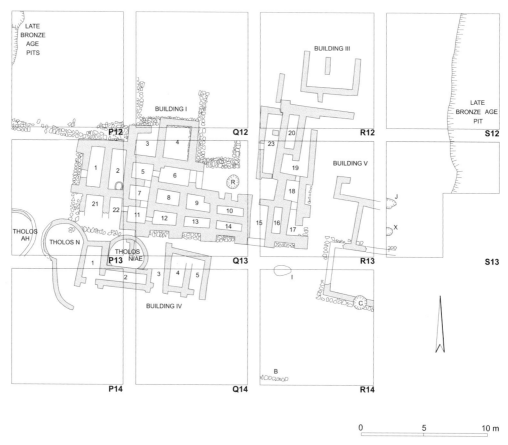

Figure 6.16. Plan of level 3A architecture.

The walls of building I still stood to a considerable height, i.e. between 0.50 and 1.20 m. The walls, each about 50-60 cm thick, were, like the walls of the other buildings, built of pisé. Most walls seem to have been founded simply on the tell surface. The building's exterior façade was supported by a series of large buttresses at wall junctions. The buttresses actually stood upon a stone foundation which ran along the proper façade of the building. This foundation wall was built of irregularly hewn gypsum boulders, laid down in two rows and one course high. The stones were joined by a loamy mortar and were covered by a mud layer. The stones carried a mud plaster, in its turn covered by a thin white coating; apparently, this stone foundation wall was not sunk. A similar plaster was found on both the exterior and interior wall faces of building I. The foundation wall and its buttresses were built along those parts of building I that were vulnerable to slope erosion; they served to consolidate and protect building I. At the same time, they gave the building a monumental, niched appearance.

Access to building I was gained in a variety of ways. The main entrance was situated at the end of the small courtyard in the northeast of square Q13. This court could be reached from the south through a passage ca. 1.30 m wide (area 15) between the western and eastern wings. The courtyard was also accessible through a narrow passage from the terrace in the north. The main doorway, ca. 1 m wide, had a low, thickly white-plastered staircase with two steps. So far, it seems that the eastern wing of building I could be entered from the east only, through two doorways situated at the far ends of the building. Some rooms seem to have had separate entrances (fig. 6.16). Circulation through the building itself was enhanced by means of narrow, ca. 40 cm wide doorways, all provided with low, mud-plastered thresholds. In some rooms no passages at floor level were found; probably these rooms were accessible from a somewhat higher elevation or even from an upper storey.

Akkermans (1993:61) has already made it clear that the main building may have had two storeys (see fig. 2.28 in Verhoeven and Kranendonk 1996), because of (1) the considerable thickness of the walls (ca. 50-60 cm), strong enough to support an upper storey (the other structures all lack such thick walls), (2) the presence of stone foundations and large buttresses along the exterior façade, able to withstand the lateral thrust of the building on the sloping tell surface, (3) the absence of doorways at floor level in several rooms, suggesting that these were accessible from above only, (4) the extremely small size of all rooms, which were hardly suitable for living in, and (5) the virtual lack of household structures (only in room 2, a *tannur*-like oven was found; cf. Akkermans 1989a:39). If, indeed, an upper storey was present in building I, this must have contained the actual areas of living, whereas the lower one may have served mainly for storage.

Building III, situated immediately north of the eastern wing of building I, basically consisted of one room, divided by a small, free-standing wall into two smaller units. Remarkably, the northern face of the building was wholly open. Most likely, building III served as a stable or barn. Building V east of building I mainly consisted of one rectangular room. Its walls were, like building I, aligned by a low stone construction consisting of one course of irregularly hewn gypsum boulders.

These rectangular buildings on the top of the tell were surrounded by a series of circular structures along the slopes (tholoi A, I, O, S, AC, N/AE; fig. 2.25 in Verhoeven and Kranendonk 1996). These tholoi were made of mudbricks and their diameters range between 1.50 to 3.10 m (Akkermans 1989a:28-35). Finally, building I and the tholoi were surrounded by a large number of pits of different shapes and dimensions (Verhoeven and Kranendonk 1996:98-101).

During level 3A the small building IV was raised immediately south of building I (fig. 6.16). As has been said, the latter structure, and the other level 3B buildings III and V, remained in use in the level 3A period. The various level 3A buildings were surrounded by fill layers, some of which were laid down during the use of these structures, others when these features had already been deserted. In the level 3A period a rather large oblong room (no. 23) was added to the western wing of building I. A buttress divided the room into two smaller units. These could be entered through a ca. 50 cm wide doorway from the small court between the western and eastern wings of building I. Building IV consisted of a series of elongated, narrow rooms oriented roughly north-south on the flanks, connected by a similarly small-sized room oriented east-west. The structure measured ca. 13.50 x 4 m. Access to the tiny rooms was gained from the south through narrow doorways each between 40 and 80 cm wide, marked on both sides by buttresses.

Building IV partly cut through a tholos (N), which was already out of use but which was still standing to a considerable height. West of tholos N, fragments of another tholos (AH) were found. Apart from tholos N, building IV cut through another level 3B circular structure, viz. tholos N/AE.

• Small Find

0   5   10 m

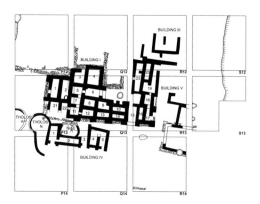

Figure 6.17. Level 3A: distribution of small finds.

9.3 Visual Inspection

In the visual inspection the distribution of the small finds, the pottery, the flint and obsidian artefacts and the animal bones in level 3A were studied. As not all areas of the level 3A settlement have been dealt with by the specialists working with the different assemblages (small finds, animal bones, etc.), the various maps (figs. 6.17-6.20) vary.

Small finds

The small finds are represented by 144 objects (excluding nine groups of sling missiles stored in pits, some of which contained hundreds of them). Sixty of the small finds (41.6%) were found in building I. Buildings III and V contained no small finds, building IV just one. So the majority of the objects stem from open areas. In fig. 6.17 it can be seen that the objects in the open areas tend to be situated in the direct vicinity of architecture. From the open spaces somewhat away from architectural structures far fewer artefacts were recovered.

In the central squares P13 to S13 a series of clusters can be distinguished. First of all there is the already mentioned cluster of sling missiles in nine pits dug into the floorlevel of the small court between buildings I and V, directly west of building V. About 1000 sling missiles were stored here (Spoor and Collet 1996:450). Clearly, they represent in-situ finds. In the immediate vicinity of the missile groups two fragments of grinding slabs, a broken polishing stone and a large fragment of a painted Halaf bowl were found. A second cluster is represented by 15 objects in room 17 of building I. Apart from two broken figurines, three complete awls were recovered from this cluster, which is remarkable, since awls are mostly broken when found.[56] A third cluster of objects was recovered from the passage to room 6 of building I. Here, in pit R, 11 tokens, a stone bowl fragment and a small clay object of unknown function were found. Between pit R and the entrance to room 6 two sling missiles and fragments of a bead, a stone bowl, a human figurine and a bone spatula were present. A fourth cluster of small finds was found in the western part of building I, i.e. rooms 1, 2, 7, 21 and 22 (n= 31, 51.6% of the small finds in building I). Within this large cluster three smaller clusters can be recognized: one in room 1, one in room 7 and one in and directly near the entrance of room 7. All these finds were recovered from floors and can be considered to be in primary context.[57] Especially the finds on and around the entrance to room 7 are interesting. They consist of a concentration (n= 101) of unbaked clay objects of all sorts, including some very stylized human and animal figurines, miniature vessels, balls, rectangular plaques, discs and cones, most likely representing tokens. One of these clay objects showed traces of a stamp seal impression (cf. Akkermans 1993:82, fig. 3.23). It is possible that room 11 served as some sort of archive, comparable with those found in the Burnt Village. Perhaps, though, the position of these objects on/near the entrance indicates a special deposit.

---

[56] The objects of the cluster in room 17 are: a fragment of a grinding slab, two stone discs (one complete), a disc of baked clay, a fragment of a polishing stone, three bone awls, a sling missile, fragments of two figurines, a token and three small objects of clay of unknown function.

[57] The finds in room 1 consist of three small ceramic painted vessels, a flattened pierced stone and a worked bone fragment (cf. Akkermans 1989b:39, 202 and fig. IV.35, nos. 255-56, fig. IV.46, nos. A-B). In room 7 three awls, one of them complete, a polishing stone and fragments of a grinder, a pestle and a grinding slab were found.

The central part of building I (i.e. rooms 4, 6, 8-10, 12-14) and the western wing (apart from room 17) contained only very few small finds (fig. 6.17). A rather large amount of objects in the level 3A village (n= 77, 53.5%) were complete.[58] The complete objects, many of which small clay items such as tokens, were concentrated in the cluster in the west of building I.[59] Likewise finds from floor contexts (n= 36, 25% of the level 3A small finds assemblage) mainly stem from the cluster in the west of building I.

Pottery

The Level 3A Halafian pottery (n= 7265) consists of various wares: Halaf Fine Ware, Orange Fine ware, Vegetal Coarse Ware, Red-Slipped Burnished Ware, Mineral Coarse Ware, Grey-Black Ware and Dark-Faced Burnished Ware (Le Mière and Nieuwenhuyse 1996: table 3.36). The majority of these ceramics are made up of Halaf Fine Ware.

Generally the pottery is evenly distributed, but small concentrations can be observed, i.e. in the north of square Q14 (n= 943) and in some rooms of building I: the south of room 23 (n= 78), the south of room 19 (n= 46), and especially in room 17 (n= 213). Within building I the majority of sherds stem from the western part (rooms 1, 2, 5, 7, 11, 21 and 22). In the open areas pottery concentrates in the passage to the entrance of room 6, building I, and in the northeast of square Q14.

Only few complete or reconstructable vessels (n= 15) were found, inside as well as outside building I. The complete vessels are represented by three small painted jars (cf. Akkermans 1989c fig. IV.35: nos. 255 and 256, fig, IV.46: a and b) and by a painted carinated bowl (Akkermans 1993: fig. 3.19: no. 29).

Bowls and jars are more or less evenly distributed, but pots tend to be present mainly outside building I (in building I only 9 out of the 47 pot fragments were found). Rim and body fragments are evenly spread, but base fragments are relatively well-represented within building I (33 out of 167 fragments: 19.8%). Within the general pattern observed, analysis of the distribution of pottery with specific characteristics, such as kind of temper, presence of decoration, nature of surface treatment, etc, did not give evidence of concentrations.

Halaf Fine Ware and Vegetal Coarse Ware, representing the majority of the level 3A sherds, were evenly distributed. At first sight the other wares seem to be mainly present outside building I (e.g. fig. 6.18). However, it has to be taken into account that building I represents ca. 25% of the excavated space. When evenly distributed, therefore, only a quarter of the finds are to be expected in this building. Indeed both Dark-Faced Burnished Ware and Grey-Black Ware indicate such a pattern, giving evidence of a rather even distribution. Of the Dark-Faced Burnished Ware sherds (n= 57) nine (15.8%) were present in building I. Of Grey-Black Ware 25 out of 115 sherds (21.7%) were recovered from building I (fig. 6.18).

---

[58] Excluding the hundreds of complete sling missiles west of building V.

[59] Presumably, the complete objects are mainly represented by small objects because these objects are less susceptible to breakage than large objects. Furthermore, small objects are more easily lost than larger ones (Schiffer 1987).

0   5   10 m

Figure 6.18. Level 3A: distribution of Grey-Black Ware.

Flint/obsidian

The flint and obsidian artefacts (n= 514) were mainly found outside building I (in this building 96 pieces, 18.7%, were found).[60] As can be seen in fig. 6.19, there are some

[60] It should be kept in mind that the artefacts of rooms 1, 2, 21 and 22 of building I have not been taken into account, since the flint/obsidian databases were not obtainable.

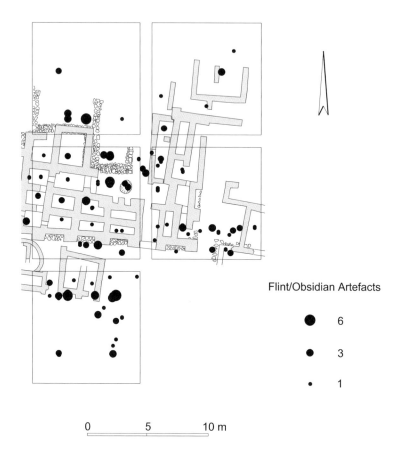

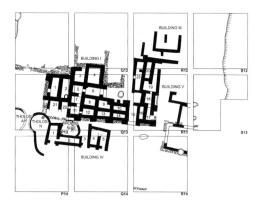

Figure 6.19. Level 3A: distribution of flint and obsidian artefacts.

concentrations of the lithic implements: (1) in square Q14, south and east of building IV (n= 153); (2) in the small court between buildings I and V (n= 25); (3) in the passageway to room 6 of building I (n= 45, and 17 in pit R); (4) on the terrace north of room 3 of building I (n= 27); (5) on the terrace east of room 3 (n= 44).

In general the different tool and debris classes follow this general distribution. Furthermore it appeared that there is no significant difference between the distribution of flint and obsidian artefacts or between tools and debris.

Figure 6.20. Level 3A: distribution of animal bones.

Animal bones

The main species identified in the level 3 village are: sheep (*Ovis aries*), goat (*Capra hircus*), cattle (*Bos taurus*), pig (*Sus domesticus*), dog (*Canis familiaris*), Equids (*Equus* sp.) and Gazelle (*Gazella cf. subgutturosa*). Apart from these species wild animals such as aurochs (*Bos primigenius*), red deer (*Cervus elaphus*), rodents, birds, molluscs and fish were identified (Cavallo 1996, 1997).

Of the 1705 bone fragments 468 (27.4%) stem from building I (fig. 6.20), again pointing to a more or less even distribution. As was the case for the small finds and flint/ obsidian implements, clear concentrations are present: (1) in room 17 of building I (n= 210); (2) in front of the doorway to room 6 (n= 154); (3) along the western wall of room 23 (n= 91); (4) in the eastern part of square Q14 (n= 441). In the central part of building I (rooms 4, 6, 9-10, 12-14) only very few animal bones were present.

Within this general pattern sheep/goat, cattle and pig (representing the majority of the animal bones) are evenly distributed. Of the few gazelle bones (n= 29) a small concentration (n= 9) was found in room 17 of building I. Of the equid bones (n= 10), seven were found outside building I, three inside it. Only three dog bones were found, all in building I. Of the general categories large mammals (n= 10) and medium-sized mammals (n= 99) 1 and 18 fragments respectively stem from building I. A distribution map of wild (3%) versus domesticated species (97%) showed no significant clustering.

9.4 The Character of the Artefact Distributions in Level 3A

From the above it appears that:

1. There is a more or less even distribution of artefacts and animal bones in the level 3A settlement;
2. Within building I a distinction between a western part (rooms 1, 2, 5, 7, 11, 21 and 22) and an eastern part can be made;
3. In many cases clusters of objects can be recognized.

How are these patterns to be explained? To deal with this question first of all the formation processes in the level 3A settlement need to be discussed. The first impression is that in level 3A we are dealing with objects mainly representing secondary refuse (see chapter 4), since (1) no fire or other catastrophe which sealed objects has occurred; (2) the majority of the objects are broken; (3) only few objects were recovered from floor contexts (most stem from fill deposits above floors). However, on floors a number of finds which are apparently in primary context were recovered: (1) a number of complete small ceramic vessels and small finds from the floor of room 1 of building I; (2) small finds in room 7 and small clay objects on and around the entrance to room 7; (3) fragments of large storage vessels appeared in room 5; (4) a complete painted bowl in room 4; (5) a complete painted bowl and a bone awl in room 16; (6) the large numbers of sling missiles in pits dug in the floorsurface of the small court between buildings I and V.

Room 1 was already excavated in 1986, and was provisionally termed room 21 (Akkermans 1989a:39, the main parts of building I were excavated in 1988: Akkermans 1993:56-66). At the time Akkermans (1989a:39) wrote: "Room 21 was marked by a hard loam floor. On this floor three complete and four fragmentarily preserved vessels were found. Near wall AB, opposite the entrance to room 21, a small heap of charred grain was found intermingled with fragments of a coarsely made bin. I suggest that this bin originally contained the cereal remains found. The highly fragile and crumbly state of preservation of the bin did not allow a proper reconstruction. The various fragments, however, indicated a low and wide shape. Next to it parts of a large storage jar appeared.

Other finds in room 21 included a flattened pierced stone and a worked bone fragment. Near the room entrance the left mandibula of a large bovine was found. All finds were concentrated in the middle of room 21, on or immediately above the floor. Do these finds represent primary contexts or not? The complete vessels and the grain bin point to a primary context but the fragmentarily recovered ceramics suggest a disturbed situation. Apparently parts of the pottery had already been lost in antiquity, either as a result of erosion or other disturbance after abandonment of the building, or as a result of incomplete deposition (i.e. refuse dumping). Although it cannot be proven yet, we are inclined to consider the finds in room 21 as being in situ."

The objects in and around building I seem to represent de facto refuse and/or abandonment stage refuse and/or provisional refuse. Apparently, the situation is less straightforward than it appears to be at first sight, since the analysis has indicated that in-situ contexts seem to be present. However, it should be kept in mind that these possible contexts in level 3A represent only a small fraction of all objects (ca. 4%). Nevertheless these artefacts are interesting.

First of all they give a clue towards room functions. Considering its relatively large size and accessibility, room 1 may have functioned as an activity area. The finds inside the room indicate that food preparation (the grain bin, the small ceramic vessels) and manufacturing/maintenance, such as spinning (the stone disc) and the making of basketry or leather tools (the bone implement?) took place in or around building I. The ceramic vessels were perhaps used for the short-term storage of food or as serving ware. The neighbouring room 2 must also have been an activity area, since here a small *tannur* was present, perhaps indicating the baking of bread in this room. The finds in room 7 and the other rooms of building I were probably stored there, since these rooms are much too small to represent activity areas. In room 7 different kinds of small finds were stored. It has already been mentioned that this room perhaps served as an archive, not unlike the level 6 archives. In room 5 large storage vessels may have been situated. In room 16 different items were apparently stored, as evidenced by the bowl and awls in this room. Considering their small size, the other rooms of building I also served as storage chambers, perhaps of objects or bulk products such as grain. The storage of sling missiles in a number of pits outside building I still remains an enigmatic procedure. Spoor and Collet (1996:450) suggest that they represent a sort of communal storage of weaponry in an 'arsenal'.

Having discussed the possible in-situ contexts and room functions something should be said about the structure and function of building I as a whole. Earlier it has been hinted at that building I mainly represents a storage building, with living areas perhaps in a second storey, and domestic activity areas in rooms 1 and 2 (Akkermans 1993:61; Verhoeven and Kranendonk 1996:94). Especially these western rooms 1 and 2, and rooms 21 and 22 are of interest, since it is here that domestic activity areas actually seem to have been present. Furthermore, these four rooms stand out because they seem to have been added to the building some time after its initial construction, because of: (1) the north-south orientation of these rooms, whereas all other rooms were oriented east-west; (2) the shifted east-west walls; leaping north with regard to the main axis of the building; (3) the separate entrance to room 1; (4) the *tannur* in room 2, absent from all other rooms. If we look at the distribution of the small finds within building I (fig. 6.17) it appears, as was already noticed, that they cluster in this western area, especially in rooms 7 and 11. On the plan it can be seen that these rooms were directly accessible from the west, via the in the level 3A period ruined room 22 (fig. 6.16). Could it be that these rooms were also functionally related to the western part of building I? If so, rooms 1 and 2, and possibly rooms 21 and 22 represented activity areas and rooms 7, 11 and perhaps room 5 marked storage areas related to the western rooms. Maybe building I consisted of living quarters and related (i.e. linked) storage chambers in the west and (communal?) storage chambers in the east.

Within the western wing of building I, a twofold division can be made: (1) the western rooms 1, 2, 5, 7, 11, 21 and 22 and (2) the central rooms 3, 4, 6, 8-10, 12-14. Apart from room 3, these rooms most likely functioned as storage rooms. This division is also suggested by the small number of finds from the central rooms (figs. 6.17 and 6.18), as opposed to the larger number of finds in the western rooms.

Having said all this, we see the possibility emerge that the difference in object quantity within building I (most objects coming from the western part) is mainly due to a functional difference. Perhaps only very few objects were stored in the central rooms, as reflected in the artefact patterning. Another possibility is that the western part of building I was still in use (apart from rooms 21 and 22), but that the central rooms, and perhaps the rooms in the western wing, were already abandoned in the level 3A period. In that case, however, it would be expected that these areas were used as dumps, giving evidence of many finds (as secondary refuse). As was noticed above, however, the dumping of refuse mainly took place in the open areas outside building I (e.g. in square Q14). This implicates that building I was abandoned and filled with debris over a relatively short period of time, thus simply not having had the oppurtunity to collect many objects. Indeed the various roomfills of the building consisted of rather homogeneous red-brown wall debris (Verhoeven and Kranendonk 1996:102), most likely indicating a rapid accumulation of debris.

If these hypotheses are correct, then the appearance of the western cluster within building I might be explained. That leaves the different object concentrations to deal with. Apart from the in-situ objects (i.e. the sling missiles west of building V) the concentrations largely consist of broken objects. Most likely, they represent discarded objects, i.e. secondary refuse. The concentrations indicate that refuse was not dumped anywhere, but mainly at specific locations. Especially the eastern part of Q14 seems to have been used as a dump, considering the large amounts of debris stemming from this area. The concentration in front of the entrance to building I (via room 6) perhaps represents a so-called door dump, i.e. debris that is swept out of buildings and remains just outside doors as provisional refuse (Binford 1983:165). Perhaps the cluster of debris in room 17 in the eastern wing of building I should be interpreted likewise, since room 17 represents the entrance to this eastern wing.

9.5 Level 3A compared to Level 6

If we compare level 3A to level 6, what appears? First of all, there is, of course, the absence of burnt deposits in level 3A. In other words, we deal with different abandonment and formation processes. The catastrophical fire in the level 6 village seems to have resulted in a rapid final abandonment (see chapter 4). In the level 3A settlement, however, a 'normal' and planned abandonment may be assumed (Cameron and Tomka, eds., 1993; Stevenson 1982, and see chapter 4).

In table 6.26 a number of general characteristics of the three major level 6 and level 3A assemblages (pottery, small finds and flint/obsidian) have been compared in order to obtain a somewhat more detailed view of the differences between both levels.[61] First of all, a number of general differences can be noted: (1) in level 6 far more objects were found; (2) the density of objects in level 6 is far greater than in level 3A; (3) in level 6 more objects stem from floor contexts; (4) in level 3A relatively more complete small finds and flint/obsidian tools were found. The main point of agreement between level 6 and level 3A is the proportion between complete and broken pottery (the majority, 99%, being broken).

9.6 Back to the Burnt Village: Which Scenario?

The way in which structures or settlements are abandoned has, of course, its effects on the composition of the objects left behind. In this respect three major factors can be distinguished: (1) was the abandonment gradual or abrupt? (2) what means of transport were available? (3) what was the distance to the new structure/settlement? (Cameroon and Tomka, eds., 1993; Schiffer 1985:26).

It has been suggested above that the level 3A village was most likely abandoned over a relatively short period of time. Apart from a number of in-situ contexts, the majority of the level 3A finds seem to represent secondary refuse. In the Burnt Village two scenarios were proposed with regard to its abandonment. In both scenarios the complete objects left behind most likely represent primary de facto refuse. In the first scenario, the level 6 village was largely functioning when the fire broke out: only building XII and the northern parts of building I were abandoned and in disuse. In these buildings and in the open areas secondary refuse was present, but inside the other buildings the broken objects mainly represent provisional refuse and/or abandonment stage refuse. In the second scenario the village was already being abandoned when the fire broke out. In this case, as in the level 3A settlement, the broken objects mainly represent secondary refuse.

In the Burnt Village many complete finds were recovered, but the majority of objects were fragments of artefacts. The proportions of complete pottery of levels 6 and 3A are more or less similar (table 6.26). In this sense, the comparison with level 3A nuances the in-situ or Pompeii character of the Burnt Village. It may be suggested that the in-situ contexts in both villages indicate that the settlements, or parts thereof, were abandoned and covered by debris over a rather short period of time. In the most straightforward case it is expected that if the settlements were gradually left, i.e. over, say, months or years, all (or nearly all) complete objects would have been removed. In view of the larger amounts of complete objects in the level 6 village (cf. table 6.26), it can be suggested that the abandonment of the level 3A village took place over a longer period of time than that of the level 6 settlement, where the fire instigated the final abandonment.[62]

[61] Pottery + small finds + flint/obsidian: n= 26,524 in level 6 and n= 7923 in level 3.
[62] Assuming that there is a positive correlation between the duration of the process of abandonment and the possibility of removal (and actual removal) of complete artefacts from a village.

	Level 6		Level 3	
	N	%	N	%
Pottery				
N	22547	100	7265	100
Density	26.5	-	10	-
Complete	59	0.26	15	0.21
Broken	22488	99.7	7250	99.8
Floor	5524	24.5	425	5.8
Fill	17023	75.5	6840	94.2
Small finds				
N	1422	100	144	100
Density	1.7	-	0.2	-
Complete	548	38.5	77	53.5
Broken	874	61.5	67	46.5
Floor	783	55.1	31	21.5
Fill	639	44.9	113	78.5
Flint/obsidian				
N	2555	100	514	100
Density	4.5	-	1.3	-
Complete (tools)	489	74.5	84	86.6
Broken (tools)	167	25.5	13	13.4
Floor	392	15.3	42	8.2
Fill	2163	84.7	472	91.8

Table 6.26. Comparison of the assemblages of levels 6 and 3.

But why did people not come back to collect complete objects? As has been indicated in chapter 1, the stratigraphy and dating point out that the prehistoric sequence at Sabi Abyad is largely uninterrupted. Only between the deep levels 11 and 10 (of the Balikh II period) a real break in settlement seems to have been present (Akkermans 1996, ed.: ix). Apart from levels 3C to 3A, all levels could clearly be separated stratigraphically and, moreover, each of these levels has its own architectural characteristics. This seems to indicate that these settlements did not grow organically over time, but instead represent separate villages. Illustrative in this respect is the fact that before the construction of the level 3A settlement level 4 and 5 remains were levelled in the central squares P13, Q13 and S13 (Verhoeven and Kranendonk 1996:73).

So there may have been an extended period of time (years) between the abandonment and decline of a village and the foundation of a new one (the stratigraphy and chronology, however, indicate that this period of time was not in the range of decades). It can be suggested that people left the site, settling somewhere else in the Balikh valley or its surroundings. In that case the means of transport most likely consisted of beasts of burden like cattle, and people would have carried things themselves.[63] The objects may have been covered by debris in the period of abandonment/decline, so that they were invisible by the time people returned, and therefore not recovered. It is also possible that people settled in another place at the tell, for instance on the northeastern mound, which shows a prehistoric sequence more or less similar to that of the southeastern mound. In that case, however, they could easily have transported their belongings to their new houses.

Summarizing and concluding, on the basis of the general comparison of levels 6 and 3A, I suggest that of both scenarios offered for the process of abandonment of the level 6 village (cf. chapter 4, section 4.1) scenario 2 is the most likely: parts of the village were already abandoned when the fire broke out.

In defence of scenario 2, the following may be put forward:

1. In level 6, as in level 3A, the character of the various deposits and assemblages indicates gradual abandonment of the village;
2. The appearance of broken objects inside the level 6 buildings is compatible with a village which is in the process of abandonment, the broken objects representing secondary refuse (and the scarcity of complete ceramic vessels was probably due to the fact that the complete ones had already largely been removed).

As was indicated in chapter 4, there are three main problems with scenario 2: (1) why doesn't the stratigraphy indicate disuse and abandonment before the fire? (lack of unburnt debris underneath the black-burnt layer); (2) why were complete objects left behind?; (3) why did a dismantled village burn so fiercely? Using the information just presented, these questions should now, finally, be dealt with:

ad. 1: Most likely the structures in the village had been out of use for a short time only when the fire broke out (i.e. accumulation of debris due to deterioration had not yet taken place);

ad. 2: Complete objects were left behind because when the fire started, instigating the final period of abandonment, people simply had no time to collect them (and they were not recovered because people settled away from the tell);

ad. 3: Apparently, the village still contained enough wooden building material (e.g. roof beams) to allow for a fierce fire in which most of the poles were reduced to ashes.

---

[63] In the Near East the earliest unambiguous evidence for the wheel, and therefore wheeled transport, comes from well after the Neolithic period, i.e. in the Early Bronze Age, at ca. 2600 B.C. (Littauer and Crouwel 1979).

As has been argued in chapter 4, it cannot be excluded that the fierce conflagration in the level 6 settlement was the result of deliberate burning, i.e. the ignition of fuel inside the buildings. In that case symbolic and ritual processes may account for the abandonment of the village. Complete artefacts, then, may not have been recovered because they were ritually 'sealed' (and there would have been a taboo on recovery), and the fierce fire would be accounted for by the ignition of deliberately deposited fuel inside structures. According to scenario 2, this possible 'ritual closing' of the village would have been executed in a village which was in the process of abandonment.[64]

---

[64] Ritual breaking/smashing of ceramic vessels (see e.g. Campbell 1992), which would account for the fact that 99% of the pottery consists of sherds, is unlikely, since almost none of these sherds could be refitted into vessels.

CHAPTER 7

ON A DUAL BASIS:
NOMADS AND RESIDENTS AT TELL SABI ABYAD

1 Introduction

This final chapter presents the synthesis of the archaeological ethnography of the level 6 community of Tell Sabi Abyad.[65] First, in section 2, the proposed spatial classification (i.e. activity areas, storage areas and discard areas) with which the analysis began will be reclassified on the basis of patterns as observed in the spatial analysis, in an attempt to present a classification which does justice to the way the level 6 inhabitants perceived space.

In section 3 an interpretation of the socio-economic structure is given. It will be argued that the level 6 society consisted of two main groups, i.e. nomads and residents. In this respect a distinction is proposed between a household economy, related to families at the site, and a village economy, related to production for the nomadic groups. In section 4 the structure and structuring character of the Burnt Village is dealt with, applying Bourdieu's *Theory of Practice*. It is argued that the round and rectangular buildings in the Burnt Village were *structuring structures*, i.e. they structured social practices, which in their turn structured architectural traditions and conventions. In this view the tholoi were related to households and residents, whereas the rectangular buildings were related to the nomads.

In sections 5 and 6 two possible mechanisms for binding together the dual society at Sabi Abyad are discussed and contextualized. In section 5 a specific ritual in the Burnt Village is discussed. Section 6 deals with the possible significance of the association of sealings, tokens and figurines for the constitution of the society. A summary and conclusions can be found in section 7.

2 From Etic to Emic: Reclassifications

In chapter 2 (section 6) a three-fold division of functional areas has been proposed, i.e. activity areas, storage areas and discard areas. By distinguishing different object contexts within these rooms (i.e. interactive, depositional and discard contexts), multifunctionality of these areas was accounted for (table 4.5). In the functional assessment in chapter 6, different kinds of activity areas, storage areas and discard areas were defined.

---

[65] In this chapter a number of ethnographic and archaeological parallels are used; for a theoretical discussion about the use of analogy see chapter 5, section 1.

Twenty activity areas were recognized (11 interior areas and 9 exterior areas). Interestingly, monofunctional areas seem to have been absent; all reconstructed activity areas gave evidence of several activities, i.e. they were multifunctional. In some areas subsistence as well as manufacturing/maintenance activities have been attested. Most areas, however, gave evidence of various manufacturing/maintenance activities. Besides activities, all activity areas witnessed the (deliberate) deposition of objects and primary discard of refuse.

Basically three kinds of storage areas can be distinguished, i.e. areas for:
1. storage of domestic equipment (n= 50);
2. storage of grain in bulk (n= 5);
3. storage of 'administrative objects' (i.e. sealings and tokens) (n= 5).

We see that storage of domestic equipment was the most widespread type of storage (cf. figs. 6.1-6.3). The domestic tools which were stored, such as pestles and bone awls, were in all cases few in number. Moreover, it appeared that of the two main functionally related sets of tools, i.e. (1) pestles and mortars, and (2) grinders and grinding slabs, one part of the set was virtually always absent. Furthermore, a large amount of the stored objects were broken.

Bulk storage of grain was attested in buildings II and X only, but possibly the very small 'cells' in buildings I and VI also were used to store grain in bulk.

Five 'archives', or rooms in which large numbers of sealings, tokens and figurines were stored, were identified in buildings II (rooms 1, 6 and 7) and V (rooms 6 and 7) (fig. 6.8). The finds in these rooms were always associated with other small finds, which were regularly broken. In room 7 of building I, moreover, grain seems to have been stored, presumably not in bulk, considering the limited quantity which was recovered.

Only one discard area, in the southeast of square P12, has been distinguished. This area was marked by a clustered occurrence of broken objects.[66]

From the above it appears that especially the storage of limited amounts of domestic items ('general storage rooms') and the 'archives' seem to represent important emic categories, as indicated by the large number of general storage rooms and apparent importance attached to the storage of broken seals, tokens and figurines. The association of these three object classes (see chapter 6) suggests another basic emic category. Apart from this, some emic concepts can be reconstructed by looking at the general structure of the level 6 village. The following two characteristics seem to be relevant: segmentation and separation (buildings consisting of many small rooms; many boundaries; basic distinction between round and rectangular architecture), closed nature of settlement (the buildings are not easily accessible, and they are closely attached).

Whether we analyse the level 6 settlement in general or in more detail, we seem to be dealing with a village, and subsequently with a community, which is structurally different from our own (modern western) villages and communities. In the following sections it is attempted to reconstruct the meaning of the emic categories mentioned, in order to reconstruct more such classifications. First, the socio-economic structure will be dealt with.

[66] Excavations in 1997, however, revealed that on the northeastern periphery of the settlement large dumps, consisting of ashes and domestic debris, were present.

3 Socio-Economic Structure

The fact that separate buildings (and six clusters of buildings, cf. chapter 6, section 5.4) do exist in the Burnt Village most likely indicates that different groups of people used these buildings. It is highly unlikely that every person at the site had access to all rooms in all buildings; this would imply a very loose, 'open', and unorganized societal structure which is at odds with the rigid structuring and segmentation of space in the village. It is far more likely (and more efficient) that specific buildings were used by specific groups, and this would also account for the (slight) functional differences between buildings which the spatial analysis has indicated. As has been argued in chapter 6, these groups did not live in the rectangular buildings, but instead used them for storage and to a lesser degree for specific domestic activities. Actual living units or houses are most likely represented by two superimposed tholoi with hearths in square R13, and a smaller tholos with a hearth in square R14. Moreover, it is likely that tholoi VI and VII also functioned as dwellings (cf. chapter 6, section 5.3). These houses would have been the 'homebases' of families or households.

Household is defined as all persons that eat, sleep and work together in a separate living unit; generally the use of such a unit is restricted to the members of the household. Consequently the household is a unit that shares a combination of production, coresidence, and reproductive and consumptive tasks (Byrd 1994:643; Netting 1982, 1990; Sahlins 1972:41-148; Wilk and Netting 1984; Wilson 1988).[67] I will designate the buildings and spaces used by the household (e.g. living rooms, storage rooms, stables, kitchen, courtyard) as the household compound (see also Flannery 1976:25, 37). A family is a group of people basically consisting of parents and children. The members of a family have ties with each other via blood, marriage or adoption (Parkin 1997:28; Shryock et al. 1976:170-173). Family is generally a biological and sociological concept, household is a socio-economic concept, household compound is an architectural concept. Note that household implies the notion of co-residence, family need not (Parkin 1997:28).

In the most straightforward cases households, families and compounds are directly related: the family is the group of people that make up the household, which has the household compound as its base (the household thus represents the family). However, as Parkin (1997:29) notes: "The family is not necessarily the property-holding unit: this may be a wider descent group or a narrower group of relatives or simply the individual. Nor is the family necessarily the domestic unit in the sense of the unit which eats, sleeps and works together: several 'families' (however defined) may form one domestic unit; a single joint or extended family may consist of more than one domestic unit; and the domestic unit may exclude or lack one parent or other relative, at least at times". In this respect, there is no direct one-to-one relationship between social organization and spatial organization: social interaction is indirectly reflected in built form (Byrd 1994:643).

With regard to households and families in the Burnt Village it has been suggested that "The rather large amount of available space may have exceeded the needs of small nuclear family groups and should perhaps be taken as an indication that the houses gave shelter to extended households" (Akkermans and Verhoeven 1995:29). The 'rather large amount

---

[67] For the origin and development of households in the Near East see Bernbeck 1995; Byrd 1994; Hodder 1990; Reinhold and Steinhof 1995, and Watkins 1990.

of available space', referred to the generally large size of the rectangular and multi-roomed buildings. As the present study has indicated, however, these buildings are not true houses, but storage buildings, some with activity areas.

How were households (with tholoi as their compounds) related to the rectangular buildings, which were mainly used for storage? It is highly unlikely that the rectangular storage buildings were only used to store the properties of the various households at Sabi Abyad: households as represented by buildings (tholoi) were probably not numerous and the amount of storage space available in the rectangular buildings (ca. 236 m²: 75% of all rooms excavated!) far exceeds that of 'normal' subsistence storage.[68] The communal storage buildings therefore seem to have served the needs not only of the settled population at Sabi Abyad, but most likely also that of people/social groups which were not physically present at the site. In the next section this off-site group wil be introduced.

3.1 Nomads

In a recent analysis of the context and meaning of the hundreds of clay sealings (n= 300) and tokens (n= 182) found in the Burnt Village, Akkermans and Duistermaat (1997) have argued that these objects facilitated the communal storage of various products and claims by a nomadic population of considerable size.[69] According to them the sealings are indicative of the symbiosis between sedentary and nomadic populations in the Late Neolithic. In their argumentation the authors first of all critizise common interpretations of sealings.

It is well-known that sealings, as devices of control, have often served the needs of elite groups in society; sealings have a large potential in terms of power and manipulation. At Sabi Abyad, however, we have no solid proof so far of any intra-site hierarchical organization or status differences. Therefore, the Sabi Abyad sealings most probably did not function in a status or prestige context, and they do not seem to have been controlled by an elite at the tell. Secondly, the sealings do not seem to have functioned in long-distance or exchange networks. This is supported by the fact that recent analyses of the clays of which the sealings were made, have indicated that they almost certainly came from the Balikh valley, quite possibly even from Sabi Abyad itself.[70] Intra-regional exchange between the few (n= 4) other contemporaneous sites is unlikely, since these sites are at a distance of 20 km of each other at the most, and exchange could proceed on a face-to-face basis.

---

[68] In Aliabad, a village in the Zagros mountains in western Iran, some 3 ha in extent and consisting of 67 houses arranged in blocks of contiguous compounds, storerooms (n= 137) represented 22 m² (or 9%) of the total village space (Kramer 1982:85 and table 4.2). In Hasanabad, another village in the Zagros mountains in Iran, ca. 2 ha large and consisting of 43 compounds, 37 storerooms were present (Watson 1979 table 5.2). No exact measurements of the storage areas are given, but it appears that in this village ca. 20% of the rooms can be identified as storage chambers.

[69] Nomads and nomadism refer to a wandering form of life in search of subsistence, both of food and of pasture for livestock, with little reliance on sedentary cultivation (Goodall 1987:325).

Ethnographically, nomadism in characterized by: (1) pastoralism (= the practice of breeding and rearing herbivorous animals (e.g. sheep, goat, cattle) for meat, milk, wool, hides, etc.); (2) periodic use of natural pasture; (3) use of tents; (4) 'tribal' social structure (Scholz and Janzen, eds., 1982:6).

[70] Moreover, it is recalled that all the sealings found at Sabi Abyad were broken, and not in-situ on containers; most likely they were kept for administrative purposes. The sealings, then, clearly did not seal products from Sabi Abyad which were to be exchanged.

Furthermore, Akkermans and Duistermaat (ibid.:26) note that the rectangular storage buildings must have been in use at the supra-household or communal level, (1) since sealings are useful only if the responsibility for one's property is transferred into the public sphere, and (2) because of the fact that numerous people dispatched sealed products to the storage buildings. It is suggested that the storage buildings also acted as distribution centres (for the seal holders only), where the products left the buildings in an unsealed state (the sealings kept as a reminder of the transaction). Like myself, Akkermans and Duistermaat also argue that it is highly unlikely that the communal storage buildings served to keep household belongings only, because the amount of storage space by far exceeds that necessary for normal storage. The collective storage of products is of course only relevant when one is unable to look after these products oneself. From this it follows that the storage buildings with their sealed containers were related to particular social groups which were not physically present at the tell.

On the basis of the above argumentation, Akkermans and Duistermaat (ibid.:27) suggest that "... the population at Sabi Abyad was not composed entirely of permanent residents, but had a considerable mobile or transhumant component which made use of the site for specific purposes at specific times". Most likely, these 'nomads' comprised entire families, since, as noted above, we seem to be dealing with particular social groups. These non-residential groups mainly relied on a pastoralist mode of subsistence in the Balikh valley and neighbouring areas. At the same time they must have been closely associated with the sedentary populations at the various sites (which relied on cultivation). It is stressed that the dichotomy between nomads and residents was probably weak only, and that both groups may easily have changed places. Akkermans's earlier analysis (1993:210ff) of the economy of later Neolithic communities in the Balikh valley seems to support the model. First, he argued that during harvest time labour requirements may easily have gone beyond the communities' capacities. Second, it seems that the food requirements of local communities could hardly be met by arable farming alone, and that other sources of food (e.g. livestock) must have been necessary. In both instances, nomads may have been of crucial importance, i.e. in providing additional labour and food.

A number of ethnographic studies of nomadic communities in arid environments in the 'Arabic World' illustrate the interactions between nomads and residents (e.g. Aurenche 1984; Bar-Yosef and Khazanov, eds., 1992; Khazanov 1994), and reinforce the model of Akkermans and Duistermaat. The 'high caves' of Jebel al-Akhdar in Libya, for instance, served as collective storage units for cereals. Each cave contained the properties of specific persons. A storekeeper protected these products (Hallaq 1994). The *agadir* of Tunesia and Marocco are large communal buildings used for communal storage. It is recalled that these extensive storage 'fortresses' consisted of numerous small rooms, which each contained the individual properties, property claims and food supplies of both sedentary and (semi-) nomadic families or other goups (Jacques-Meunié 1949; Montagne 1930; Suter 1964). Here, too, there was a caretaker who lived in or near the building and who supervised the private belongings.[71] Both the 'high caves' and the *agadir* were in use until recently.

---

[71] The large 'fortress'of the Berber tribe of Nalut (Suter 1964) consists of a large building (ca. 40 x 26 m), made of limestones and held together by gypsum mortar, and is located on a hilltop. There is only one narrow entrance; the high walls are only marked by small ventilation holes. The ca. 500 storage rooms (*ghorfas*) are arranged in eight storeys. The rooms, chaotically arranged next to and upon each other, are

In western Iran the Berovand tribes of the Lur represent sedentary agriculturalists who still retain long-range nomadism. The social structure mainly consists of family corporations which are commonly made up of men who are brothers. Most often, one brother stays 'at home' and farms, while his partner is on the move with the sheep, migrating to Khuzistan in the winter, and returning the following spring. The year after roles are reversed (Rosman and Rubel 1976).

Köhler-Rollefson (1987, 1992) has observed that the Marrai'e of the Huweitat tribe in Jordan, who combined pastoral and agricultural subsistence activities, had three main patterns of residence, which are directly related to subsistence modes. She particularly studied a community related to the village of Suweimrah. First there are people who have permanent houses in villages. The men earn a living in the army, as bus drivers, etc, and are often away from home. The wives and children, however, reside in Suweimrah. Families often own one or two goats for milk. These animals are kept nearby. Second, other families only live in the village for a part of the year. These families maintain medium-sized herds of sheep and goats (ca. 30-70 head), and usually stay in the village in spring, when grazing is possible on the fallow fields. While one or two family members are occupied with the herds, the other family members may pursue wage labour. Third, there are the families who own large herds (of more than 70 head) and who live in tents throughout the year. About every month tents are relocated. Virtually the entire family is engaged in herding and related activities. Köhler-Rollefson (1987:526) has noted that "Although these families are permanent tent dwellers, many of them own houses in the village simply to use these for storage (of grain, animal food, wool, tools) or to get access to tap water". In fact about 40% of the buildings in Suweimrah are owned by these year-round nomads. These structures are never inhabited, and are used for storage only (ibid.:538). Circa 20% of the houses are only occupied during certain times of the year, it appears mainly in the summer. The remaining 40% of the buildings are occupied

accessible through narrow passages. Sticks and stones protruding from the wall act as steps. The doorways to the rooms are narrow and low. The wooden doors are closed with padlocks. Often the rooms are subdivided into three to six compartments by low walls. In many instances, a large ceramic vessel is incorporated into the walls of one or two compartments (for storing food products such as olive oil, butter, dates, figs, flour or meat, and valuable goods, i.e. carpets, wool, clothes, jewellery, money, etc.).

Almost every family of the Nalut community which uses the storage fortress owns two to five rooms. The rooms are arranged over different storeys; the heavy products are stored in the lower rooms, the lighter stuff in the upper areas. Grain is often stored in the upper rooms; in the lower chambers the grain rots due to moisture. The rooms are private property and inheritable.

The Nalut fortress is guarded by an old man. This man is responsible for the proper arrangement of supply and demand of goods, and for the general order. The guard gets paid in money or in products such as flour and olive oil. The fortress is closed during the night.

Because of their central location, the fortresses had many functions other than storage, In Cabao, for instance, important community and many non-official social meetings took place in the inner court. Domestic animals which were found wandering around were brought to the fortress; they could subsequently be returned to the owner by the guard. The fortresses with their valuable inventories were, of course, the focus of many raids by hostile tribes. Due to their strategic location and their architectural qualities, however, they could be well-defended. During the hostilities the fortresses served as places of refuge for the community (but see Suter 1964:251). In Cabao, finally, religious practices centred around the capel of the holy *Hadj Abdallah el Baruni*, which was situated in the inner court (ibid.:250-251).

In Nalut and Cabao the fortresses are surrounded by a domestic settlement consisting of dwellings; when the semi-nomadic population temporarily left the settlement their goods were stored and looked after in a well-protected place. Some storage fortresses, such as at Médenine, however, are without a domestic 'counterpart'.

throughout the year, but near many of these houses tents are set up, which are primarily used for relaxing and entertaining the men.

What at first sight appears as a 'normal' year-round occupied sedentary settlement, even if this village had been uncovered in an archaeological excavation, thus in fact represents a highly dynamic settlement. In Suweimrah the whole range of residence patterns, from permanent habitation in stone houses in the village to year-round habitation in tents outside the village, is attested. Archaeologists dealing with tell settlements mostly assume that they are dealing with year-round occupied villages, but the above examples (and many more could be given) indicate that settlements may represent a much more dynamic society, consisting of interacting sedentary and mobile components.

On the basis of other ethnographic and historical evidence Akkermans and Duistermaat suggest that Sabi Abyad, and other large sites such as Mounbatah (contemporaneous with the Burnt Village, and located in the central Balikh valley) may have acted as winter camps for the mobile pastoralists, providing food, shelter and security.[72] These 'central' sites may have acted as "... points of exchange, storage and distribution centres and as the scenes of marriage contracts, communal festivities and ceremonies" (Akkermans and Duistermaat 1997:28). When the weather improved and crops started to grow, the nomads moved into the steppe (see also Buccellati 1990:96-98). In eastern Anatolia and the central Jordan valley, for instance, the lower-lying valleys are used as winter stations by mobile pastoralists, because there is water, pasture and fuel, and agricultural surplus is stored in order to survive the lean months (Van Der Steen 1995). In southeastern Anatolia the nomads often return to the same village, and the contacts with the villagers are formal rather than incidental. The nomads are under the protection of village chiefs, and use services and land resources of the villagers (Cribb 1991:198).

In her study of the animal bones from Neolithic Sabi Abyad, Cavallo (1997) found evidence supporting the view that pastoral nomadism played an important role at Sabi Abyad. Detailed analysis of the ovicaprid remains from the Late Neolithic levels indicates a process of transformation that involved a shift from a generalized pattern of meat production towards a more selective culling for secondary products. Furthermore, a shift towards a more conscious control of the herds of sheep and goats was deducted on the basis of a reduction of the frequency of ovicaprid remains with pathological alterations (ibid.:65, fig. 6.10). With these results Cavallo suggests that seasonal pastoral movements or mobile pastoralism were relevant subsistence strategies at Sabi Abyad.

3.2 Residents and Nomads

This section deals with the socio-economic relations between the nomads and the residents in the Burnt Village. Before proceeding, it should be mentioned that it is most likely that the village had a year-round occupation. This seems to be indicated by the animal remains (i.e. the high percentage of domestic animals (ca. 95%), such as pigs, a controlled cattle husbandry (Cavallo 1997:151-153)), crop cultivation, and the domestic character of the artefact assemblages (e.g. pottery).

---

[72] In fact, the authors refer to pastoralism, which is the seasonal practice among settled pastoral farmers of moving domesticated animals between two areas of different climatic conditions (Goodall 1987:478).

As has been argued, it is highly unlikely that the rectangular buildings were used as houses (cf. chapter 6, section 5.3). What about the activity areas and domestic installations, such as ovens, which were present in the rectangular buildings? In what context did they function?

Undoubtedly, families at the site took care of and guarded the products of the nomads in the rectangular storage buildings. Probably the residents (living in tholoi) carried out the attested activities in the rectangular structures. Considering the close connection between the location of these activities and the storage chambers, it can be suggested that these activities were directly related to the main function of the rectangular buildings, i.e. storage. On the other hand, it is likely that in these buildings the residents also made products for immediate consumption. Ovens (e.g. tannurs), for instance, were present in the rectangular buildings only.

In this respect a distinction can be made between a 'household economy', related to the resident households, and a 'village economy', indicating the production of goods at and around Sabi Abyad (e.g. grain) for the nomads (with a 'pastoral economy') (fig. 7.1). The village economy was in the hands of the resident families, which were perhaps occasionally grouped into cooperating economic units. The domestic (agricultural/food preparation) equipment stored in the various rooms (see section 2) would then have functioned in the village and pastoral economies.

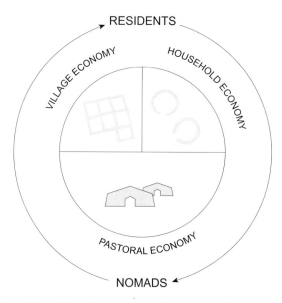

Figure 7.1. Level 6: relations between social groups, architecture and economy.

The distinctions between the distributions of artefacts, and especially between the sealings noted in chapter 6, seem to indicate that the various rectangular buildings were used by different groups and sealing agencies (Akkermans and Duistermaat 1997:26; Duistermaat 1996). It could be argued that the unequal distribution of sealings indicates a marked social hierarchy in the Burnt Village. This seems unlikely, however, since other possible indications of status differences have been wholly lacking so far: no concentrations of valuable objects have been found, and neither does the architecture show differences which might be explained as the result of competition.[73] In this respect it should also be noted that 67 sealing agencies seem to have been present at the site[74], indicating that a large number of persons were involved in the sealing and handling of commodities.

3.3 Population Size

How large was the population? As has been indicated, at least 67 sealing agents used the storage buildings at Sabi Abyad. Suggesting that each of these agents represents a family of 6-10 persons, Akkermans and Duistermaat (ibid.:27) argue that the storage facilities at the site were used by a non-residential group (i.e. the nomads) of 400 to 670 people.[75] It is further maintained that, even if this number is considered too high and therefore halved, yet hundreds of non-residents must have relied upon the settlement at Sabi Abyad.

How many people, then, were present at the site? In addressing this question I have used the concept of household rather than families, since families, which are so difficult to define, are, one way or another, organized in households, which archaeologically can be grasped with somewhat more ease. So far, possibly three households (based in tholos VI, and tholoi in R13 and R14) have been indicated in the Burnt Village, giving a minimum of perhaps 30 persons. Clearly more people must have been present in the settlement. Including the 1997 excavations, as yet ca. 1300 m^2 of level 6 has been excavated. Taking the large dump in square T12 and the open areas in the north and east (in squares R11 to T11, S12) into account, it seems that the northeastern periphery of the settlement has been reached. The extension to the south and west, however, has not yet been excavated (fig. 3.1), and more buildings are to be expected. The excavated part of the settlement seems to be representative of the entire village, considering the presence of different types of buildings, courtyards and large dumps. If half of the village has been excavated, the population, therefore, may be doubled to 60 persons. If, however, a larger settlement of a 0.5 hectare (= ca. 4 times the excavated area) is proposed we arrive at a settled population of about 120 persons.

---

[73] Based upon a survey of settlement distribution, architecture, burials and artefacts, Akkermans has argued that Halaf society was generally non-hierarchical, i.e. Halaf social organization was based on kinship relations, and decision-making relied upon concensus rather than formal authority (Akkermans 1993:288-318). At Sabi Abyad, Halaf layers of occupation immediately follow the 'transitional' levels, which level 6 is part of.

[74] At least 67 different stamp seals were used for sealing purposes (Akkermans and Duistermaat 1997:26).

[75] From ethnographic accounts it is well-known that nomadic tribes can be of considerable size. According to a count in 1918/1919, as reported in Ehlers 1980, tribes in the highlands of Iran ranged in number from 1500 (*Lashani*) to 33,045 families (*Qashqa'i*). Subtribes of the *Qashqa'i* range from 1000 to 10,000 people (Beck 1986: table 2; 1991:14). In 1961 the *Basseri* tribe near Shiraz in southwestern Iran consisted of ca. 16,000 people (Barth 1961). In the Kerman area of central Iran nomadic tribes and subtribes ranged from 10 to 3000 families (according to counts from 1922 to 1963; Stöber 1978). In the 1920s and 30s the *Ayas* in the Silifke region of southern Anatolia consisted of 750 people, living in around 100 tents (Cribb 1991:117). The nomadic population (of the *Huweitat* tribe) of Suweimrah and El-Qurein, two Jordanian villages used for storage by nomads, ranged from ca. 300 to ca. 1000 (Köhler-Rollefson 1987:538).

For the Halaf stages of occupation at Sabi Abyad Akkermans (1993:166) has suggested that the site, which was of an open nature with buildings widely separated from each other, may have been occupied by four or five households only, together perhaps comprising a population ranging between 30 to 50 persons (i.e. 6-10 persons per household). Contrary to the Halaf settlements, level 6 consisted of many buildings, all closely attached. Therefore, more households and subsequently more persons can be suggested. The above calculations, although admittedly highly speculative, suggest that in the level 6 village 60 to 120 persons may have been living.

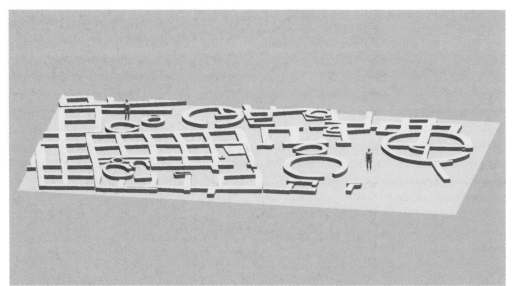

A

B

Figure 7.2. A: level 6: people in their surroundings. Virtual reconstruction of the central portion of the Burnt Village as excavated. Depicted are buildings I, II, VI to X, and XII. This figure serves as a reference for the following figures. The people in the figures are 1.60 m in length. B: people in buildings II and IX and open area 3.

Taking together the nomadic and residential components the community related to Sabi Abyad may consequently have consisted of 460 to 790 persons.

Summarizing, the presence of an (important) nomadic population related to Sabi Abyad seems to be indicated by the present spatial analysis, by the analysis of the context and meaning of the sealings, and by the osteological analysis of the Late Neolithic animal bones. Moreover, many ethnographic studies illustrate the cooperation between residents and nomads. Therefore, as indicated in the title of this chapter, it is suggested that the social structure had a dual basis, and concomitantly Tell Sabi Abyad served as a basis for nomads and residents (fig. 7.1).

Of course, these groups were interrelated; they are not separate isolated units, but social segments that probably maintained kinship ties and other kinds of contact. The suggested social and spatial partition should not be regarded as very strict and necessarily durable. The on-site groups may easily have changed places with the off-site groups and vice versa.

In the next section the relationships between the nomads, the residents and the architecture in the Burnt Village is further explored.

4 Round and Rectangular: Structuring Structures

As has been argued, the level 6 community[76] seems to represent a highly structured society, as already appears from the almost obsessive segmentation and demarcation of space, and from the obvious structuring of space. According to Bourdieu's *Theory of Practice*, which was introduced in chapter 2, this structuring can be illustrated as pairs of related oppositions (or structural dualities), the most important and basic of which are:[77]

round	:	rectangular
household	:	storage
residents	:	nomads

[76] By community the whole level 6 society, i.e. the nomads and residents is meant.

[77] These 'oppositions' are to be conceived not so much as contrasts, but as 'fused dualities', the one element of which cannot exist without the other.

Dumont (1980:239-245) makes a basic distinction between complementary or symmetrical oppositions (in which two complementary elements are opposed), and hierarchical oppositions. In hierarchical oppositions X is indentical to the universe and Y belongs to this universe but at the same time Y is inferior to X; at the superior level there is unity, but at the inferior level there is a marked difference (e.g. in Western Christian tradition man and woman, both part of man, are opposed, with woman being inferior to man: ibid.:239-340, see also Allen 1985).

The oppositions isolated in this study are non-hierarchical (each of the two elements is first dealt with separately, and later they are interrelated in order to reconstruct the whole system). The symmetry of the oppositions, however, is due to the fact that *archaeologically* there is no evidence for hierarchical classifications in the level 6 community. However, ethnographical parallels indicate that in the Near East nomads are often perceived as inferior to residents (e.g. Kalter 1992:126), and it cannot be excluded that opinions of this nature also existed in the level 6 community. If indeed this was the case, we would be dealing with hierarchy.

The dichotomy round versus rectangular refers to the tholoi versus the rectangular buildings. In fact, the distinction between round and rectangular architecture is one of the most conspicuous features of the Burnt Village (cf. figs. 3.1 and 7.1). The difference between both types of architecture, reproduced by *practice,* may have been an important part of people's habitus. Apparently, people felt the need to build two wholly different types of buildings, i.e. to make a distinction. Obviously, such a distinction must have been a meaningful element in the community, otherwise all buildings would have looked more or less the same (figures 7.2 to 7.5 schematically depict people in and around the buildings of the Burnt Village). In this section I will attempt to give an explanation for this interesting phenomenon.

As has been argued, the rectangular buildings were largely used for communal storage, and to a lesser degree for executing activities. The tholoi have been related to households. At least some of the tholoi were presumably used as houses. The other tholoi served as storage places and/or activity areas. It could thus be argued that the dichotomy round : rectangular does not indicate functional differences, since both types of buildings were used for storage and activities. However, that would be a rather general statement, not taking into account the more precise functions and meanings of the buildings. The amount of storage facilities in the rectangular buildings, for example, was far larger than that of the tholoi ($236 m^2$ in the rectangular buildings versus $47 m^2$ - at the most - in the tholoi). Furthermore, that functional differences between round and rectangular buildings do exist has been argued in chapter 6. From a comparative statistical analysis it appeared that with regard to inventories the tholoi differed from the rectangular buildings. Therefore, tholoi VI to VIII and tholos IX were grouped in two separate clusters (nos. 4 and 5). So there seem to have been functional differences between round and rectangular buildings in the level 6 village.

I suggest that these functional differences and the contrast in shape indicate that both types of buildings played a role in two different social contexts, or *fields* in Bourdieu's terms. A 'private' household context related to residents, and a communal storage context related to nomads is suggested. The round : rectangular distinction, therefore, reflected and structured functional as well as social differences. At the same time, architectural, functional and social dimensions were inextricably linked, i.e. tholoi were related to residents and households, and rectangular buildings to nomads and communal storage. By creating a difference between circular and rectangular buildings the social structure was materialized and 'objectified'; it was given meaning and substance in the most literal sense.

Sabi Abyad may be seen as a kind of microcosm of the society at large. By living in and moving around the level 6 village one would have been made aware (largely unconsciously) of the social divisions, which were materialized in the built environment. Tholoi, storage buildings, ovens, etc. in principle had preferred locations and uses, but through *practice* alterations came about, which are dealt with in the following section.

4.1 Structure, Habitus and Practice in the Burnt Village

The oppositions round : rectangular; household : storage; residents : nomads seem to have been the main organizing principles in the Burnt Village. Now, these three basic sets of oppositions can be extended, adding formal, functional and social characteristics:

	tholoi	:	**rectangular buildings**
shape	round	:	rectangular
	some segmentation	:	extreme segmentation
activities	household	:	storage
	arable farming	:	animal husbandry
social groups	residents	:	nomads

In my view this structure at least partly represents the system of generative principles relevant for the level 6 community. In the following the various characteristics that have not yet been given attention will be discussed.

The extreme segmentation of the rectangular buildings is among the most conspicuous features of the Burnt Village (e.g. figs. 7.2 and 7.3). Apparently, the need was felt for an extreme division and 'domestication' (i.e. control) of space. This segmentation of space can be partially explained by the functional difference of storage chambers; it is recalled that general storage rooms (for the storage of domestic equipment), grain storage rooms and archives can be distinguished. Even taking these functional divisions into account, the segmentation is extreme; why were there 60 very small rooms with (more or less) the same function? It could be argued that many interior walls were necessary in order to carry the roof. As was indicated in chapter 3, the roofs consisted of wooden poles covered by a layer of reeds, which in its turn was covered by a thick layer of mud (it is recalled that a second storey on the level 6 buildings is not to be expected; cf. chapter 3). The buildings are rather wide (e.g. the central building II has a width of 7.75 m), and probably the wooden beams were not long enough to cover the space entirely. Most likely, two opposing sets of beams, each set resting with one end on the exterior walls, and with the other end on an interior wall, were necessary to span the distance. Therefore, the interior walls may indeed be regarded as constructural features. However, this only partly explains the segmentation in the buildings. In order to support two sets of beams, one interior wall (over the length of the entire building), or two at the most, would have been enough. The many interior walls, apart from separating functionally different areas and from being construction elements, may have served a third purpose. I suggest that an important function of the segmentation was to separate the belongings of specific persons or groups. If indeed the non-residential groups related to Sabi Abyad consisted of some 400 to 670 people, and these numbers are divided over the 60 thus far discovered storage chambers then only 0.15 to 0.09 storage chambers per person would be available. If the non-residential groups consisted of families of 6-10 persons ca. 67 families may have used the storage facilities. If the storage rooms were evenly distributed, this would mean that hardly one chamber per family was available. Even if we double the amount of storage chambers to 120 (in view of unexcavated storage buildings) only about two store-rooms per family would be available.

Figure 7.3. A: people in building II and open area 3; B: people in building II and open area 5.

Therefore it may be suggested that the segmentation of the storage buildings was largely due to the fact that many families made use of these buildings. The above calculations have shown that what at first sight seems to be an overkill in storage facilities in fact may reflect a 'normal' pattern of 1 to 2 storerooms owned by families.

Considering their subsistence economy, the nomads have furthermore been associated with animal husbandry, and the residents to arable farming. In this way I have reduced space and socio-economic structure to basic sets of oppositions, related to the central round : rectangular dichotomy. The *practice* of people was informed but not dictated by these oppositions and 'rules'.

Habitus not only allows participants to act, but also offers the possibilities for change within the limits it sets. It reproduces itself in a manner which all participants perceive as meaningful. So it contains the means for social change and variations in social practices. Tradition, which sets the limits within which acts are socially acceptable, can thus be re-interpreted. Habitus has to be produced and reproduced in everyday practice, enabling and resulting in cultural change. The actions of people are informed by conceptual schemes which organize space, but the actual meaning given to the spatial order is in its turn dependent on the actions of people. Through action in space, persons interpret that space and produce a spatial representation.In this respect I wish to distinguish a long-term and a short-term perspective of habitus (Fontijn 1995:12). The long-term perspective of habitus refers to changes which take place over generations. In the case of Sabi Abyad this means changes between the various settlements (i.e. between levels). The short-term perspective refers to changes within settlements. Let us begin with the short-term perspective, i.e. with the dynamic use of space in the Burnt Village.

Apart from storage, the rectangular storage buildings were also used for conducting activities. Activity areas were encountered in buildings I (area 3, rooms 1, 2 and 12), II (area 15 and room 18), IV (room 2) and V (rooms 5 and 8). Perhaps the various ovens in the rectangular buildings were used for the preparation of products which were subse-quently stored in the storage buildings (cf. the village economy). Most likely, a part of the processed products was reserved for the households, either to be stored or to be directly consumed in the tholoi (cf. the household economy).

Moreover, some tholoi were, like the rectangular buildings, segmented (i.e. nos. VI and IX), and also used for storage, and apart from arable farming households at the site would also have engaged in animal husbandry.

These examples indicate that the use of space in the Burnt Village was dynamic, and while guided by the schemes, it was not dictated by them. In this respect I like to stress here that the suggested oppositions are only basic schemes.

In order to analyse the impact of the short-term aspect of habitus, we have to look at the long-term perspective of habitus. Therefore level 6 is compared with the Halafian level 3 (ca. 5100 B.C., 5910 cal BC) at Sabi Abyad. This level (fig. 6.16) was, similar to level 6, excavated over a relatively large area (ca. 875 m^2), and a general spatial analysis is available (cf. chapter 6, section 9), allowing a comparison. Moreover, it is felt that the chronological distance between these levels (ca. 100 years) is large enough to observe structural changes in settlement structure and function.

At first sight, the same structural round : rectangular pattern emerges in the level 3 Halaf settlement as in level 6. In level 3 a large central and multi-roomed building (I), situated on the top of the tell, is surrounded by a series of tholoi on the slopes. These tholoi were probably used as granaries (Akkermans 1989a:66). The small rooms of the large central building were most likely also used for storage (cf. chapter 6). The western rooms 1 and 2 were domestic areas. A second storey may well have been present, and perhaps living quarters were situated there. If these interpretations are correct, it would seem that, with regard to level 6, the settlement structure remained the same (rectangular : round; extreme segmentation : some segmentation), but that the function of the structural units of the settlement had changed. The tholoi in this phase most likely largely functioned as storage buildings, while the rectangular building I may have served for storage and dwelling at the same time.

It may be suggested that the habitation shifted from tholoi to rectangular buildings, the Early Halaf tholoi mainly (but not only) serving as granaries[78], and the rectangular buildings as houses and storage buildings at the same time. In fact, in level 5, which was built directly upon the level 6 remains, the multi-roomed rectangular architecture gave evidence of many ovens and hearths in various rooms (ibid.:63-73). This level 5 archi-tecture, then, already seems to be of a more domestic nature than the level 6 storage buildings.[79]

The basis of these functional changes can probably be found in the 'domestic use', i.e. the construction and use of occasional activity areas, of the storage buildings in level 6. The following scheme illustrates this structural change.

Level 6
(ca. 5200 B.C.)

rectangular	:	round
extreme segmentation	:	some segmentation
storage	:	household

Level 3
(ca. 5100 B.C.)

rectangular	:	round
extreme segmentation	:	some segmentation
household/storage	:	storage/household

Most likely, nomads also played an important role in the level 3 community. In fact, for the Early Halaf period Cavallo (1997:65, fig. 6.10) has found evidence for selective culling of ovicaprids, a possible indication of seasonal pastoral movements or mobile pas-toralism. As in level 6, this would also explain the large amount of storage facilities, which seem to exceed storage for normal domestic requirements (Akkermans 1993:59-61).

The above analysis of the short and long-term aspects of habitus has suggested that while the basic structure of the level 6 and 3 communities remained more or less the same, the meaning of space was altered by social practice. This practice was informed by struc-tural schemes, but the actual meanings given to this spatial order were dependent on interpretations and actions of individuals. As was argued above, this meaning was context-dependent, and therefore probably different in various *fields*. The reasons for the observed changes are difficult to reconstruct. Perhaps the suggested 'merging' of household and possible communal storage in level 3 (building I) indicates that the fields of the nomads and residents were becoming more interwoven. This may have represented a more stable socio-economic system, based on nomadic and residential subsistence strat-egies, which continue to the present day.

---

[78] In level 3 tholos O in square P14 a small fireplace and a posthole was present, and in tholos P, built directly upon tholos O, three postholes and a stone platform were found. These features may indicate that these tholoi (with interior diameters of ca. 3 m), were used as small dwellings or perhaps as kitchens instead of as storage buildings (Akkermans 1989:28-33).

[79] Since only limited portions of level 5 have been excavated and we probably have an incomplete view of settlement structure, a more detailed comparison of settlement structure of levels 6 and 5 is not presented. The same holds for levels 4, 2 and 1.

Figure 7.4. A: people in buildings I, VII and IX; B: people in building I.

   In her thesis Breniquet (1996) has also noted the basic distinction between round
and rectangular Halaf architecture (pp. 80-96). On the basis of archaeological and
ethnographical evidence she suggests that the small tholoi (1-3 m^2) served for storage,
while the larger ones (often with antechambers and annexes) were dwellings (see also
Seeden 1982, and section 5.3 in chapter 6). A typical small Halaf village, consisting
for a large part of tholoi, may have been inhabited by one extended family, which
consisted of two or three generations, and which centred around the head of the
family. The large rectangular buildings, such as found at Yarim Tepe II (level VI),
Arpachiyah (the TT6 'Burnt House'), Tell Songor B (levels II and IV) and Tell Hassan
(levels 1 and 2), combined various activities (storage, food preparation, dwelling)

according to Breniquet. Their generally large size suggests that they were inhabited by an extended family (ibid.:95).

Breniquet considers rectangular buildings to be a feature of the later Halaf period, and she argues that they combined activities which were earlier separated, i.e. carried out in several tholoi. The excavations at Sabi Abyad, however, have revealed that already in the earliest Halaf (i.e. level 3) rectangular architecture played an important role (Verhoeven and Kranendonk 1996:84-114). Nevertheless her suggestion is relevant, since it has been argued that at Sabi Abyad dwelling may have shifted from (mainly) round houses to (mainly) rectangular buildings, which were used for storage, food preparation and possibly dwelling. Changes in layout and function of architecture and associated social changes at Sabi Abyad seem to have been (not surprisingly) part of a wider trend, as indicated by Breniquet's analysis.

Nomads and residents acted in different *fields*, each of them having its own logic and rules, i.e. its own ideology. Of course, these fields were closely related; all groups operating in them were part of the same community and were probably related through kinship. Moreover, nomads and residents may regularly have changed places. The socio-spatial organization of the settlement was related to the society as a whole. Society was reproduced through domestic activities related to tholoi and communal activities related to rectangular buildings. Buildings and society constituted a dynamic complex in which changes in one were mirrored yet transformed in the other (Robben 1989:572). Because of socio-economic interactions, the fields were tied up and the society was made coherent.

Notwithstanding these interactions, the fact remains that residents and nomads may have been separated for extended periods of time, and therefore formal mechanisms for linking and binding these elements of society are to be expected. In the following sections 5 and 6 two such 'strategies' are discussed.

5 'Betwixt and Between': Ritual Practice in the Burnt Village

In building V in the level 6 settlement nine large and heavy clay objects were found, and in addition two were found directly along the northern wall of this building. The objects were recovered from the fill of the various rooms (nos. 3, 6 and 7 in building V, and 11 and 13 in building IV). Most often they were situated high above the floor and amidst the charred roof beams; in view of this position the objects must originally have stood on the flat roof along the northern and eastern façades of building V. When the building collapsed (due to the fire) they must have fallen down. Apart from one, the objects were all oval and loaf-shaped, i.e. they had a flat base and a rounded, convex body (fig. 3.5, see also fig. 8.7 in Spoor and Collet 1996). The objects were rather large; their length varied between 45 and 62 cm, their width between 35 and 41 cm, and their height between 21 and 27 cm (Spoor and Collet 1996:443). All objects had one or two shallow and elongated holes (9-10 cm long, 3-4 cm wide and 5-6 cm deep) along each of the long sides. Another hole was often found at the top. In two of the objects the hole at the top carried the horn of a wild sheep (*Ovis orientalis*), which was largely but not wholly hidden from view (fig. 8 in Akkermans and Verhoeven 1995). In another of the objects a proximal part of a femur of a bovid (*Bos*) and a fragment of a bovid rib were found (Cavallo in press). Possibly these remains are also of wild animals. The animal remains were put in the objects when the clay was still wet. Apart from the oval objects another small (28.5 x 26 x 16 cm) and more or less rectangular clay object with a saddle-like top was found in room 13 of building IV. It had a flattened base on which it could stand upright. The back was flat, but the front was strongly convex. Near the base two 'grips' were placed lengthwise.

Between the objects, high in the fill of room 7 of building V, the remains of two skeletons were found, one of a male and one of a female. Their position indicates that they were also originally present on the roof of building V. The bones were completely crushed and burnt. The female was lying in anatomical order on her left side, the legs tightly flexed and the head facing south. The clearly intentionally flexed position must have been fixed, either by a strong *rigor mortis* or by a bondage. The female was over 30 years old (Aten 1996). Of the male skeleton most, if not all, of the bones were also lying in anatomical order. The exact position, however, could not be reconstructed. For the male likewise an age of over 30 has been suggested.

It has been proposed that the clay objects may have served as prehistoric seats, or as rollers, used for compacting the roofs after rains, otherwise cracks would develop while drying. These rollers, however, are still used in the Near East, and they are made of stone, and not of clay. In fact they are heavy stone cylinders fitted with wooden handles for pushing (Watson 1979:119; Wulff 1966:114). Both suggested functions seem rather unlikely. I suggest that the objects served in a ritual context, particularly when taking into account the horns and bones inside the objects and their position on the roof around the two skeletons. The clay objects may have represented stylized animals or mythical creatures. Possibly the holes along the sides may have carried wooden sticks or the like, which represented legs. Or perhaps they served as grips to allow transport (Spoor and Collet 1996:444). Furthermore, part of the horns inside the objects must have been visible from the hole at the top (Cavallo in press). Apart from the oval clay objects, no artefacts could be associated with the skeletons on the roof.

5.1 Ritual Theory and Practice

In a recent publication about rituals Bell (1992) points out that ritual is *practice* (à la Bourdieu 1977), i.e. that ritual activity is not a secondary aspect of religion (secondary to 'beliefs', which are primary), but that through ritual action religion/beliefs acquire meaning. Ritual should not be opposed to beliefs or cosmology, but instead should be seen as an integration of cosmology and social practice, the one defining the other in a dialectical fashion.

Focussing on ritual as a form of social action, Bell introduces the term 'ritualization', which is meant to indicate that ritual is deeply embedded in social systems, and that ritual actions derive their significance from the interplay and contrasts with other practices. Ritual, therefore, should be analysed not as a very specific context of activities, but as a form of social practice which is indeed different, but also closely connected to other activities.

The two main strategies of ritualization are: (1) the generation of a privileged position among other activities, and (2) the production of ritualized agents through the generation of a structured environment experienced as bodies act in it. These 'external' strategies involve three 'internal' operations: (1) the construction of schemes of binary oppositions; (2) hierarchization of these schemes (some dominate others); (3) the generation of a loosely integrated whole (ibid.:101). This way a 'semantic universe' is created. In fact this universe is part of Bourdieu's habitus: it consists of the ideas, notions, experiences, etc. which enable a person to act in the ritual environment or field. The active and dialectical relationship between habitus and ritual practice is a continuous process of interpretation, re-interpretation and change.

So ritual communicates, and apart from verbal messages it contains non-verbal messages about how to act in the world. In a sense it is a memory system (see also section 6). However, it does not merely send messages; it also creates situations. The messages are expressed in a structured and symbolic environment characterized by e.g.: the use of a space to which access is restricted; a special time; restricted codes of communication; special persons and objects; special verbal and gestural combinations; particular physical or mental states; and especially formalization (Bell 1992:204-205). As has been argued, sets of binary oppositions underlie action in rituals. These oppositions inform but do not control social action. Barth (1975) and Tilley (1991:123-126) have argued furthermore that meaning does not derive from opposition or contrast alone, but also from the use of metaphors: "An analogic logic depends on a digital logic but cannot be reduced to this level" (Tilley 1991:125). Symbols such as used in rituals have important metaphorical qualities, i.e. they are meaningful elements (e.g. the cross in Christian rituals or the sacred host).

The deep structure in rituals consists of strategies in which the social world is integrated and differentiated at the same time. Rituals are marked by ambiguity, and the structural spatial and related conceptual oppositions are confirmed and negated at the same time. Metaphorically speaking limits are being crossed and are brought under discussion. Dealing with these limits Van Gennep (1960) has shown that *rites de passage* (or rites of passage; rites which accompany every change of place, state, social position and age (which almost all rituals do)) are marked by three phases: (1) a preliminal phase, or rites of separation; (2) a liminal phase, or rites of transition, and (3) a postliminal phase, or rites of incorporation. Beneath a multiplicity of forms, either consciously expressed or merely implied, this typical three-staged pattern occurs very frequently (ibid.:191). A funeral, for instance, is marked by the separation of body and soul, the transition from the world of the living to the world of the dead, and the incorporation of the soul in the world of the dead (Van Gennep 1960:146-165; Metcalf and Huntington 1991:29-36).

In the *Ritual Process* Victor Turner (1969 and see e.g. Turner 1986 and 1990)[80] has explicitly analysed the liminal phase of Van Gennep's scheme. Mainly by a detailed study of women's rituals of the Ndembu of Central Africa, Turner demonstrates how the analysis of ritual behaviour and symbolism may be used as a key to understanding social structure and processes (for a prehistoric example see Tilley 1991:139-148). "The attributes of liminality or of liminal *personae* ("threshold people") are necessarily ambiguous, since this condition and these persons elude or slip through the network of

---

[80] In *The Ritual Process* Turner has laid the foundations of his theory of liminality and communitas; it is in this book that his theory is dealt with most extensively, and therefore this book, instead of subsequent works in which the ritual proces is described and analyzed in terms of liminality (e.g. Turner 1986), has been used here.

classifications that normally locate states and positions in cultural space. Liminal entities are neither here nor there; they are betwixt and beween the positions assigned and arrayed by law, custom, convention and ceremonial. As such, their ambiguous and indeterminate attributes are expressed by a rich variety of symbols in the many societies that ritualize social and cultural transitions" (ibid.:95). Furthermore, the ambiguous liminal status is often expressed as contradictory behaviour (i.e. inversions take place), e.g. people behaving like animals, forbidden food is consumed, homosexual contacts, etc. "The ideal that is present in ritual is often the inverse of patterns in daily life, giving those patterns meaning and coherence" (Hodder 1982c:167).

A

B

Figure 7.5. A: person in builing VI; B: people in open area 3.

Turner argues that there are two models of social interaction. First a society can be perceived as a structured, differentiated and often hierarchical social, economic and political system. Second, in a ritual liminal period, a society can emerge as an unstructured and relatively undifferentiated *communitas*. In ritual liminal periods the social world is bound and differentiated at the same time. Differentiation may find expression in restricted access to rituals, difference between actors and audience. A main purpose of rituals may be to show that social differences (which are not necessarily hierarchical) do exist by using metaphors which indicate particular social groups or practices. These symbols, however, are ambiguous and represent a fusion of opposing categories; they differentiate and bind at the same time. The binding of the social world lies in "... giving recognition to an essential and generic human bond, without which there could be no society" (Turner 1969:97). Different 'worlds' or fields are given attention and are confronted with and related to each other. "...social life is a type of dialectical process that involves successive experience of high and low, communitas and structure, homogeneity and differentiation, equality and inequality" (ibid.:97). The ritual 'chaos' or 'anti-structure' during liminal periods or *communitas* is temporal, and in fact serves to reinforce the social order, which generally is not fundamentally threatened. Liminal situations and roles are almost everywhere related to rituals, since they represent dangerous and anarchical periods, which have to be surrounded by the prescriptions, prohibitions and conditions which rituals as formal occasions offer. Douglas (1966), too, has noted that that which falls between traditional criteria of classification is potentially dangerous and polluting.

In liminal periods fundamental societal values are symbolically expressed, often by ambiguous metaphorical persons and objects which are 'betwixt and between' (e.g. Corbey 1994:298-305). These objects are contradictory; they are, as it were, a denial of the oppositions which they represent.

The possibility of change is always there in rituals. In fact it is in these periods that fundamental issues can be approached and changed, since the ritual allows for it and the relevent personae are present and aware of the possibilities. These changes will have influence on the other fields, since the ritual brings together the different fields in communitas. Rituals thus are important occasions in a society's life cycle, presenting focal points for elucidating and questioning societal norms and values.

5.2 A *rite de passage* in the Burnt Village

Now, let us return to the enigmatic clay objects and the suggested ritual in the Burnt Village at Sabi Abyad.

In my view the association of the clay objects with the two skeletons indicates that here we deal with the remains of a rite of passage related to a mortuary ritual on the roof of building V. As has been argued in chapter 4 (section 4.1) this mortuary ritual was possibly associated with abandonment rituals. In this context two human bodies awaiting burial, or perhaps decomposing, were surrounded by loaf-shaped mythical creatures or monsters (i.e. strange beings) of clay. The deceased were in the liminal, transitional, phase of Van Gennep's scheme: they were situated between the world of the living and the world of the dead (Van Gennep 1960:148). According to Turner (1967:98) the process of rotting and dissolution of form provides a metaphor for social and moral transition (and see Metcalf and Huntington 1991:71-74). Furthermore, the clay 'monsters' surrounding the bodies separated these two worlds (Turner (1969:95) has argued that monsters are typical liminal objects, see also Schieffelin 1985:722).

With regard to the decomposition, a famous mural in the so-called Vulture Shrine at Late Neolithic Çatal Hüyük (ca. 6200 B.C., 7130 cal BC) should be mentioned. In the shrine, murals indicate vultures with outspread wings and open beaks above headless human beings. The scenes have been interpreted as representations of the ritual scavenging of human bodies by vultures (Mellaart 1967: 167-168, fig. 47). According to Mellaart another wall painting (ibid.: pl. 8) depicts a charnel house of reeds and matting, in which the dead were laid out for excarnation. Moreover, many burials at Çatal Hüyük have provided evidence of secondary burial; burial after the flesh had at least partially decomposed: sometimes bones were missing, traces of paint were occasionally applied after the flesh had decayed, and in one skull a piece of textile was found.

Most interestingly, the horns inside the clay objects were of *wild* sheep. As Cavallo (1997, in press) has noted, wild sheep is only sporadically found in the faunal spectrum of Sabi Abyad. Domesticated sheep, on the other hand, has been found in large quantities. Wild sheep may have had a special meaning, and horns of this species were probably deliberately chosen as a metaphorical symbol to be used in a ritual context. Wild sheep may be regarded as a transitional element, referring to and mediating between wild and domesticated animals, as an (in etic terms) 'undomesticated domesticate', as a liminal being.

Douglas, in her famous *Purity and Danger* (1966), has pointed out that that which is physically marginal or liminal is most often symbolically central. She refers to elements which are marginal, ambiguous, liminal, 'betwixt and between', e.g. faeces, urine, sweat, pus, snot, ear mucus, menstrual blood, cut hair and nails, etc. Worldwide these 'dirty' and 'impure' things have a central and symbolic role. Traditional ritual Zuni jesters, for instance, drank urine, ate faeces and living mice, etc. In essence these ritual clowns, which are present in many traditional cultures, mediated between profane and sacred worlds, i.e. between humans and spirits (Corbey 1994:302). These dirty things play a central role in liminal rituals; because of their ambiguous nature they are well-suited to 'think with', to express meanings, especially those with regard to moralities and social norms and values (ibid.:306). Horns, like hair and nails, can be interpreted as impure, dirty and liminal elements, physically marginal, but symbolically central. Apart from being betwixt and between wild and domestic animals, they were also marginal and liminal with regard to the animal body. The horns operated on at least two different but associated ritual liminal levels.

It is well-known that animal horns have had, and still have, an important ritual meaning in Near Eastern societies. A remarkable instance of the use of horns in a liminal and possibly ritual context, has been found in Tell Aswad, located in the Balikh valley at a distance of only 15 km northwest of Sabi Abyad. Tell Aswad is a Neolithic mound dating from the later seventh and early sixth millennium B.C.. In Mallowan's initial excavations at the mound (Mallowan 1946:123-126) parts of a small rectangular building consisting of several narrow and oblong rooms were unearthed. Across the threshold of the doorway to one of the rooms the skull of an ox with the horns still attached was found (ibid.: fig.

2). On the basis of this discovery Mallowan suggested that the building served as a shrine. Two low mud pedestals were interpreted as offering tables. The skull with horns represents a liminal object *par exellence*, as it was literally upon a threshold, possibly dividing sacred and profane areas.

Of course the use of the horns of wild bulls in Neolithic Çatal Hüyük (ca. 6500-5750 B.C., 7490-6470 cal BC) should also be mentioned here (Mellaart 1967; Todd 1976). In the early levels X to V the bull is the most commonly represented animal among the plastered animal heads (otherwise of rams and stags) in the cult buildings or 'shrines'. These buildings are marked by attractive murals (depicting animals, people and geometric designs), reliefs, cut-out figures, statuettes and bulls' horns (Mellaart 1967:97). In the early levels the horns are most often of clay and plaster, while in the later levels VII-VI actual horns were used extensively. Furthermore, horned pillars and benches are commonly found in the shrines (of levels VII-II). In most cases the benches are located near the south end of the buildings, and set against the east wall. In a number of instances opposing pairs (sometimes as many as seven) of bulls' horns were mounted on each side of the bench (ibid.: fig. 4). The horned pillars, consisting of a rectangular mud brick pillar ca. 50 cm high with a pair of bulls' horns set at the top, occurred more frequently. They were found singly as well as in groups (ibid.: fig. 9).

In other less known Neolithic sites the horns of animals seem to have figured in ritual contexts as well. For example, in Ganj Dareh in the central Zagros of western Iran, level D (late eighth and early seventh millennia B.C.), two skulls with horn cores of wild sheep were attached to the walls of a small niche in a sub-floor cubicle (Smith 1976: fig. 5, 1990: fig. 1). In the late seventh to early sixth millennium B.C. site Zaghe, located in the Qazvin plain in northern Iran, numerous mountain goat skulls and horns had originally been attached to the walls of the so-called painted building. This building is an unmistakable ritual structure with richly decorated walls not unlike more or less contemporaneous Çatal Hüyük (Negahban 1979, see also Malek-Shahmirzadi 1977).

So it seems that in the seventh and sixth millennia B.C. the horns of wild animals were regularly used for cultic purposes (e.g. Cauvin 1972b, 1994; Hodder 1990:3-19). I feel that it is significant that the horns are of wild and not domesticated species. It can be suggested that these objects played a role similar to that suggested for the Sabi Abyad horns, i.e. they may have referred to an undomesticated and 'wild' or 'natural' (i.e. nomadic) way of life as opposed to the domesticated and cultural (i.e. resident) contexts in which they had a ritual meaning. Moreover, as has been argued above, horns may be perceived as meaningful liminal and ambiguous objects, which are characteristically used in rituals (e.g. Hodder 1990:11).

Furthermore, the materiality and visual effect of horns (i.e. the experience the object evokes; see e.g. Tilley 1994:11-17) should be taken into account. Horns are generally perceived as impressive and powerful and visually attractive objects. Moreover, they most often refer to concepts such as strength and dominance. Through their high visibility and their power to arouse feelings (much more than, say, a rib) they are symbols that are very appropriate to be used in rituals, which are usually executed in a highly structured and symbolic environment. The powerful visual image of horns, therefore, may have been especially used to evoke referential meanings which were transmitted in the ritual (e.g. Hodder 1994:73-74; Schieffelin 1985:722). Perhaps the oval clay objects with included horns should be regarded as 'dominant symbols' (Turner 1967), i.e. symbols which are compelling because they are represented in emotionally arousing imagery and contexts. As Schieffelin (1985:723) states: "The meaning of the ritual is discovered by unpacking the densely interwoven messages hidden within the dominant symbols and showing how their presentation in symbolic form obscures, combines, or mediates fundamental social and cultural issues, and thus opens the participants to new perspectives on their life situations".

In later times horns and objects representing horns also seem to have been used regularly as sacred objects in the Near East (Diamant and Rutter 1969). For example, from the 'Ubaid levels (XII-IX) at Tepe Gawra 'double-horned clay objects' (ca. 25 x 17 cm) were found, interpreted as cult objects by Tobler (1950:173). Most interesting are the 'horned objects' in 'shrines' from the Early Bronze Age levels (XVI-XIV, dated ca. 2600-2300 B.C.) at Beyçesultan in Anatolia. The clay horns, comparable to the well-known Minoan horns, were associated with hearths in what were undoubtedly religious structures, given the elaborate hearths or 'altars' surrounded by fine 'votive pottery' (Lloyd and Mellaart 1957, 1958; Yakar 1974). The horned objects thus functioned as - sacred - hearth furniture, and it has been suggested that they served as pot-stands (Diamant and Rutter 1969).

A modern parallel for the possible ritual use of horns can be found in Luristan, where it has been observed that two pairs of moufflon skulls with long horns were placed on poles at the side of the flat roof of a rectangular clay building (Mortensen 1993:117, fig. 6.52).

With regard to the horns in the Sabi Abyad clay objects it is furthermore interesting to note that horns of sheep, and not of cattle, were chosen as symbolic elements. Presumably sheep horns were selected because they referred to the herds of the nomads, instead of the cattle of the residents. In this view, symbols mainly related to the nomads were brought within the domestic context of the residents; in a sense, two fields were tied together.

The oval shape of the clay objects represents a transitional form between round and rectangular. From the material culture from the Burnt Village we know that the level 6 people were skilled artisans in several ways (cf. chapter 3). Undoubtedly, if they had wanted they could have produced naturalistic clay animals, with heads, legs, etc. Nevertheless, in the case of the clay objects they chose to make rather coarse and undiagnostic objects, most likely just because these objects had to be ambiguous. In this respect Turner (1969:102, 105) has noted that sexlessness and anonymity and mysticality are highly characteristic of liminality. As has been argued extensively, the circular and rectangular buildings seem to have embodied important structuring principles, related to nomads and residents. An oval shape may indicate a fusion between these and related categories.

The location of the clay objects and bodies upon the flat roof of building V may perhaps also be interpreted in liminal terms: they are situated on the border (or threshold) between outside and inside, above and below, and light and dark.

If mortuary and abandonment rituals were associated (see above), it would be expected that the dead persons were 'buried' on the roof.[81] If not, the two bodies were to be buried

---

[81] It cannot be excluded that the persons and objects were sacrificed. Sacrifices are to be viewed as rites of passage (from the profane to the sacred; Hubert and Mauss 1964), and therefore these rituals may have liminal aspects as well; i.e. my discussion of liminality and the meaning of the mortuary ritual also holds if we are dealing with a sacrifice.

after the ritual. As yet, no Neolithic cemetery has been recovered at or near Sabi Abyad. However, three burials were recovered from level 6 at Sabi Abyad: one of an infant, and two of children. From level 5 another child burial was recovered. All these graves were situated inside rectangular buildings (cf. chapter 3, section 3). In view of the absence of their graves, it seems that adults were buried outside the settlement, perhaps outside the tell. Apparently, mortuary rituals related to children/infants were different from those related to adults. It may be the case that through burial children were symbolically related to buildings and the settlement, whereas adults were related to the landscape. Perhaps when more burials are recovered, such a pattern can be substantiated.

The suggested liminal dimensions in the reconstructed rite of passage were 'betwixt and between' some major and meaningful societal, ideological, categories, which were structured by and at the same time structured the community: the oval shape of the objects intermediates between the round and rectangular shape of the architecture, and the wild sheep's horn is seen as transitional between wild and domesticated animal species. Both liminal categories thus link symbols related to residents (round buildings and domestic animals) on the one hand and nomads (rectangular storehouses and wild animals) on the other hand.

It can be expected that the death of two individuals was a major occassion in a relatively small Neolithic community. Mortuary ritual was most likely not a separated activity isolated from other social practices (e.g. Bell 1992; Campbell 1995:33). Therefore, it can be argued that at a ritual which was connected with deaths of persons (and perhaps of the settlement; see above), members of the whole community, i.e. the nomads and the residents, were present. The ritual was a performance, a kind of drama; a dialogic interaction between the people present and between the people and the death persons (or ancestors), and perhaps spirits and gods (Schieffelin 1985). The meanings embodied in the symbolic materials were actively used in a social drama. The group of people supposedly present at ritual performances (e.g. Schechner 1994) related to the rite of passage formed a communitas, a liminal, relatively undifferentiated group of people consisting of both nomads and residents. The ritual, as a period of communitas, served "... to underline the essential and generic human bond, without which there could be no society" (Turner 1969:97). There was a dialectic "... for the immediacy of communitas gives way to the mediacy of structure, while, in rites of passage, men are released from structure into communitas only to return to structure revitalized by their experience of communitas. What is certain is that no society can function adequately without this dialectic" (ibid.:129). Nomads and residents, who may not have had close contacts for extended periods of time (i.e. months, or seasons), were brought together and reminded of the unity of the society of which they were a part. A sense of identity was transmitted. Furthermore, ritual regulated social behaviour and it sanctified the social structure (Rappaport 1971).

During the ritual, through ritualization, the social world was bound and differentiated at the same time through the fusion of disparate groups and symbolic social categories. That which was physically marginal or liminal (i.e. the clay 'monsters', the included horns and the communitas itself) was symbolically central (Douglas 1966). Structural social norms and values were being symbolically expressed, and communicated through the ritual (see Turner 1969:106-108). This would have resulted in a reinforcement of ideas and rules shared by the society as a whole. Through ritual the social order was legitimized, and group solidarity and identity were maintained. Following Bell, the Sabi Abyad ritual is perceived here as an integration of cosmology and social practice, both defining each other in a dialectical manner. The ritual was deeply embedded in the social system and by the structured and symbolic environment it created, it communicated symbolic messages concerning proper behaviour. At the same time, through the active participation of people in the ritual performance, i.e. through the continuous process of interpretation and re-interpretation, the ritual practice and concomittantly the social structure could be transformed (or re-constructed).

The ritual functioned at several interconnected levels. Not only the dead and the transitoriness of life, but also relations in life itself and fundamental social norms and values were given attention through the metaphorical symbols as used in the ritual. Through the ritual habitus and *practice* of the people was thus being structured, and, because of the opportunity of change inherent in liminal ritual periods, perhaps re-structured.

6 Breaking and Binding

In this section the second 'binding strategy' is discussed. The small figurines (of unbaked clay) found in the Burnt Village (anthropomorphic, n= 31, as well as zoomorphic, n= 16) were virtually all without heads (cf. fig. 3.6). So far, only three human heads and three animal heads have been recovered. Presumably the heads were intentionally broken off (Collet 1996:403). The intentional destruction of material culture is a phenomenon which merits further discussion, and which, as will be shown, can be related to the dual nature of the level 6 society at Sabi Abyad.

In an article about religion and ritual in sixth millennium B.C. Mesopotamia, Oates (1978:119-120) has devoted some attention to 'ritual breakage'. She mentions Halafian Yarim Tepe II (level VII) where a cremation of a 12-13-year-old girl was found in a specially constructed oven. It seems that at the time of cremation six clay and three stone vessels were intentionally broken and deposited in the oven, together with some small objects. It is also suggested that the smashed Halaf pottery in the TT6 Burnt House at Arpachiyah (Mallowan and Rose 1935:99, 106), associated with 'cult objects', had perhaps also been ritually broken (see Campbell 1992). Ethnographic parallels furthermore indicate a variety of possible reasons for the breaking of objects as a mortuary rite, e.g. the releasing of the spirit of the object to accompany the deceased to ritual toasts, the prevention of quarrels among surviving relatives, or the destruction of the enemies of the deceased (Grinsell 1961).

The potlatch ritual is the most famous instance of the intentional destruction of material culture. The potlatch was characteristic of traditional totemic Northwest Coast Indian groups, i.e. the Tlingit, Haida, Bella Coola, Nootka and Kwakiutl (e.g. Rosman and Rubel 1971; Vertovec 1983). In these societies a potlatch ritual, initiated by family heads, was part of a rite of passage, e.g. birth, death, or initiation in a 'secret society' of community members.

With regard to the characteristics of the potlatch ritual, the intentional breaking of figurines in the level 6 village does not seem to indicate a similar procedure. The objects destroyed in potlatch rituals are most often precious commodities which were highly valued as objects of exchange. The simple clay figurines of level 6 can hardly be regarded as valuable objects which served in gift exchange. I therefore suggest that the deliberate destruction of the figurines served another purpose, which I will now come to.

Ethnohistorical and ethnographical evidence has indicated that the intentional breaking of objects often serves as a form of identification or as an agreement between two people or parties. The practice has a long history in a variety of contexts (Talalay 1987). The Classical author Herodotus, for example, reports 'split tokens' which sealed an agreement between Miletos and Glaukos of Sparta. Examples of such 'documents', excavated in the Athenian Agora, are probably represented by small clay plaques cut in half along an irregular line, the shape of which is such that only the original other half can be fitted to the plaque (ibid.:165). Often economic transactions were and occasionally still are symbolized in this way. Each party would keep one piece in order to have proof of the identity of the owner of the other (ibid.:164). Discussing clay figurine legs from Neolithic Greece (first half of the fifth millennium B.C.) Talalay (ibid.) convincingly argues that the leg fragments formed parts of originally attached pairs and that they functioned as special contracts or identifying tokens symbolizing social and economic bonds among Neolithic communities in Greece. The broken figurines of Sabi Abyad may perhaps be interpreted in a similar vein.

It is recalled that these figurines were found in association with the clay sealings and tokens (cf. figs. 6.7 and 6.8). It has already been argued that this seems to indicate that they functioned in administrative contexts. In this respect it has been suggested that the human figurines represented services, and the animal figurines (and tokens) goods (Matthews 1989:94-95; Schmandt-Besserat 1992:178). Expanding on this view it is a possibilty that the figurines were intentionally broken in order to signify economic trans-actions and social bonds. On the basis of the above-mentioned association and Akkermans and Duistermaat's suggestions, I hypothesize that the figurines figured in transactions of sealed products. It is possible that the (ritual) breaking of figurines accompanied and sanctioned such transactions. As has been argued, the sealed products probably belonged to the nomads related to Sabi Abyad, and the residents were responsible for the storage at the site. It then follows that the transactions were between these two groups. Apart from producing two 'documents', the destruction of figurines captured the moment of inter-action between nomads and residents, and activated a mental image of union and cohesion. In a sense, then, they were not a view towards the past, but towards the future, towards the continuation and reproduction of society (cf. Kuechler 1987; Rowlands 1993, both discussing the role of memory in the transmission of culture). Presumably figurines (and not e.g. pierced discs) were chosen for the above suggested practice because they symbolized living beings, i.e. humans and animals. They therefore may have referred to life itself[82], and concomitantly to social organization. The choice of breaking human or animal figurines was perhaps related to different kinds of transactions.

---

[82] In this respect it should be noted that the human figurines all depict females, who perhaps symbolized life-giving power and fertility.

As Strathern (1988:161) has argued, the exchange of commodities in 'primitive' economies also includes the exchange of qualities of persons and parts of things, which merge and interpenetrate. In this respect Mauss (1990) has argued that exchange in 'traditional' non-modern societies is not only an economic activity. Exchange in these societies is part of a holistic worldview in which humans, ancestors, spirits and gods, or the cultural, natural and supernatural worlds are interlinked. In such systems the subject (the participant) and object (the goods) of exchange are inextricably linked. In traditional societies persons are not only dependent upon relations with other persons, but also on relations with the supernatural world (e.g. Bazelmans 1996:91-106). Especially in exchange the various relations between the human and supernatural world are activated. In this holistic process of exchange, the society as a whole, as a unity, is continually constructed and reconstructed.

When viewed from this perspective, not only the breaking of figurines, but also the exchange or transaction of products between nomads and residents itself, may indeed have constituted the society. Finally, it can be suggested that the breaking of the figurines released the 'life force', which was reactivated and rechannelled to the living. Perhaps they were ritually 'killed' for the sake of giving life (Bell 1992:176).

7 Summary and Conclusions

In this final chapter of this archaeological ethnography, an interpretation and reconstruction of the level 6 community at Sabi Abyad has been presented. It has been argued that this (non-hierarchical) community was organized on a dual basis of nomads and residents. A nomadic population related to Sabi Abyad has been based upon three archaeological indications: (1) the analysis of the use of the sealings, functioning in a system of distribution rather than in an intra-regional exchange context; (2) a spatial analysis, indicating large scale communal storage, exceeding that of household requirements; (3) analysis of the animal bones, which suggests that seasonal pastoral movements or mobile pastoralism were relevant subsistence strategies at Sabi Abyad.

The nomadic population may have consisted of ca. 400 to 670 people, the residential group possibly was made up of ca. 60 to 120 persons. The architecture in the Burnt Village can be divided into two classes, i.e. round and rectangular buildings, which have been related to the residents and nomads respectively. A distinction has been made between a 'household economy', related to the resident households, and a 'village economy', indicating the production of goods at and around Sabi Abyad (e.g. grain) for the nomads (with a 'pastoral economy'). The village economy was in the hands of the resident families, which were perhaps occasionally grouped into cooperating economic units.

The apparent architectural order was not only the outcome, but also the cause of human action: the tholoi and rectangular buildings were meaningful *structuring structures*. The level 6 settlement and its buildings shaped people's cultural being and the way they related to one another outside the settlement. The settlement and its buildings were given meaning by social practice, which reproduced their habitus. This practice would have been informed by childhood education, the architectural structure of the surrounding houses and buildings, the nature of the activities, and social interaction outside the domestic world. Buildings and society continually generated and regenerated one another in a structuring dynamic. In this sense, Tell Sabi Abyad acted as a microcosm for the society as a whole, the buildings and related material culture referring to different social groups and practices, i.e. to the different fields (i.e. the fields of nomads, residential families, and a ritual group, or communitas).

Two mechanisms for binding together the dual society and creating a sense of identity were isolated, and contextualized. First, and probably foremost, a ritual (a rite of passage) related to the death of two persons deposited on the roof of building V and surrounded by a number of clay 'monsters' was discussed in detail (and possibly this ritual was associated with abandonment rituals). It has been argued that a series of liminal, 'betwixt and between', categories were used (especially the horns of wild sheep) in the ritual. These categories (which are typical of rituals) served to express, and at the same time to question, fundamental social rules and values, mainly related to the nomads and residents. Second, the deliberate destruction of figurines has been interpreted as an act of symbolizing economic transactions, which moreover was a moment of interaction between nomads and residents, activating a mental image of union and cohesion between both groups.

Apart from these mechanisms, it should be taken into account that persons were related to different fields, each of them linked and the members closely interacting, resulting in cohesion rather than division. Sabi Abyad probably functioned as a central place where different but related groups of people, ideas and objects were brought together at certain times. This would have resulted in a reinforcement of ideas and rules shared by the society as a whole, of a shared identity, ideology and social solidarity.

Although the society seems to have consisted of two main elements, a real dichotomy between nomads and residents should not be postulated. In all probability pastoralist groups easily changed places with sedentary groups and vica versa. So the people's habitus was marked by an ambiguous worldview in which tholoi and storage buildings, nomads and residents, fields and herds, houses and tents, and village and landscape were distinguished but at the same time inextricably linked.

APPENDIX 1

FUNCTIONS OF ARTEFACTS OF THE BURNT VILLAGE

In this appendix the functions of the different artefacts found in the level 6 settlement will be discussed. It is the background for the models as presented in chapter 5. The objects have been described in Akkermans, ed. 1996, chapters 3 (pottery), 4 (flint and obsidian industries), 5 (seals and sealings), 6 (figurines), 7 (ground-stone industry), and 8 (other small finds).

1 SMALL FINDS

Sealings

Sealings consist of lumps of clay either pressed on the fastening of a container or closing this container entirely. Many sealings carry stamp-seal impressions. Evidently sealings secure against unauthorized opening of containers, and at the same time they may carry information about contents, destination and ownership (e.g. Duistermaat 1996:342). In chapter 7 the functions of sealings are discussed in detail, and no further mention of them will be made here.

Anthropomorphic and zoomorphic figurines

The function of prehistoric figurines has been dealt with by various authors, particularly by Ucko 1968, Voigt (1983:186-195), Rollefson 1986 and Ochsenschlager (1974). From such reports it follows that figurines probably had various functions, i.e. they may have served as: (1) cult figures; (2) vehicles of magic; (3) didactic or teaching figures, or initiation figures; (4) toys; (5) representations of deceased persons, and (6) burial gifts (Voigt 1983:186-195). For the specific function and meaning of the figurines from the Burnt Village the reader is referred to section 6 in chapter 7.

Grinding slabs, mortars, grinders and pestles
Ethnographic accounts

Lumps of buttermilk (*kashk*) can be pounded with water in a wooden mortar to make a *dugh* substitute that may also be eaten hot as a soup (Watson 1979:110). In Tell Toqaan metal mortars and pestles were used to crush meat and spices (Sweet 1960:130). Watson (1979:162, 169 and fig. 5.40) mentions that wooden pounders in Hasanabad were sometimes used for pulverizing meat in a wooden mortar. Salt was crushed with a small limestone rock which was used as a pestle (ibid.: fig. 5.6). The pastoral Tuareg also pound meat in a large wooden mortar (Nicolaisen 1963:233). Furthermore, they use a large wooden mortar and a heavy pestle for pounding millet and sorghum, and wild seeds, dates, cheese and locusts (ibid.:235). Smaller mortars which rest on a foot are used for pounding rock-salt, dried tomatoes, tobacco, and other plants or materials to be used in small quantities. A grinding slab (or 'saddle quern') is used for grinding wheat, barley, lentils and wild seeds (ibid.:235; Beazly and Harverson 1982:81 and fig. 98). When saddle querns are used, a piece of cotton or leather is put under the slab to catch the flour (ibid.:236 and figs. 180 and 182). In Baghestan mortars and pestles served for crushing and grinding grain, seeds, herbs, sugar and salt (Horne 1983:18). In Luristan, the nomads use wooden mortars, and pestles served for pounding herbs and spices and for removing husks from the rice (Mortensen 1993:245). In the city of Sana (Yemen) grinders and slabs were commonly used for crushing wheat and spices and herbs, and pestles and mortars for pounding spices (Serjeant and Lewcock 1983:543-544). All these grinding implements were used in the kitchens of the houses.

Seeden (1982:56) has suggested that the absence of larger grinding slabs at Shams ed-Din Tannira might be explained by the observation that at present much of the crushing of bigger quantities of soft substances such as 'whitewash' is simply done on a working floor in or near the settlement. Furthermore, broken ground-stone tools are often secondarily used in building and other activities.

Interesting with regard to the above is the analysis of Molleson (1994) of the human bones from Neolithic Abu Hureyra. Molleson suggests that deformations occuring on many of the Abu Hureyra bones (i.e. the big toe, the spine and the leg) were due to the strains of grinding grain on querns. It is suggested that prior to the grinding the grains were dehusked by pounding them with a pestle in a mortar. The operation of grinding is pictured as a woman (as indicated by some foot bones and ethnographic parallels) kneeling before one of the ends of the slab and rubbing a grinder with both hands. Kneeling for long periods deforms the toes and knees, and the action of grinding puts pressure on the hips and particularly the lower back.

Archaeological functional interpretations

Tools which are designated 'handstones', 'querns', 'saddle querns', 'slabs', 'mortars', 'pestles', and 'grinders' *metate'* (= base-stone), *'mano'* (= upper stone) (Wright 1992) are generally interpreted as instruments which were used for grinding, pounding and cracking of foodstuffs. Grinding slabs (or querns) and grinders (or handstones) are mostly related to the grinding of grain. With regard to mortars and pestles it is commonly suggested that smaller amounts of foodstuffs other than grain (i.e. seeds, nuts, herbs) have been 'crushed', 'pounded', and 'grounded' with these tools (see e.g. Ellison 1984). Apart from foodstuffs substances such as ochre are sometimes mentioned (e.g. Hattula 1983:291-292; Semenov 1964:134). At Hajji Firuz, red and black, probably red ochre and carbon black, adhered to the grinding stones. Voigt (1983:261) therefore suggests that the grinding stones were used for the preparation of paint for use in burials and on floors and walls. Grinders (or grinding stones) appear, like grinding slabs, throughout all periods in the Near East. In numerous excavation reports they are interpreted as tools for the processing of grains (e.g. Dorrel 1983:527; Merpert and Munchaev 1993a:91). Mortars are commonly assumed to have been used for the preparation of seeds, nuts, herbs, etc. (e.g. Howe 1983:60).

'Boulder mortars' (at Sabi Abyad one was found in room 5 of level 6 building V) are immovable features set into the ground. Voigt (1983:261) suggests that such mortars (according to her used to grind cereals and other food plants) would have been located in roofed areas. She also points out, however, that the stones were moved seasonally, so that they were located in exterior areas during the warm dry months (see also David 1971:116; Kidder 1932:67; Reynolds 1968:84).

Forbes (1955:57-149) makes the important distinction between crushing by impact (or pounding) and grinding by rubbing. Mortars and pestles are said to have been the general tools for the crushing and pounding of e.g. cereals, herbs and pigments. According to Forbes, cereals were dehusked by crushing or pounding by impact, whereas grinding by rubbing was used for making flour. It is mentioned that many of the dehusking mortars and pestles were made of wood, and the mortar was sometimes placed on a high stand (ibid.:148 and fig. 38). In the Near East, saddle querns and grinders were used to ground grain, corn, sesame or dates (ibid.:58). The use of querns and grinders in ancient Egypt is illustrated by a number of statues and depictions on papyri and on walls (e.g. Hodges 1970: fig. 23). Furthermore, on a number of cylinder seals from Mesopotamia the use of grinders on slabs is also depicted (ibid.: fig. 23).

The grinding slabs, mortars, grinders and pestles of Sabi Abyad have been divided into 13 classes or types, mainly on the basis of shape (Collet and Spoor 1996:416-417). Collet and Spoor mention that "Sometimes the shape is directly related to the function of the object but usually items with a more or less identical shape served different or multiple purposes" (ibid.:416). The different types and their functional interpretations are indicated in table A.1.

Stone vessels

In the literature studied I have encountered no information regarding the function of stone vessels. Obviously, stone vessels were used as containers for food (dry as well as liquid) and other products such as herbs and pigments. Food may have been stored and served in such vessels. Furthermore, it cannot be excluded that stone vessels were used as mortars, i.e. to grind substances. The often scratched interior of the recovered vessels at Sabi Abyad may in fact indicate such a use. In this respect, Seeden (1982:56) notes the rarity and shape of the small stone bowls at Shams ed-Din Tannira suggest that they were used as mortars.

Type	Description and suggested function
1	Rather large cylindrical pestles or grinders; most objects have traces of use along the long sides (grinders); short ends occasionally show traces of pounding and grinding (pestles)
2	Like 1, but conical; more pestles than grinders
3	Like 1 and 2, but more irregular in shape; frequent re-use; unworked stones used as abrasion tools included in this class
4	Small spherical or cubical tools, used as pounders, hammerstones, grinders and/or pestles
5	Mortars
6	Flattened grinding slabs and associated flattened grinders
7	Small cylindrical grinders or pestles with smoothened top and/or base; the ends used for grinding and pounding
8	Small conical pestles
9	Unidentifiable pieces
10	Rectangular grinders
11	Flat, more or less rectangular or slightly rounded objects made of extremely light and porous basalt; likely that these tools were used for scraping, rasping or grinding of soft materials
12	Large spheres, smooth on one or more sides
13	Large bulbous or pear-shaped tools with often roughly-hewn or damaged working ends; most likely used as pestles

Table A.1. Typology and possible functions of ground-stone tools of Sabi Abyad (after Collet and Spoor 1996:416-417).

Celts (axes and adzes), chisels and maceheads[83]

Ethnographic accounts

Obviously, axes are used for felling trees and for working the wood which is to be used in construction of houses, etc. (e.g. Hodges 1989:115; Mellaart 1967:215). The periodic resharpening is done by rubbing them against sandstone while the surface is kept wet (Carneiro 1979; Townsend 1969). Sandstone pebbles, found in archaeological excavations, may have been used for the resharpening of axes. Watson (1979:108) reports that butchers in Kermanshah use metal axes and adzes to chop the bones, cartilage and meat of slaughtered animals in a number of portions. The Tuareg use axes to make domestic implements such as spoons and wooden bowls (Nicolaisen 1963:261, and fig. 206).

Archaeological functional interpretations

Axes and adzes are commonly held to have functioned as wood-cutting instruments (e.g. Braidwood and Howe 1960:45; Mellaart 1967:215; Voigt 1983:262). Generally it is suggested that axes were used for felling trees, cutting off the branches, and for splitting logs and planks. Adzes were used for hollowing out

[83] Collet and Spoor 1996 (pp. 424-425) use the term *celt*, and make no distinction between axes and adzes. In the literature, however, this distinction is widely applied. The difference between axes and adzes is in the method of hafting (and use): axes are hafted with their working edges parallel to the haft; adzes are hafted transversely to the haft (e.g. Semenov 1964:126-127). In the present study I have chosen to distinguish between axes and adzes.

trunks and smaller pieces of wood, and for flattening planks (see e.g. figs. 49-51 in Drenkhahn 1976). The wood would mainly have been used for construction purposes. On one of the fully ground axes from Arpachiyah the impression of the wooden haft was still discernible (Mallowan and Rose 1935:102 and fig. 52:12). Howe (1983:55) suggests that axes and adzes which show no edge scarring may have been used for the working of soft materials such as skins, hides, reeds and rushes. Hattula (1983:296) suggests that the small axes at Jarmo were used for carving wood and bone, whereas the larger ones were choppers for clearing the land of wood (see also Dorrel 1983:509).

Seeden (1982:58) reports that the axes of Shams ed-Din Tannira were not hafted, which would make their use as axes rather unlikely. On the basis of wear traces and percussion marks she suggests that the axes were used for knapping and breaking up hard materials, such as rock, bone and nuts. Therefore she regards axes as multipurpose tools which were used for chopping, cutting, breaking, grinding, rubbing, polishing and whetting.

Dorrel (1983:509) has observed that the PPNA axes of Jericho were shaped with a punch or pick. Subsequently, the axes were finished by grinding and polishing; the peck marks were smoothened out and the cutting edge was sharpened. According to a use-wear analysis of Coqueugniot (1983) the flake-adzes from Mureybit were (like some scrapers) used to chop and plane wood. Roodenberg (1983) has suggested that the ground stone 'hatchets' from PPNB Bouqras were used as felling axes, and as wedges and adzes. The flint axes from Arjoune, a fifth millennium B.C. site in the Homs Basin, Syria, were, according to a microwear analysis, most likely used to chop wood (Unger-Hamilton 1988:167).

Collet and Spoor (1996:424) have suggested that the miniature axes of Sabi Abyad served in luxury or ritual contexts instead of in domestic activities. The carefully made maceheads may have been used as symbols of power, as ritual instruments or as weapons.

Stone discs

Archaeological functional interpretations

On the basis of tiny striations, scratches and traces of red ochre, the circular or oval stone discs at Sabi Abyad have been interpreted as palettes or pigment grinders, and as polishing stones or vessel lids (Collet and Spoor 1996:425).

Unworked stones/flat stones/pebbles

Ethnographic accounts

The Tuareg use flat stones for smoothing leather (Nicolaisen 1963:275). The Pueblo potters use smooth pebbles for burnishing pottery (Kidder 1932:63 and figs. 39-40). These pebbles are often highly valued (e.g. De Boer and Lathrap 1979:115). As a general remark, Campana (1989:25) mentions that 'abrasive stones' (and sand used for grinding) are used to shape and polish bone tools, such as awls. In the process of hide-working the North-American Plains Indians worked hides during tanning by rubbing them with smooth stones (Lowie 1954). The Tanaina in Alaska sharpened their awls (i.e. bone points for their fish spears) by rubbing them with a stone (Osgood 1937).

Archaeological functional interpretations

In many archaeological reports it is suggested that unworked stones, such as 'stream-worn' pebbles (with gloss/polish/striations) may have been used as rubbers, whetstones (for the sharpening of artefacts of stone, bone and shell, see e.g. Drenkhahn 1976: fig. 50), polishers, anvils, 'working stones', knapping tools, hammerstones, abraders, and in general for pounding and pecking (e.g. Dorrel 1983:505; Hattula 1983:291-294; Howe 1983:53-54; Wright, ed., 1981:146). The smooth pebbles of Hajji Firuz Tepe are interpreted as implements which were used to burnish pottery and other small artefacts of clay (Voigt 1983:262). The large smooth pebbles are interpreted as smootheners for large flat surfaces, e.g. floors, hearths and bins.

Collet and Spoor (1996:426) suggest that at Sabi Abyad the limestone, sandstone, quartzite and dolorite cobbles that were found may have been used as polishing stones or, perhaps, grinders.

Palettes

Archaeological functional interpretations

Palettes (flat stones, often with a shallow depression) have been connected with painting (e.g. Hattula 1983:293; Howe 1983:61). The palettes at Hajji Firuz Tepe showed traces of pigment. Voigt suggests that the palettes were used for grinding small amounts of ochre and carbon which were used for decorating ceramics, textiles, leather and people (1983:261). The grinding stones which would have been used with the palettes were not identified. Traces of pigment have repeatedly been observed on palettes from a large variety of Near Eastern sites (e.g. at Arpachiyah; Mallowan and Rose 1935:100, see also Wright, ed., 1981:146). Very fine stone palettes with two rims preventing spillage have been recovered from Arpachiyah and Yarim Tepe III (Mallowan and Rose 1935:100; Merpert and Munchaev 1993c: fig. 9.14:1).

At Sabi Abyad flattened pebbles with one or more shallow concavities and either circular, oval or more or less rectangular in shape, and often with traces of polishing, have been interpreted as palettes (Collet and Spoor 1996:425). One of these implements was found in association with a small spherical grinding stone. Other flat stones have been found with traces of red ochre.

Stone spatulas

Archaeological functional interpretations

Some 'pebbles' found at Sabi Abyad have tiny scratches on one of the ends. Perhaps these thin and oblong, unworked objects were used as spatulas (Collet and Spoor 1996:426) or as retouchoirs (e.g. Semenov 1964: fig. 14, no. 1).

Whetstones

Archaeological functional interpretations ˙

At Sabi Abyad, flat objects made of tabular sandstone or limestone, some with tiny striations, others with highly polished surfaces, have been interpreted as whetstones (Collet and Spoor 1996:426). These implements may have been used to sharpen stone tools which were used for cutting.

Working platforms

Archaeological functional interpretations

At Sabi Abyad, fragmented limestone and gypsum objects of considerable size (over 28 cm in length and up to 9 cm thick; Collet and Spoor 1996:426) may have been used as worktops for carrying out certain activities (e.g. making stone tools, hammering, etc.).

Perforated stones

Archaeological functional interpretations

Perforated stones which might represent weights for digging sticks that were used for preparing the ground for agriculture (e.g. Wright, ed., 1981:49), have been found from the Hassunan period onwards on many sites (e.g. Braidwood et al. 1952:21; Braidwood and Howe 1960:45; Lloyd and Safar 1945: figs. 19-20). Perforated stones may also have functioned as maceheads (Moorey 1994:71-73), net sinkers and loom weights (e.g. Braidwood and Howe 1960; Hattula 1983:294; Holland 1983; Howe 1983:53; Watson and LeBlanc 1990:103; Wright, ed., 1981:49).

Some irregularly-shaped and coarse perforated objects of limestone at Sabi Abyad may have been used as weights (Collet and Spoor 1996:426).

Stone balls/hammerstones

Ethnographic accounts

It has been observed that hammerstones which were used for shaping axes should be harder than the axes themselves, furthermore hammerstones must be shaped in such a manner that they fit the hand comfortably (Kozak 1972:18). Hammerstones are most often selected as angular pieces of rock which become rounded through use (Kidder 1932:60; Moorey 1994:71-73; Reynolds 1968:84). Hammerstones are often found among the tool kits of hunters-fishers-gatherers; most often they are used as multi-purpose tools (McCarthy 1967:56, 59).

Archaeological functional interpretations

Voigt (1983:261) suggests that stone balls or hammerstones were multipurpose tools, used for pounding and/or rubbing (e.g. of pigments; Semenov 1964:137) and for the manufacture and maintenance of ground stone artefacts. The Jericho hammerstones, discussed by Dorrel (1983:514) were presumably used for flint-working, as weights, as mauls for stone dressing and as 'pot boilers' (i.e. cooking stones).

Arrowshaft straighteners
Archaeological functional interpretations

Grooved stones with or without decoration, such as found at e.g. eighth millennium B.C. Çayönu in Anatolia, seventh millennium B.C. Kültepe in northern Iraq, and sixth millennium B.C. Tell Judaidah in the Amuq (Bader 1993b:59 and fig. 4.5; Braidwood and Braidwood 1960:90 and fig. 66; Davis 1982:110-111 and figs. 3.12-3.13) have generally been interpreted as objects which served for the production of some kinds of stone or bone tools or for the straightening, i.e. the removal of irregularities at the surface, of arrowshafts (e.g. Collet and Spoor 1996:426).

Howe (1983), discussing the artefacts of Karim Shahir, a hunter-gatherers' camp in the Central Zagros, suggests that these grooved stones may, apart from the straightening and removal of irregularities (i.e. thorns) of arrowshafts (e.g. Hattula 1983:294; Voigt 1983:203), have been used for the polishing or grinding of beads, rings and other artefacts of stone, bone, shell or wood (see also Solecki and Solecki 1970).

At Tepe Yahya, a mound in South Iran dated to ca. 4900-3300 B.C. (5680-4010 cal BC), 14 stone (mainly chlorite) shaft straighteners were found (Lamberg-Karlovsky and Beale 1986). Two specimens were found in association with a foundation deposit. Perhaps this tool kit gives a clue to the function of the straighteners. The deposit consisted of a stone female figurine, two sickle or knife handles of bone, a bone spatula, and 63 flint sickle blades (ibid.: fig. 7.16). The excavators suggest that the association of artefacts may indicate that the shaft straighteners were used for producing bone tools, or that they may have been used in conjunction with the bone tools. The wear traces on the Tepe Yahya straighteners suggest that soft materials such as wood and bone were processed. The shape of the groove (mainly U-shaped, occasionally V-shaped) furthermore suggests that the worked objects already had a circular or U-shape (ibid.:185).

Spindle whorls/pierced discs/loom weights
Ethnographic accounts

The use of spindles and whorls for weaving in Hasanabad has been discussed in detail by Watson (1979:174-180, but see also: Barber 1991:39-51; Bazin and Bromberger 1982:60-75; Champault 1969; Dalman 1964). Both spindles and whorls were made of wood (ibid.: fig. 5.51); stone or ceramic whorls have not been attested. The length of the spindles varied from 27 to 38 cm; the spindle whorls, however, were remarkably uniform in diameter (5.5 to 6 cm) and thickness (generally 2.5 to 3 cm) (ibid.:178). The main fibres which can be spun are flax, cotton, wool and silk (Forbes 1956:149).

The process of spinning is described as follows by Watson (see also Forbes 1956:152-156; plates 5.7 and 5.8 in Watson 1979; Costa 1991:145-189; Hodges 1989:128-129; Wulff 1966:185-186, for Syria: Van Der Kooij 1976:90):

"In the village, spinning is done by the women using a very simple grooved, wooden spindle with a wooden whorl. These spindles are usually bought in Kermanshah or from the Kwali. The wool is first cleaned by picking out the larger impurities (dung, twigs, weed seeds, etc.), then washing it. When the wool is dry it must be fluffed out, and then part of it will be drawn out and shaped by rolling between the palms into a flat rope. The rope is coiled into a basket and is ready to be spun. A filament is drawn from the end of the roll and wound around the spindle inside the groove. The spindle is strongly twirled with the fingers of the right hand, the whorl acting to balance the spindle and prolong the spin. The base of the spindle often rests on the floor or upon some solid object but this is not essential. The wool (or goatshair, or cotton; the same process is used for all) filament is twisted into yarn, and, as it becomes yarn, is wound onto the spindle just above the whorl; the next section is wound into the groove, and the procedure is repeated. The fingers of the left hand may be used to guide the wool. An experienced spinner working with clean wool rapidly produces an unbroken length of yarn whose diameter is quite consistent" (ibid.:174-175, for a description of the use of whorls: see also e.g. Curtis 1984:38).

The yarn is subsequently woven into balls. To produce rugs, tent cloth, salt bags, spoon bags[84], goatshair feed bags and horse coverings (ibid.:182), looms are used (see also Bazin and Bromberger 1982).

A loom is a weaving instrument which is designed to keep the warp-threads under tension. Hodges (1989:133): "Weaving demands the use of two distinct groups of parallel threads. One set, the warp, chain or ends, is kept under tension during weaving, while the second set, normally referred to as the weft or woof (...), is passed over and under the threads of the warp, at right angles to it".

---

[84] A spoon bag is a storage place for the large wooden spoons used by the Hasanabad women (Watson 1979:184, see also Mortensen 1993:252).

Archaeological functional interpretations

Spindle whorls and pierced discs (the use of potsherds as whorls is widely attested; Braidwood and Howe 1960:45; Lamberg-Karlovsky and Beale 1981:188; Watson 1979:178) have been found in a large number of Neolithic and other prehistoric excavations. Commonly they are interpreted as flywheels for spindles, which were used for spinning (Akkermans 1989b:288; Broman-Morales 1983:390; Voigt 1983:202, 262), and sometimes for drills (e.g. Hattula 1983:294). Schmandt-Besserat (1992:17, 77-84, 108), however, has suggested that pierced discs (and she refers to small objects of unbaked clay as well as to reworked sherds) were used as tokens. She presumes that the perforations served to hold the perforated tokens on a string. Another interpretation is Bader's idea that the perforated stone discs at Tell Maghzaliyah (northern Iraq) served as pendants (Bader 1993a: fig. 2.15). Seeden (1982:61) has suggested that some of the more rregular 'perforated disc sherds' at Shams ed-Din Tannira had been perforated in order to be hung, possibly as weights. And at Arpachiyah, finally, a pierced disc was found in a primary context on the top of a jar, where it had been fastened by means of bitumen. Mallowan and Rose (1935:90 and fig. 49:22-23) suggest that the hole in this 'vase lid' was for attachment with a string.

The earliest evidence for woven thread comes from the Near East. Impressions of textile dated at ca. 7000 B.C. (8020 cal BC) were found at Jarmo, actual remains of cloth were found at Çatal Hüyük (ca. 6000 B.C., 6860 cal BC), and other evidence of weaving was found in early sixth millennium B.C. Shimshara (Iraq) and Tepe Yayha (SW Iran) (Barber 1991:51). In Mesopotamia we have some pictorial evidence for the use of spindle whorls from the late fourth millennium B.C. onwards. For example: on a cylinder seal from Choga Mish (Khuzistan) a spinning woman is depicted seated on a low stool, and on a fragment of a mosaic panel from Mari (early second millennium B.C.) a number of women with spindle whorls are shown.

Perforated clay balls, commonly interpreted as loom weights, have appeared at many other sites in the Near East. In Israel, for example, they have been found in nearly every Iron Age site (Sheffer 1981). At Iron Age Lachish (Israel) groups of clay balls were found near a burnt still pole still standing upright; here a warp-weighted loom can be postulated as well (Starkey 1936:88). A seal impression from Susa (ca. 3000 B.C.) depicts two weavers at work on a horizontal double-beam loom. Two vertical posts next to this loom may depict a vertical loom (Ochsenschlager 1993:55). Ochsenschlager (ibid.:55) suggests that the existence of horizontal and vertical looms perhaps points to a division of labour (as today) with women weaving on horizontal looms and men on vertical looms.

The earliest representation of the vertical two-beam loom comes from wall paintings in Egyptian tombs from the New Kingdom (1550-1070 B.C.) (Barber 1991: figs. 3.29 and 3.30).

Early evidence for band looms is so far lacking, but from 6000 B.C. (6860 cal BC) onwards fragments of textile bands have been found (e.g. at Çatal Hüyük).

Tokens

Archaeological functional interpretations

Schmandt-Besserat is well-known for her pioneering work about all aspects, particularly the use, of tokens. Tokens are described as "small clay counters of many shapes such as cones, spheres, discs, and cylinders, which served for accounting in prehistory" (Schmandt-Besserat 1992:6). Her book *Before Writing - From Counting to Cuneiform* (1992) was the first systematic study of tokens. In short, she proposes that tokens were the immediate precursors of cuneiform writing (in her view, tokens were already in use at about 8000 B.C., 9110 cal BC). Furthermore, she suggests that tokens reflected an archaic mode of 'concrete' counting prior to the invention of abstract numbers. According to her tokens were used in a one-to-one correspondence (ibid.:190), and particular counters were needed to account for particular goods.

In table A.2 I have summarized the information which Schmandt-Besserat presents about the contexts of tokens. She deals with prototypes of the late fourth millennium B.C. and she tentatively suggests that the meaning of these types may perhaps be extrapolated backwards in time (ibid.:152).

Regarding the meaning of plain and complex tokens, it is concluded that both categories varied in shape, markings, handling, storage and the type of commodities they represented. It is suggested that plain tokens were used for 'products of the country', whereas complex tokens stood for goods manufactured in cities. As yet cannot be proved, but Schmandt-Besserat is inclined to believe that there is a continuity in use of tokens from the eighth millennium B.C. to the third millennium B.C.. She thinks this continuity might be due to the general conservatism (seen worldwide) in the use of signs and symbols.

At Tepe Sarab in northern Iraq small geometric clay objects such as balls, flattened discs, cones, tetrahedrons, blocks and oblongs, rod fragments and rolled pieces were discovered. Broman-Morales (1990:22-25) interprets these objects as 'counters' or 'marbles', or gaming pieces. She suggests that the counters were used to aid in the control of flocks. The marbles, in her view, were perhaps used by shepherds while tending the herds. The cones and the tetrahedrons, however, are related to abstract human figurine forms.

Following Broman-Morales (1983) and Schmandt-Besserat (1977), Voigt suggests that small geometric objects, tokens, are abstract figurines and/or elements in a recording system (1983:195-202).

At Sabi Abyad, in addition to tokens, several miniature vessels of unbaked clay were found (Spoor and Collet 1996:443). It has been suggested that these little objects also represent tokens.

Context	Description
Types of settlements	Ca. 8000 B.C. (9110 cal BC): semi-permanent and sedentary open air compounds with round houses; 7th - 3th mill. B.C.: caves; sedentary villages of rectangular buildings; cities with monumental public architecture
Distribution within settlements	In cities tokens more frequent in official rather than secular quarters; recurrence of tokens among refuse in vacant lots, suggesting that they served for record keeping rather than reckoning
Structures	Tokens often located in storage structures; since 6th mill. B.C. tokens also occur in nondomestic architecture (e.g. temples, citadels, gates)
Token clusters	Recovered token clusters indicate that tokens dealt with relatively small quantities of different kind of commodities
Containers holding tokens	Pottery bowls and jars, baskets, wooden boxes, leather or textile sacks
Associated assemblages	8th mill. B.C.: tokens coincide with cultivation and storage of cereals; 7-4th mill. B.C.: tokens part of typical agricultural assemblages: pestles and mortars, spindle whorls, figurines, seals, etc.

Table A.2. Generalized information about contexts in which tokens have been recovered (after Schmandt-Besserat 1992 (94-101).

Oval clay objects
Archaeological functional interpretations

The oval, or loaf-shaped, clay objects of Sabi Abyad (level 6) are as yet without a parallel. In chapter 7 the function of these objects is discussed in detail.

Beads and pendants
Ethnographic accounts

Women's ornaments at Hasanabad mainly consist of rings, bracelets, necklaces, anklets and pendants (Watson 1979:194). Men do not wear ornaments or jewelry. These items are sometimes bought in Kermanshah. Ornaments of gold and semi-precious stones are available at the bazar, but only the rich people can afford these items. Necklaces and anklets consisting of beads (and cowry shells) are often made by the village women. By using different kinds and colours of beads, complex patterns may be created. Watson (1979:195) noticed that often some blue beads are tied to a woman's turban as decoration, for good luck, and for protection against the Evil Eye.

Archaeological functional interpretations

Pendants and beads are commonly interpreted as (personal) 'ornamental' and 'decorative' objects (e.g. Moorey 1994:106-110; Watson 1983a:573; Voigt 1983:262; Lamberg-Karlovsky and Beale 1986:167).

These objects may have been used as bracelets, anklets, necklaces (e.g. a beautiful bead necklace found at Hassunan Yarim Tepe I, level 12; Merpert and Munchaev 1993a: fig. 6.28), hair-ornaments, ear-ornaments, sew-on ornaments, clothing fasteners, etc. (i.e. Hattula 1983:297). At Tepe Gawra beads, mostly found in graves, were used in necklaces, as bracelets, anklets and possibly girdles (Tobler 1950:192).

Labrets
Archaeological functional interpretations
The best evidence for the use of labrets stems from an early sixth millennium B.C. male burial at Ali Kosh in the Deh Luran plain. In the grave a tersed labret was found in primary context upon the lower jaw, suggesting that labrets served as lip (and ear) ornaments (Hole et al. 1969:235-236).

Sling missiles
Ethnographic accounts
Based on ethnographic analogies, Korfmann (1972:4) produces some useful, but rather general information, on the use of sling missiles. These projectiles, propelled by means of a sling, are generally round, oval or bi-conical. Not only clay, but also lead, stone and even hard fruits are used as ammunition. The careful smoothening of the missiles enlarges the velocity, range and the accuracy of the projectiles. Slings are generally made of wool, leather, intestines, hair or plant fibres. A width of 3 cm and a length of ca. 1 m seem to be the minimal dimensions of successful slings. Generally the slings were used as follows: one end of the sling was tied around the little finger (or held between this finger and the ringfinger), the other end was held between the thumb and the forefinger. Subsequently the sling missile was put into the strap, which was then swung three or four times above the head before the end between the thumb and forefinger was released (see also Bonnet 1926:115).

Sling missiles in Hasanabad consist of round pebbles. Watson (1979:186-187) reports that the slings are made by plaiting yarn and goatshair strands for the straps. The yarn was wound several times in a loop around the big toe and finger for knotting the pocket. Across this loop a series of half-hitches was made with the end of the yarn, and the loop was knotted to each end of the loop strands. Subsequently, the half-hitches were continued back and forth until the pocket was complete. Eyes were left at either end for attachment to the straps. In Hasanabad the slings were used by shepherd boys only. At Al-Hiba in southern Iraq, the "... most common sling is made of small egg-shaped balls of mud dried in the sun, but small stones, when available, and pot sherds can also be used. The sling is used by both boys and girls primarily for edible birds but occasionally for small animals as well" (Ochsenschlager 1993:47 and fig. 4).
Archaeological functional interpretations
Sling missiles have been found at numerous prehistoric and historic sites, e.g. Tell es-Sawwan, Hassuna, Matarrah, Tepe Gawra, Girikihaçyan, Mersin (Watson 1979:187). Based on ethnographic analogies the majority of researchers interpret sling missiles as weapons (e.g. Bonnet 1926:117; Korfmann 1972). The Assyrians, for example, used special slingers in their armies (Korfmann 1973). Most likely, they were also used for hunting.

Some have proposed other functions for sling missiles. Starr (1937/1939) interpreted the biconical objects of unbaked clay from Nuzi as 'counters'. He considered their use as sling missiles unlikely; they would be too fragile for this purpose (it was thought improbable that animals larger than birds could be hunted with these implements), and there would be no need for them because suitable stones could be found in the direct vicinity of Nuzi. The large quantity of the 'clay pellets', and their association with household objects makes it rather likely, according to Starr, that they functioned as counters. The sling missiles at Tepe Yayha have also been interpreted as possible counting devices (Lamberg-Karlovsky and Beale 1986:190).

The sling missiles at Tepe Gawra (Tobler 1950:173ff), mainly stemming from 'Ubaid and Uruk levels of occupation, resemble those of Sabi Abyad (Spoor and Collet 1996:448-450) in some respects. First, different types have been attested at Gawra: ovoid sling missiles (type 1 at Sabi Abyad), cigar-shaped sling missiles (type 3 at Sabi Abyad)85, and long-drawn pieces which were formed in the hand (resembling type 2 at Sabi Abyad). Secondly, the sling missiles at Gawra were usually found in caches composed of 20 to 150 specimens. In level 3B at Sabi Abyad virtually all sling missiles were stored in circular or oval pits (one pit contained up to 682 missiles). These pits were concentrated in and around level 3 buidling V (Verhoeven and Kranendonk 1996:96 and fig. 2.25). In the other levels at Sabi Abyad pits with sling missiles were not encountered. It has been suggested (ibid.:450) that the cluster of pits related to building V points to a communal storage of weaponry in an 'arsenal'. Perhaps this interpretation also is significant for the Tepe Gawra missiles.

Mortensen (1982:214), discussing early villages in Mesopotamia, suggests that sling missiles were used for hunting. He suggests that the bow and arrow and sling pellets at Umm Dabaghiyah could have been used for killing large animals (hyenas, wolves and swine), and smaller animals such as foxes, hares and birds. Sling missiles are also used to make stray animals return to the herd (pers. comm. S. Bottema).

Stands
Archaeological functional interpretations

Three large and conical perforated clay objects of level 6 may have served as stands (Spoor and Collet 1996:451).

Enigmatic clay objects

At Sabi Abyad a number of clay objects cannot readily be interpreted functionally as yet. First there are disc fragments of unbaked clay (all found in level 6). These flattened discs are circular or oval and are rather small, with diameters ranging between 6 and 11 cm (Spoor and Collet 1996:451). Secondly, a few 'mushroom-shaped' objects, up to ca. 6 cm high, are as yet unexplained.

Whistles
Ethnographic accounts

Clearly, a whistle is made for making music. In Hasanabad a reed pipe (Watson 1979:202 and fig. 5.65) is used on feast-days (weddings, or circumcision ceremonies).
Archaeological functional interpretations

At Hassunan Yarim Tepe I a hollow and cigar-shaped object with a hole has been interpreted as a whistle (Merpert and Munchaev 1971:16-17 and fig. 5b, pl. V). At Tepe Gawra bone 'playing-pipes' (i.e. bone tubes) have been found (Tobler 1950:214). Shorter bone tubes have been interpreted as mouthpieces for playing-pipes (ibid.:214).

White ware
Archaeological functional interpretations

It is evident that 'white ware' represents a class of gypsum containers. At Sabi Abyad 'white ware' rim fragments have shown that bowls were made of this material (Spoor and Collet 1996:451-452). Some fragments, however, indicate that they may also have served for the interior coating of large ceramic vessels.

Copper ore
Archaeological functional interpretations

The two small pieces of copper ore (malachite) found in level 6 at Sabi Abyad (Spoor and Collet 1996:452) seem to indicate copper processing (probably on a very modest scale). Perhaps copper was used for weapons (e.g. arrowheads), or for luxury or status items such as jewelry.

Awls/needles
Ethnographic accounts

Commonly, awls are described as piercing or drilling tools (e.g. Hodges 1989:116). Awls and large needles were used in Hasanabad to produce slippers (of rubber). An awl with an eye in it is used to punch holes in the sole and to thread the string through (Watson 1979:186 and fig. 5.60, see also Nicolaisen 1963:278). A needle is used to build up the slipper by a kind of crocheting. Awls are also used for making

---

[85] According to Korfmann (1972:204), the cigar-shaped sling missiles at Tepe Gawra represent a "Zwischenform" in the production sequence of sling missiles. At Sabi Abyad, however, these types of missiles (type 3) were found together (in pits) with the more common type 2 missiles (flattened cylinders). Clearly the type 2 missiles were finished products; it seems highly unlikely that finished and unfinished missiles would have been stored in the same pit. Therefore it is suggested here that the cigar-shaped objects were projectiles which were ready for use.

basketry, clothing and items of leather. Slender awls, or needles, may have been used to make small perforations in skins (Semenov 1964:100). Larger awls were probably used to make larger holes in e.g. leather (Hodges 1989:152).

The coiled Tuareg basketry is sewn with a small awl (Nicolaisen 1963:269). Coils may be joined either by wrapping or by sewing. The requirements for an awl which is to be used in basketmaking are: thick butt, strong haft, pointed tip (e.g. Voigt 1983:215). The awl is pushed through the foundation element of the basket in order to make a hole, subsequently the binding element or splint is passed through the hole, forming a stitch that joins adjacent coils (Voigt 1983:215, see also Costa 1991:145-189). In Brekelmans 1979 the tools and techniques related to making basketry are discussed. Two types of awls are mentioned: (1) a 'normal' awl, made of wood; (2) a hollow awl: an awl with a wooden handle and an iron point which is hollow in section (ibid.: fig. 12: e, fig. 18). This hollow awl is almost identical to the awls such as found at Sabi Abyad and many other Neolithic sites in the Near East. The hollow awl is like the normal awl used to make an opening for the splint. A difference in use is not indicated in Brekelmans 1979.

The Tuareg also use awls to sew leather bags (with leather threads; Nicolaisen 1963:273). Needles are rarely used; the Tuareg prefer to sew their skin articles by means of little awls. With the awl the women pierce holes close to each other, through which a thin leather thread is drawn (ibid.:278). After having been heated in a fire large awls are used for wood and horn working.

Watson (1979:196-198) has observed that in Hasanabad bone awls were used as applicators for eye make-up (ibid.: figs. 5.11 and 5.63). The make-up consisted of carbonized fat. The awl, together with a skin bag which contained the make-up (carbon), were kept in a cloth bag. The tips of the applicators can vary from sharp to blunt (Voigt 1983:216).

Archaeological functional interpretations

Bone awls and needles are found in many Neolithic (and later) sites. In the Levant they have been attested as early as the Natufian period (Bar-Yosef 1983). Commonly, bone awls (and spatulas) are interpreted as skin-working and sewing tools, (ibid.:19; Mellaart 1967:215).[86]

Campana (1989) has carried out a microwear analysis of Natufian (and "Protoneolithic") bone tools. The 'Natufian pointed implements' (i.e. awls) were, not surprisingly, most likely used as perforaters, presumably for hides. The tools may also have been used in the production of matting and basketry. It is mentioned that points used for basketry need not be so sharp; blunt points are preferable (ibid.:131). Campana suggests that the large tips were used to penetrate open structures such as basketry, the smallest tips were used to pierce hides, and the intermediate tips for either or both purposes (ibid.:118, 132). The fine and slender bone awls, or needles, were probably used for weaving textiles according to Campana.

Watson (1983b:362) discusses the functions of the bone objects found in Jarmo. She recognizes four groups:

1. Hide-working tools: awls, fleshers and burnishers. The awls were most likely used for 'working up' the hides to make garments, bedding, tents and containers. Watson suggests that hides were removed by obsidian blades and flakes. Scrapers may have been used to scrape the hair off the skin;
2. Cloth-working tools: awls and needles. It is suggested that awls were used for coarse cloth such as burlap, and needles for fine cloth;
3. Decorative items: jewelry. Jewelry most likely consisted of personal ornaments such as rings, beads, pendants and bracelets;
4. Eating utensils: spoons, pins (if used as skewers).

[86] Before hide-working tools like bone awls, spatulas, etc. can be used, hides have to go through three main stages of leather manufacture: (1) preparation for tanning: cleaning of the skin, opening up of structure to absorb the tanning agent; (2) tanning: the impregnation of leather in a tanning agent (such as smoke and water containing oak galls and cypress leaves) to make and to keep it water-resistant, strong and flexible; (3) finishing: rolling, dyeing, embossing, glazing, etc. (Aten, Faraday Innes and Knew 1955:41-72; Forbes 1957:3; Mann 1962:80-137; Reed 1972).

Reed (1972:88) lists the following applications for Mesopotamian leather: doors, shields, boats, chairs, chariots, thrones, containers for solids and liquids, bags, sandals, musical instruments, writing materials, labelling tags, scabbards, quivers, harness, saddles, armour. To this, of course, cloth should be added.

At Arpachiyah some of the bone awls showed a handle of bitumen (Mallowan and Rose 1935:103, pl. xiia). These handles, which have also been attested at e.g. Hassuna and Uruk (Ess and Pedde 1992:205; Lloyd and Safar 1945, pl. x:2, no. 23), allowed the awl to be conveniently pushed and turned into e.g. leather with the palm of the hand. The somewhat irregular proximal end of the awl is likely to cut into the hand if no handle is used; therefore it can be expected that handles of bitumen, or perhaps other substances, were much more common than presently assumed.

At Halafian Girikihaçyan two types of bone awls were found: short blunt points for 'heavy duties' and long thin points, for 'light duties' (Watson and LeBlanc 1990:90). At Ilipinar in northwestern Turkey there was also a distinction in long slender and short awls, probably pointing to at least two activities with regard to bone awls (Marinelli 1995:130).

Spatulas
Ethnographic accounts

Bone tools with a broad spatulate tip with an edge at right angles to the shaft were used by the Ojibwa Indians to loosen the excess tissue from the inside of hides (Campana 1989:59). These tools were used over long periods of time, therefore the number of tools in use at any time was small. Perhaps the (few) Sabi Abyad spatulas were used as leather burnishers. Burnishing leather is neccesary to make it tough and more impermeable (e.g. Semenov 1964:178). A burnisher should be a hard and smooth tool (e.g. Wulff 1966:232). Smooth stones, but bone tools as well can be used for this, as has been shown by Semenov, who studied wear patterns on Upper Paleolithic bone tools which he interprets as burnishers. These tools showed curved bevels and fine striations, which attributes were both assigned to use as burnishers (1964:175-167).
Experimental evidence

According to Campana (1989:59), 'spatula-tipped implements' have two general functions: the broad surface is used as a smoothing tool (e.g. for hides; Semenov 1964:175-179); the edge of the tool can be used as a wedge, chisel or pry. Campana (1989:59-71) made and used a number of experimental spatulas. On the basis of the microwear patterns it was concluded that the spatulas were pushed into the worked material and considerable force was indicated. About 80% of the spatulas could have been used to penetrate hides, the tips of the other 20% were too large for piercing. The high proportion of deep scratches suggests that spatulas were used for the production of basketry and matting rather than for the working of hides.
Archaeological functional interpretations

Campana's experimental study on Natufian and "Protoneolithic" bone tools has pointed out that the Natufian spatulas were commonly used as 'hide rubbbers'. These rubbers were used to compress, thin and smoothen leather surfaces (Campana 1989:93, 132). Watson, too, (1983b:354-356 and fig. 144:23-27), considers these tools to represent burnishers or hide smoothers (see also e.g. Bader 1993a and fig. 2.10).

Voigt suggests that spatulas could have been used to smoothen a relatively soft surface such as leather and clay (1983:215; Semenov 1964:185). Large spatulas have been interpreted as 'sword beaters', which are weaving tools (e.g. Curtis 1984:45; Marshall 1982; Voigt 1983). Sword beaters are long flat tools which are used to compact the weft threads and to keep the countershed open, i.e. the weft is straightened and pressed into position at right angles to the warp (e.g. Adams 1977: fig. 6; Jopling 1977: fig. 7; Wulff 1966:200-202). According to Voigt (1983:215), smaller spatula can have been used on very narrow looms which were used to make straps or belts. Flat tools made from ribs at Jericho are also interpreted as 'weaving tools': shuttles (Marshall 1982:571-572). Wright suggests that spatulas could have been used for scraping the fatty deposits from the inside of hides (ed., 1981:145). According to Moorey (1994:113) spatulas could have been used as a sort of fork for eating and also for cotton manufacture.

The spatulas of Halafian Girikihaçyan showed two types of wear: a high polish and deep striations (Watson and LeBlanc 1990:89-90). These patterns suggest at least two different uses of these implements. On the polished specimens the polish was concentrated at the rounded ends, which were often worn thin. At least one flat surface, however, was also polished. The striated spatulas predominantly showed striations which were transverse to the long axis of the object.

Gouges
Ethnographic accounts

On the basis of ethnographic parallels of North American groups, Voigt (1983:214) identifies the bone gouges of Hajji Firuz Tepe as fleshing tools. Longe bone fleshers with straight, slanted or convex edges were used by Indian and Eskimo peoples. Voigt mentions, for instance, that the Sioux used long bone fleshers with a deeply notched edge (Mason 1891:569-570). The Ojibwa used a deer or moose tibia with a slanting working edge which was finely serrated (Lyford 1943:97, pl. 55).

Archaeological functional interpretations

Campana's microwear study included the analysis of bone implements with broad chisel-shaped tips (i.e. 'gouges') from the Zagros Protoneolithic sample. The study suggests that the broader-bladed tools were perhaps used as hide scrapers, whereas the small-bladed tools were used in the 'tooling and decorating of leather' (Campana 1989:132).

The gouges which were found at Jericho have been interpreted, like the used scapulas, as skin-working, i.e. scraping, tools (Marshall 1982; Wright, ed., 1981:150).

Incised bones

Archaeological functional interpretations

Incised bones (or notched bones) have been found at a number of sixth and fifth millennium B.C. sites in the Near East, e.g. at Hassunan Yarim Tepe I (Merpert and Munchaev 1993a:112), the Halafian sites of Arpachiyah (Mallowan and Rose 1935: pl. xii, 716 b), Girikihaçiyan (Watson and LeBlanc 1990: fig. 6.6) and Sakçe Gözü (Du Plat Taylor et al. 1949: fig. 30). Watson and LeBlanc (ibid.:93) suggest that notched bones were used as tools or perhaps as musical instruments (i.e. rasps).

The 'grooved bones' found at sixth millennium B.C. Hajji Firuz Tepe have been interpreted as counts or tallies or musical rasps (Voigt 1983:216; Redman 1973:258). Rasps or scraping sticks have two components: a grooved or notched rod which is usually made of wood, but may be of bone; and a smooth rod of wood or bone. The instrument is played by rubbing the smooth rod over the notched one. In some cases the notched piece rests on a resonator, for example a gourd or a basket (Voigt 1983:216, see references on this page). Wear, such as transverse striations across the grooves, and polish over the notched area of an incised bone may indicate use as a rasp.

Recently, Schmandt-Besserat (1992:160-161) has proposed that the notches on the incised bones were symbols "... promoting the accumulation of knowledge for specific ends". In other words, she interprets the notched bones as tallies. Schmandt-Besserat sees the tallies as the first attempt at data processing in the Near East. However, she argues that these tallies were still rather rudimentary devices: (1) the notches were abstract and unspecific; they could have been interpreted in a number of different ways; (2) notched bones could store only quantative information; (3) only one type of data could be handled (i.e. only one type of symbol - a notch - was being used).

Prime knives

Archaeological functional interpretations

The single 'prime knife' at Sabi Abyad ('triangular object' according to Spoor and Collet 1996:453) may perhaps be regarded as a so-called scapula scraper (Masson and Sarianidi 1972:41; Voigt 1983: 214). Scapula scrapers were used to clean skins in the initial stage of processing (Masson and Sarianidi 1972:41). With regard to metal scrapers Forbes (1957:4) and Reed (1972:53) have pointed out that sharp edges are used for fleshing, and blunt edges for dehairing. Since the wide end of the Sabi Abyad 'prime knife' is sharp, it may have been used as a fleshing instrument.

Basketry

Ethnographic accounts

Basketry is generally made from unspun fibres, e.g. reeds, rushes, sedges and grasses (Forbes 1956:176; Wendrich 1991). Forbes distinguishes six types of basketry: (1) coiled basketry; a core (a bundle of grass, rushes or fibres) is coiled spirally in the shape required, the different layers are fastened by a sewing-strip of similar material; (2) twined work; single rushes, or bundles of rushes and flax are laid side by side and are interlaced by two threads which intertwine between each read or bundle; (3) wrapped work; the wrapping strand passes round bundles of reeds, passing over two and under one; (4) matting work; a sort of weaving; a series of strings (the weft) is woven through another series (the warp); (5) plaited work: plaits are made separately and then sewn into the required shape (6) wickerwork; strands woven in and out of a frame of wooden stakes (ibid.:177-178).

Basketry in Hasanabad is made by some of the men during winter time. Making baskets is a seasonal activity because the pliable willow shoots which are used are too brittle in summer (Watson 1979:189, see also Champault 1969).[87] The technique used in Hasanabad is marked by "... close twining, in which a framework of twig bundles is interlaced by a series of withes that go over-two-under-two of the bundles while intertwining among themselves" (ibid.:189). Coiled basketry was not made. Baskets were used to transport dung and to cover dough for bread making.

In Tell Toqaan baskets and trays are made by the women. The only tool needed is an awl. Contrary to Hasanabad, the coil technique is used exclusively (Sweet 1960:135).

Voigt (1983:160) reports that in Azerbaijan large and shallow coiled straw baskets are used to dry apricots.

Archaeological functional interpretations

Coiled basketry has been recovered at a number of prehistoric sites in the Near East, e.g. at Çatal Hüyük, Haçilar, Jarmo and Baghouz (Kleindienst 1960; Mellaart 1967, 1970). Mellaart, for example, has interpreted the Çatal Hüyük baskets as containers which were used for the collection and storage of food and as grain bins. They were furthermore used in a ritual context; some of the baskets contained human skeletons (ibid.:219). Hole (1983:259) has suggested that baskets were used for winnowing grain. Mats found in archaeological contexts may have been used for covering roofs, walls and floors (e.g. Ess and Pedde 1992:253).

At Sabi Abyad coiled as well as plaited basketry has been attested in the Burnt Village (Duistermaat 1994, 1996). This basketry may have been used for storing dry products such as cereals and other staple products, legumes, bread, flour, cheese, dried fruit, dried vegetables, dried meat, herbs, nuts, household tools, textiles (ibid:67 and see table A.3 in this study).

Ochre

Ethnographic accounts

Ochre may be regarded as a pigment, used for pottery and wall paintings (e.g. Forbes 1955:207). In Hasanabad pigments are used for bodily ornamentation of women (Watson 1979:196). The material used for tattooing was soot mixed with milk. The tattooing was done by the Kawli tribe, not by the village members themselves. Occasionally, tattooing was used for relief of headache, arthritis, and other diseases (for tattooing against disease: Sweet 1960:217; Van Der Kooij 1976:59). As in many villages in Syria, henna was used to colour the hair, and for festivities the soles of feet, the palms of hands, and finger and toe nails may be coloured with henna (e.g. Van Der Kooij 1976:59). Black eye make-up was made from fat of goat or cattle, mixed with soot. The make-up is applied with a blunt awl (ibid.: fig. 5.11). This awl is most often made by a man from the bone of a wild sheep or an ox (ibid.:197). The awl was cut from the bone with a metal knife. In prehistoric villages, ochre has often been found in a ritual context: in burials (e.g. Hole et al. 1969:248, 254; Mellaart 1967:207). Kramer (1982:93) has noted that in a few villages around Aliabad, red ochre was used as a decoration of houses.

Archaeological functional interpretations

At Jarmo one of the female figurines was coloured with red ochre. This 'ochre lady' was special because other figurines were generally smaller and not coloured with ochre (Broman-Morales 1983:384).

Collet and Spoor (1996:415) suggest that the common occurence at Sabi Abyad of ochre traces on pestles, mortars and grinding slabs and its presence in small fragments on the floors of some rooms seems to indicate that ochre was widely used at Sabi Abyad, perhaps for decoration of the body or for decoration of ceramics and other artefacts.

---

[87] In fact, the major handicrafts at Hasanabad (rug-making, weaving, felting, basket-making, construction of fish traps) are seasonal occupations (Watson 1979:193).

Container	Dry or liquid products	Possible contents
Basketry	Dry	Cereals and other staple products, legumes, bread, flour, cheese, dried fruit, dried vegetables, dried meat, herbs, nuts, household tools, textiles
Pottery	Dry and liquid	Cereals and other staple products, flour, dried fruit, dried vegetables, dried meat, salt, honey, milk, cheese, yoghurt, butter and other dairy products, water, beer, pigments, herbs/medicine, nuts, herbs, oil/fat
Plaited mat	Dry	Bundled products like textiles, hides, rope, wool; dates
Stone vessel	Dry and liquid	Small substances, precious commodities: precious raw materials, ornaments, pigments, make-up/perfume?
Bag	Dry	Cereals and other staple products, legumes, flour, bread, dried vegetables, dried meat, dried fruit, herbs, household tools, textile
Leather bag	Liquid, semi-liquid	Water, beer, milk, yoghurt, cheese, oil

Table A.3. Possible contents of sealed containers at Sabi Abyad.

Loamers

Archaeological functional interpretations

With regard to loamers Tobler (1950:173) reports that at Tepe Gawra these implements appear as sherds cut into oval or elliptical shapes. Tobler designates these objects as buffers or smoothers, suggesting that they were used to smoothen ceramics. The loamers of Hajji Firuz have also been interpreted as 'sherd rubbing tools' which were used to shape pots and to finish their surfaces (Voigt 1983:203).

2 CERAMIC CONTAINERS

Bowls

Ethnographic accounts

According to Henrickson and McDonald (1983:632), serving and eating vessels are mainly represented by open bowls with flat bases. Often these vessels are decorated. They distinguish between individual and family-sized bowls. Both sizes of serving/eating bowls generally have a maximum diameter of one to almost six times the height (typically two to three times). The maximum diameter is typically equivalent to the rim diameter, resulting in open and 'unrestricted' bowls. The family-sized bowls are mostly three times larger than the individual-sized bowls.

In Luristan the nomads use small metal bowls for dishing out and drinking water and buttermilk (Mortensen 1993:245). In the Sahara, the pastoral Tuareg use wooden bowls for making cheese; the milk to be used is poured into a wooden bowl (Nicolaisen 1963:225). Tuareg wooden bowls are also used as containers for water and milk for greasing the hair (ibid.: fig. 166). From the region of the Alawites (western Syria), Hansen (1976:23, plate 7:a) reports that simple undecorated hemispherical bowls (diameter: 20 cm, height: 15 cm) were used for dishes made with fat or oil. A somewhat larger pot of the same type (ibid.: plate 7:b) was used for boiling water. N.b.: both bowls have rounded bases. In Sana (Yemen) shallow bowls were used as sauce pans, little bowls for drinking water.

Archaeological functional interpretations

Voigt (1983:159), who has suggested functions for ceramic vessels at Hajji Firuz Tepe, argues that large and deep open bowls with thick straight sides (Voigt's type 10 and 9 bowls) were used for cooking. The flat bases of these vessels were often dark coloured on the exterior, perhaps due to heating over a fire. The burnished interiors would have facilitated the cooking of soups or stews. Furthermore, the vessels are undecorated, they have a low centre of gravity, thick walls, a large basal surface and a large mouth. According to Henrickson and McDonald's ethnographic survey of vessel function these characteristics are typical of cooking ware. It is suggested that the size of the type 10 bowls was related to the kind of dish prepared and the number of people to be fed (1983:159).

Medium or large closed bowls (types 14 and 15) may have served for yoghurt making; these containers would have prevented spilling, they could be easily covered, and the contents were quite accessible.

Large open vessels at Hajji Firuz (types 6, 7 and 11) could have been used to dry food, such as yoghurt (turning into *kashk*), wild fruits and nuts; the large open surface in these vessels permits evaporation (ibid.:159).

Wide and relatively shallow open bowls which had been carefully burnished and which sometimes showed elaborate bases (types 3-8) are interpreted as serving utensils (ibid.:160). The small closed type 13 bowls are, by analogy to such modern vessels, interpreted as individual serving vessels. The smallest bowls and jars (types 2 and 16) may have been personal drinking cups.

The typical Halafian 'cream bowls' have been interpreted as milk bowls by Mallowan and Rose (1935:131). They have suggested that the groove on the inside of the vessels allowed the drinker to swirl the liquid freely round the base of the pot. Similar metal milk bowls were observed in the modern village of Arpachiyah.

Pots
Ethnographic accounts
Henrickson and McDonald (1983:631) found out that the great majority of cooking pots are low and squat, have a large basal surface for efficient heat transfer, and usually have a restricted mouth to prevent rapid evaporation from cooking foods. Cooking vessels have relatively thick walls and are generally unpainted. Handles and lugs are optional rather than essential.
Archaeological functional interpretations
According to Voigt (1983:159), medium or large closed bowls (which would have been termed 'pots' at Sabi Abyad) at Hajji Firuz (types 14 and 15), were used for several kinds of food processing which did not involve heating. Such vessels prevent spilling, are easy to cover and allow stirring and scooping out. Small closed bowls (type 13) are interpreted by Voigt as individual serving vessels.

Jars
Ethnographic accounts
Henrickson and McDonald (1983:632-633) distinguish between dry-storage and liquid-storage vessels. Dry-storage vessels are further divided into long-term (i.e. for weeks or months) and temporary (i.e. for hours or days) containers. Lids are seldom reported, but the generally rolled-over or everted rims possibly facilitated a pliable cover over the opening. The long-term vessels are usually tall and proportionally rather thin. The temporary storage vessels tend to have a maximum diameter which is larger than the maximum height.

The liquid storage vessels show a considerable morphological variation. The general size and form ranges for long-term liquid storage vessels suggest that they are on the whole taller and thinner than dry storage vessels. The temporary liquid storage vessels are smaller than either of the long-term types. Both types of liquid storage vessels are rather tall and thin in general form, and usually they have rounded or everted rims.

Water-transport vessels are generally globular (with or without necks) or biglobular in shape and have a small orifice. Handles and lugs are rare on long-distance water jars; the short-distance vessels, however, are often equipped with two or even three handles. Henrickson and McDonald (1983:634) conclude that: "... the size and form of water-transport vessels greatly depend on several key factors, including the topography and distance over which water must be routinely transported, the means of transport, and the number of people being supplied from the vessel".
Archaeological functional interpretations
Voigt (1983:160) assumes that jars were used for transporting and storing liquids and granular solids such as grains, pulses and wild legumes. She suggests that medium and large collared jars (type 18 at Hajji

Firuz) are the most suitable for storing liquids such as water and milk. Medium or large jars (types 18-20) with porous walls were probably used for dry storage, for example of grain. The jars were perhaps covered with wooden or basketry lids (ibid.:160). Smaller collared jars (type 17) were most likely used to store liquids and dry goods in small quantities, such as honey, seasonings, dyes or pigments.

Recently the residue in a biconical jar with a short narrow neck from Hajji Firuz has been chemically analysed (McGovern et al. 1996). The analysis showed that the jar had contained wine, with a *Pistacia* tree resin additive. Most likely the resin was added to the alcohol to disturb and inhibit the growth of bacteria that convert wine into vinegar. As yet the Hajji Firuz jar presents the earliest chemical evidence of wine (ca. 5400-5000 B.C., 6170-5760 cal BC).

Trays/dishes
Ethnographic accounts

Henrickson and McDonald (1983:632) report that cooking trays have a large basal surface, are generally squat and only occasionally have lugs or handles. The majority of cooking trays are unpainted.

In Luristan the nomads use large copper dishes for preparing rice and for serving cooked rice and meat (Mortensen 1993:245).

In Busra (Syria) round bread moulds strongly resemble the so-called husking trays. These moulds, 15 cm high and 18 cm in diameter, are, like husking trays, always incised with a design on the inner flat surface (Bresenham 1985:9 and figs. 9-12). The vessel wall prevents the dough from spreading off the flat base. The moulds are special purpose features which are used on festive occasions. The design patterns on these clay moulds are also found on wooden moulds, which are also used for baking special bread in many areas of the Near East. Bresenham furthermore reports that ancient 'cake moulds' have been recovered from the kitchens of Bronze Age palaces (ibid.:95).

Archaeological functional interpretations

Lloyd and Safar (1945:278) have suggested that husking trays were perhaps used to separate grain from husks. Voigt (1983:159) suggests that the Hajji Firuz type 11 and 12 trays which were less completely oxidized at the base may have been used for cooking. According to her, husking trays (type 12) were not used for husking (Lloyd and Safar 1945:278), for they seem too large and heavy, and baskets were available for tossing grain into the wind. Voigt suggests that perhaps they were used to bake flat bread. According to her, the textured interior would have facilitated the easy cooking and removal of flat bread. She supports her hypothesis by pointing out that (1) "... husking trays would have provided a baking surface similar to that of present-day *sangak* ovens, which have a bed of hot pebbles on which bread is cooked" (ibid.:159, see also Wulff 1966:294-295); (2) trays may have served as portable ovens (in the lower levels husking trays were found, domed ovens only occurred in the latest levels); (3) the smoked bases do suggest exposure to fire.

2.1 The Association of Ware and Vessel Shape

Here I present a comparative statistical analysis of the vessel shapes of the various ceramic wares of level 6: Standard Ware, Standard Fine Ware, Dark-Faced Burnished Ware, Grey-Black Ware, Fine Painted Ware and Orange Fine Ware. The purpose of this analysis is to define and to compare the different vessel shapes within waregroups, since it is assumed that there is a relationship between vessel shapes and wares. It is expected that the different wares are not only technologically (temper, surface treatment, etc.; Le Mière and Nieuwenhuyse 1996), but also morphologically differentiated. Morphological variation (i.e. variation in vessel shape) would point to functional differences.

Of bowls, pots and jars of each waregroup the variables height (of rims: the distance between rim-endpoint and first carination), diameter (of mouth) and thickness (of vessel wall) were studied. The statistics included are: number of sherds (n), minimum (in mm), maximum (in mm), mean (in mm and rounded), standard deviation, and the coefficient of variation. The coefficient of variation provides a comparative measure which expresses the spread of values within distributions; the coefficient is a convenient way to express the differences between means and standard deviations. The coefficient of variation is usually denoted as V and it is found by simply dividing the standard deviation by the mean. A value towards 0 denotes a very narrow spread, whereas a value towards 1 indicates a wide spread (Fletcher and Lock 1994:46).

By means of X^2 tests it has been assessed whether there is an association between vessel shape and waregroup. As has been indicated earlier, only small numbers of ceramics could be assigned to specific vessel shapes (the majority of the pottery is represented by body sherds). Therefore the number of sherds compared may vary largely; very small numbers are compared with relatively large numbers. Clearly this affects the various outcomes. The interpretations based upon these outcomes have to be regarded as first tentative hypotheses, which have to be tested if more data become available.

First it was assessed whether there is a significant difference between the dimensions of bowls of the various waregroups.[88] Compared were the minimum and maximum rim heights, diameters and thicknesses of the bowls and jars. The numbers of bowls were far too low to be used in statistical tests.

The rim heights of bowls vary between 20 and 130 mm, the means between 52 and 60 mm. The coefficients of variation show that rim heights of Fine Painted Ware and Orange Fine Ware are less varied than the rim heights of the other wares. A X^2 of 38.89 (df= 4) has been calculated, indicating that there is an association between rim heights and wares. A Cramer's V of 0.24, however, shows that this association is not particularly strong. The diameters of the bowls range between 50 and 480 mm, and the means between 113 and 230 mm. Fine Painted Ware shows a relatively high coefficient of variation, perhaps indicating that diameters varied largely within this ware. It should be taken into account, however, that only three Fine Painted Ware bowls were identified! Orange Fine Ware shows the lowest coefficient, perhaps pointing to limited variation. The other wares were more varied, according to the coefficients. The X^2 of 30.47 (df= 4) demonstrates that there is an association between waregroup and diameters. The rather low Cramer's V (0.13), however, indicates that the association is not very strong. The thickness of bowls varies between 5 and 28 cm. Again, Fine Painted Ware and Orange Fine Ware indicate the lowest variation. A X^2 of 3.67 (df= 4) allows the 0 hypothesis to be confirmed: there is no association between thickness and ware.

The rim heights of jars ranges between 20 and 155 mm, with means between 33 and 65 mm. The coefficient of variation of Fine Painted Ware is very low (0.08), indicating that rim heights were standardized, i.e. between 30 and 35 cm high. The other wares show more variation in this respect, especially Standard Ware and Standard Fine Ware. A X^2 test indicates that there is an association between rim heights and wares ($X^2 = 39.76$, df = 4). A Cramer's V of 0.25 indicates that this association is not very strong. The same goes for the relation between jars and diameters; a X^2 of 109.85 (df = 4) and a Cramer's V of 0.26 indicate a not particularly strong relationship. Grey-Black Ware and Fine Painted Ware prove to be the least varied wares. An association between thickness (ranging between 5 and 25 cm) and ware was not supported, as a X^2 of only 4.68 (df = 5) indicated. However, this result seems to be due to sample size, since Le Mière and Nieuwenhuyse (1996) have indicated that such an association is present (the finer wares having thinner walls than the coarser wares).

In conclusion it can be said that there are indeed associations between vessel shapes and wares. Clearly Standard Ware represents the largest vessels, but this ware seems to be closely related to Standard Fine Ware. Both wares show the largest rim heights and diameters, as well as the widest variation in shapes. The bowls of both wares vary from small (20-25 cm high, diameter of 70-80 cm) to large (13 cm high, 36-48 cm wide). The jars of both wares have very low (2 cm high) or 10-15 cm high necks and show very narrow (4-6 cm) to very wide (270-370 cm) neck diameters. Both wares seem to be used for a variety of purposes. Perhaps the rather coarse Standard Ware was mainly used for the actual preparation and storage of the daily food (e.g. bowls used for cooking and temporarily holding food, and jars for storing ordinary dry goods and liquids). The finer Standard Fine Ware may have been used for some food preparation as well as for the serving of various foodstuffs and meals. In the various Standard Fine Ware jars special dry goods and liquids (e.g. special herbs and beer (?)) were perhaps stored.

A second class of morphologically related vessels is represented by Fine Painted Ware and Orange Fine Ware. Of these two Orange Fine Ware shows the largest vessels, but as the following will indicate, both wares are indeed comparable. Both of these 'finest of wares' (Le Mière and Nieuwenhuyse 1996) show the smallest bowls and jars. Compared with Standard Ware and Standard Fine Ware the Fine Painted Ware and Orange Fine Ware bowls are lower and smaller. The minimum rim heights (4 and 5 cm), however, are larger

---

[88] Dark-Faced Burnished Ware has been excluded from the various computations; as can be seen in the tables only few vessels of this ware have been assigned to shape categories, and, moreover, the attributes rim height, diameter and thickness could not be recorded in many instances.

than those of Standard Ware and Standard Fine Ware. However, the maximum rim height (6 to 9 cm) is much lower than the maximum height of Standard Ware and Standard Fine Ware, and the diameters are also much smaller (50-180 cm). As has already been indicated, the Fine Painted Ware jars in particular seem to be rather standardized (mean rim height of 33 cm, mean diameter of 125 cm). The Fine Painted Ware jars are somewhat smaller.

So the vessel shape analysis has indicated that both Fine Painted Ware and Orange Fine Ware represent the finest and probably the most luxurious wares in use. Both wares may have had a serving and 'display' function, and perhaps small quantities of valuable products were stored in jars of both wares (if so, smaller quantities of more valuable products than in Standard Fine Ware jars). Grey-Black Ware (and perhaps Dark-Faced Burnished Ware, which could not be analysed) seems to fall between both classes of wares just mentioned. The diameters of Grey-Black Ware jars show little variation (mean of 96 mm), perhaps indicating that these were designed for a special purpose. Most likely, Grey-Black Ware was used, like Fine Painted Ware and Orange Fine Ware, as a serving and display ware.

3 FLINT AND OBSIDIAN ARTEFACTS

Pressure flaked pieces (F, O)
Archaeological functional interpretations

Copeland (1996:293) suggests that these well-made and relatively rare objects were used as javelins and bifacial knives. The amount of energy and time spent on these carefully made objects perhaps suggests that they were valuable tools which had a special function.

Lustred sickle elements/"shape defined elements" (F)
Ethnographic accounts

In Hasanabad grain is cut with an iron sickle (cf. Watson 1979: fig. 3.4, see also Van Gijn 1990: figs. 20-23; Van Der Kooij 1976:21, photos 6, 7; Wulff 1966:271-272 and figs. 375 and 376). The stalks of grain are gathered with one hand, with the other hand the sickle is applied near the roots and drawn across the stems. When the bunch of grain which is held in the hand becomes too large to hold it is put on the ground. The deposited heaps behind the cutter are gathered by the other workers in the field. Before being transported to the threshing floor (in goatshair nets on the back of a donkey), the grain stored in heaps is left in the field for some time (ibid.:80).

Reconstructions of the shape of prehistoric sickles in the Near East have suggested that they were more or less used like their metal counterparts (Van Gijn 1990:37). The grain could have been cut in three ways. First, the stems can be cut or broken just below the ears. Van Gijn reports that this method is still applied in Syria (ibid.:37). Second, stems can be cut a little below the ear, somewhere half-way. According to Van Gijn this is not a very practical procedure, since in fields with stems of different heights one has to cut further down to collect the ears from the stems of lower height as well. Besides, this method results in straw of variable length, which is not useful for e.g. roofing. The third method consists of cutting the stem just above the ground.
Experiments

Sickle blades have received attention since the beginning of functional analyses (e.g Semenov 1964:19, 113-122). The literature on sickle blades and ancient agriculture is overwhelming (see e.g. Anderson, ed., 1992; Grace 1989; Juel Jensen 1994:105-113); here I will limit myself to presenting some of the most interesting results of three of the best known experimental efficiency studies (see Unger-Hamilton 1988:171). Well known are the experiments of Steensberg (1943), who harvested barley and oats with sickles of the European Stone, Bronze and Iron Ages. It appeared, not surprisingly, that crescentic flint sickle blades with serrated straight edges were less efficient than iron sickle blades. Furthermore it was found that a flint flake mounted perpendicular to the handle was most useful for cutting weed and thistles, grain was more difficult to cut with these implements.

From her extensive work on use-traces and function of flint implements Anderson (1992:62-63) has come to three main conclusions: (1) the harvesting of silicious plants results after ca. two hours in a gloss which is visible to the naked eye; (2) traces which are virtually indistinguishable by the naked eye from the above traces are produced by the working on other soft or humid mineral materials such as clay; (3) microscopic analysis provides criteria to distinguish between uses in most cases.

The experiments of Unger-Hamilton (1988:181-187) with flint sickle blades, secured in rectilinear and curved hafts, gave the following results. It appeared that non-cereal species could easily be harvested with single unhafted blades; these blades, however, had to be large and had to have backing retouch. The best composite sickle for plants in dense stands (e.g. reeds) proved to be a short curved sickle, which was most useful for reaping rather than cutting or sawing. The unretouched edges were best suited for cutting in all instances. On hard-stemmed plants, however (e.g. reeds), unretouched edges became damaged within seconds. Denticulated edges were in this respect much more efficient. For the cereals single unhafted blades could be used, but it appeared that composite sickles were more efficient. A long rectilinear sickle turned out to be more useful than the short curved sickle. Toothed or denticulated blades were unsuited, because the culms got stuck in the denticulations. Sickles consisting of blades with fine denticulation set in rectilinear wooden shafts were the most efficient and they lasted longest.

Archaeological functional interpretations

In many excavation reports blade segments are considered to have been parts of composite tools, and to have been hafted in compound sickle tools which were used for harvesting cereals, grasses and reeds (e.g. Mortensen 1983:214; Stager 1985:12). Commonly these blade segments are called 'sickle blades'. Many of the blades are lustred on one (unretouched) edge. This lustre derives from silica contained in the stalks of the plants, during the cutting and dragging motions of the reaper. Reeds may have been used in crafts such as mat making, basket making and thatching. The gloss often observed on sickle blades is commonly assumed to have resulted from the cutting and processing of a variety of plants, especially the harvesting of cereal grains (e.g. Hole 1983:257; Voigt 1983:244, for a discussion of gloss see e.g. Del Bene 1979; Diamond 1979; Kamminga 1979; Keely 1980).

Unger-Hamilton (1988:168-205) has dealt in some detail with sickle blades from Arjoune and other sites in the Levant (see also Copeland in press). These implements, blades and bladelets with or without lustre along one or both edges, have been found in Palestine and Syria from the Epi-Paleolithic (ca. 18000 B.C.) onwards. At the beginning of the Neolithic (eighth millennium B.C.) the amount of sickles increases and they become larger in size. Commonly sickles are related to plant cultivation and cereal gathering (ibid.:168). Cereals in particular are mentioned repeatedly in relation to sickles. Unger-Hamilton (1988:168-169), however, points out that this relation rests upon at least two false assumptions. First, gloss is not necessarily caused by contact with grasses and cereals: "Not only grasses, such as cereals, but also other plants, such as reeds and rushes, have siliceous stems; these were common in the Levant and had their uses in daily life. They all produce gloss when rubbed against a flint surface for a length of time" (ibid.:168). In this respect it may be mentioned that on the basis of typological and microwear analyses M.-C. Cauvin (1983) has, suggested that the earliest sickles in the Near East were used for cutting plants other than cereals (and see also Van Gijn 1992). Second, cereals do not always have to be harvested by cutting: harvest by beating (inefficient because grain is lost this way), picking or stripping the seed heads without tools (Anderson 1992:188) and uprooting are other practised methods. However, it is very probable that cereals were cut (with stone sickle blades) for at least the following reasons: (1) metal sickles of today seem to indicate stone precursors; (2) up till recently stone sickles were in active use; (3) uprooting of cereals is not ideal; soil from the roots gets mixed into the grain. Anderson (1992:206) notes in this respect that at the beginning of the Neolithic in the Near East there appears to have been a more intensive use of harvesting tools. Furthermore, the harvest seems to consist of cutting off many stems at a time close to the ground (see also Anderson 1994:63). This suggests that in the Neolithic the fields were denser than in the preceding Paleolithic in the Levant.

With regard to the unlustred 'sickle elements', or 'shape defined elements' (Unger-Hamilton 1983), Copeland (1989:241) suggests that the fine-grained pieces without lustre at Sabi Abyad were used on plant material softer than cereals (leaving no macroscopically visible traces). The longer pieces were perhaps used as knives for cutting meat or for whittling wood or bone.

Sickle hafts in the Near East were made of various materials such as bone, wood, antler, clay and horn, and they were rectilinear or curved in shape (Camps-Fabrer and Courtin 1982; Anderson, ed., 1992: figs. 2-4; Juel Jensen 1994: figs. 33-35, 43, 52). Remains of curved and straight sickle hafts have been found at a number of Neolithic sites, e.g. Tell Hassuna and Tell es-Sawwan (M.- C. Cauvin 1983). So far, no sickle hafts have been reported from Syria. According to Unger-Hamilton (1988:182) this may indicate that in

Syria sickle hafts were made of wood. On the basis of the size, shape and length of sickle blades from Neolithic Byblos, J. Cauvin (1968) has suggested that in the *Neolithique Ancien* rectilinear sickles were in use, whereas in the *Neolithique Récent* both rectilinear and curvilinear sickles were used.

At Sabi Abyad, the evidence does not yet allow any definite conclusions as to the procedure of harvesting (i.e. reaping high on the straw, leaving stubbles on the field, or reaping low on the straw). Akkermans (1993:218-219) is inclined towards the option of reaping high on the straw, for the following arguments: (1) as a source of fuel straw is inefficient (and animal dung could have been used); (2) there was probably little need for straw for thatching, considering the small size of the settlements and the presence of reeds near watercourses; (3) the majority of ceramics were mineral-tempered; straw for tempering would have been needed in limited quantities only; (4) use as fodder is unlikely, since the herds were probably small and could graze on vast areas of natural pasture around the site.

Truncated pieces (F, O)
Archaeological functional interpretations

Unger-Hamilton (1988:147) reports that at Arjoune soft materials (e.g. hides, meat, horns, feathers) and fibrous material (e.g. wool) seem to have been worked with 'truncated elements shaped like sickle-elements'. Copeland suggests that truncated pieces at Sabi Abyad may have been used as sickle elements, because it is known that truncations were typically used to fashion sickle elements on PPNB sites in the Balikh valley (Copeland 1979, 1982, 1989, 1996:297; M.-C. Cauvin 1973).

Truncated blades from the region around the Caspian sea were used to scrape wood and bone, according to Beyries and Inizan (1982).
Experiments

Experiments by Grace (1989) have suggested that truncated blades and bladelets were used for cutting fresh saplings, and for whittling wood.

Backed knives (F)
Archaeological functional interpretations

Steeply retouched edges on blades are commonly interpreted as the backs of blades; the hafted or hand-held part. However, the retouched back of blades may also have been used as rasps. The sharp edge could have been inserted into a haft (Unger-Hamilton 1988:139).

Microwear analysis of backed bladelets at Paleolithic Pincevent (France) suggested that they were used as barbs and projectile points (Moss 1983b:115). Copeland (1996:293) suggests that the backed knives of Sabi Abyad were utilized on the lateral edge which is opposed to the edge with abrupt retouch.

Scrapers (F, O)
Ethnographic accounts

Scrapers are commonly used to prepare hides (scraping and softening), for the skinning of small animals, for shredding plant fibres, for working wood, and for shaving and incising during the production of bone tools (e.g. Campana 1989:25; Hole 1983:257; Moss 1983b:38). End-scrapers are generally interpreted as implements which were used for the processing of animal skins (e.g. Voigt 1983:245). Semenov (1964:87) distinguishes two types of end-scrapers. Scrapers with a wide and sharp working edge are most useful for fleshing, whereas scrapers with blunt edges were used for softening the skin by rubbing it. Hodges mentions that round-ended flint tools such as scrapers, would be most effective for cutting leather (Hodges 1989:151).

Van Gijn (1990:27) has pointed out that, although scrapers are often used for skin-processing, they are not necessarily related to this activity. Fresh hide scrapers (used for removing fat before the skin can be dried), for instance, may not exist because skin can come off clean enough for processing. Moreover, the scraping of dry hides (for softening them) can, instead of by using dry hide scrapers, be done by pulling a hide over a wooden beam, as has been observed with the North American Plains Indians.
Experiments

Unger-Hamilton's experiments (1988) showed that for working wood, bone and hides, hafted scrapers are preferable to unhafted scrapers (ibid.:123-127). Hafting with resin, wax and sinew seemed more efficient than wedging; the wedges losened and the dental floss broke easily. Scrapers which were used for fleshing hides had to have a sharp angle (ca. 45°) and an overhang to grip the hide with. Steep scrapers without such an overhang were wholly unsuited for defleshing. Steep scrapers were also unsuited for working shell.

On the basis of use traces and shape of the working edge, Semenov (1964:85-87) rejects the notion that narrow end-scrapers were used as knives. Basing himself on usewear analysis, he supposes that they were used for scraping and softening skin. End-scrapers with a wide and sharp working edge may have been used for removing flesh from hides. Blunt end-scrapers were intended for rubbing skins to make them soft.

Semenov (1964: fig. 19: no. 2) reports a 'piece of tabular flint' which is virtually identical to the 'Tile Knives' of Sabi Abyad. According to him this implement (from the Neolithic in the Baikal area in the Soviet Union) was used for sawing stone.

Archaeological functional interpretations

In functional analyses scrapers have received a good deal of attention. Microwear studies in general have pointed out that scrapers were used as wood choppers (Mureybit in Syria), as burins and most often as tools to scrape and soften hides (Unger-Hamilton 1988:122).

The scrapers from Arjoune seem to have been used on hide, wood and possibly bone and stone (ibid.:130). Due to sample size, definite conclusions cannot as yet be given, but it appeared that there are significant form-function correlations. First, large scrapers were apparently used on wood and hide, whereas small scrapers were used on bone and stone. Second, round tabular scrapers were mainly used on wood, end-scrapers seem to have been used on hide, and straight-sided scrapers were apparently used on bone. Third, scrapers with steep edges were used to scrape hard materials. Scrapers with overhang seem to have been used on soft materials such as hide. The thin butt on many scrapers suggests that they were originally hafted (hafting would have increased the efficiency and ease of working).

From microwear studies dealing with western European assemblages, it appears that in general end-scrapers were used for processing skin. Wood and plants were also important, and end-scrapers were also used for mineral substances such as ochre and schist (Juel Jensen 1988:67). Juel Jensen (ibid.:68) has pointed out that skin-processing is a complicated process consisting of many stages (e.g. removal of epidermal structures, thinning of hides, etc.). Furthermore, hides can be worked in different conditions: fresh, moist, hard-dry and soft-dry. Various substances such as fat, ash and ochre may be added during the process. The processing procedures also influence the use of scrapers: thickness and fat of the skin, climatic conditions, cultural traditions, etc. Nevertheless, various examples have shown that microwear studies can determine whether fresh, moist, hard-dry or soft-dry hide was worked. 'Fresh-hide' polish on scrapers indicates the defleshing of hides. Scrapers with polish of moist and dry hides indicate the epilation, thinning and softening of hides (ibid.:69). Beuker (1983:110) has suggested that scrapers could have been used for rubbing ochre into hides.

Copeland (1989:241) has suggested that the racloirs at Sabi Abyad could have been used to scrape bone or hide material. The tabular knives or 'Tile Knives' (e.g. Du Plat Taylor et al. 1949: fig. 37: no. 11) have thin bifacial working edges which suggest a cutting/slicing rather than scraping function. On a number of Transitional and Halaf tabular knives plaster adhered to the cortex. Copeland (1996:294) therefore suggests that these implements may have been used for applying plaster on walls, floors and the inside of pits. Another possibility is that they were used for pulverizing plaster nodules which formed during plaster making (ibid.:255). End-scrapers are interpreted as skinning tools; they separate the skin from the flesh without cutting the hide.

Burins (F, O)

Ethnographic accounts

It is well-known that burins are used for producing bone tools such as awls (e.g. Campana 1989:25; Semenov 1964:19, 155-158).

Experiments

Unger-Hamilton (1988:158) found that most materials can be grooved easily with burins. Soaking of material (e.g. wood) increased the use-life of burins considerably. For fine and/or deep grooves in bone, shell or stone borers, flakes and narrow-angled scrapers were better suited than burins. The experiments by Grace (1989) showed that carving bone and shell, scraping hide and bone, piercing holes in wood could all be done with burins.

According to Semenov (1964:94-100) burins are used for cutting skins and meat and carving (e.g. for decorative purposes) stone, bone and wood.

Archaeological functional interpretations

In his extensive study on prehistoric burins from Upper Paleolithic Ksar Akil (Lebanon), Newcomer (1972) suggested that burins could have been used as groovers for hard organic materials, scrapers, borers/ piercers, and as stone engraving tools (see also Beuker 1983:108). At Tepe Hissar burin spalls were used as borers (Tosi and Piperno 1973). Microwear studies of burins from early Neolithic Abu Hureyra in Syria (Moss 1983a), have suggested that here they were used to work reeds or wood. The burins from Pincevent were used for a much larger spectrum of activities: working bone, scraping hides, piercing and cutting, butchering, and wood-working.

Microwear analyses in western Europe have pointed out that the 'burinated section' of burins, the active work edge, was used on bone and antler, but also on hide and occasionally on wood, shell and minerals (Juel Jensen 1988:72). The carving and engraving of antler and bone (i.e. for making bone awls and needles) with the burin bit was most common. The edge of the burin facet, however, was also used regularly to plane and shave hard substances (most often bone and antler). Burins were also used for piercing and boring holes in bone. On many burins microwear is lacking. The reasons for this are still not well understood, but it has been suggested that the lack is due to unsuccessful production attempts, accidental burination, and it is possible that burin spalls were the primary means of production, leaving burins as waste products. Burins were used as borers, gravers and scrapers; they have to be perceived as several classes of tools which are united by a common technical feature (ibid.:73).

Mortensen (1982) regards burins (like notched and retouched knives and scrapers) as tools used for cutting hides. He suggests that other scraping and cutting tools for hides are notched and retouched flakes and blades, knives, and a variety of scrapers.

According to microwear analysis the burins from Arjoune were used to incise relatively hard materials, perhaps bone, shell or limestone, (Unger-Hamilton 1988:160). One burin was probably used for splitting reeds.

Borers (F, O)

Experiments

Unger-Hamilton reports that carving with piercers is an easy task (1988:133). Fresh watersoaked hide proved more difficult to pierce than dry hide. Fresh wood was more easily worked than seasoned wood. For wood a relatively broad tip was most useful. Bone could be drilled easily, but the tip of the borer often got crushed or broke (e.g. Semenov 1964:18, 74-83). On quartzite the tip of the borer broke, but on limestone boring was not so difficult. When used to perforate, coarse-grained flint broke more easily than fine-grained flint.

Archaeological functional interpretations

Of the borers at Arjoune most seem to have been used on wood. Furthermore, stone, pottery and perhaps hide were perforated. The use-wear traces strongly suggest that mechanical drills, possibly a bow-drill, were used at Arjoune (Unger-Hamilton 1988:137).

Generally it is assumed that borers and drills were used for piercing, boring and drilling stone, bone, wood and leather (e.g. Hodges 1989:106, 116; Howe 1983:71; Voigt 1983:243; Wright, ed., 1981:43). Hole (1983:258) determined through experiments that the grinding on many drills at Jarmo could have been produced by boring into hard material (most likely beads, since the drills fit into the holes of the beads; Moorey 1994:106-110).

Tosi and Piperno (1973) found residues of lapis lazuli on small perforators at Tepe Hissar, pointing out that borers were indeed used on stone. Borers from Abu Salabikh were used on shell and stone (Unger-Hamilton 1988:132). The microwear analysis further suggested that sand was used as an abrasive during the process of drilling. The borers from Abu Hureyra (Keeley 1983b) were used to drill wood (the small borers for drilling the holes, the larger ones for enlarging them).

Semenov (1964:18-19) has discussed the use of piercing and reaming tools in some detail. Borers and drills may be turned by hand, by a stone drill inserted in a dowel and rotated between the palms of the hands, or with a bow (ibid.:78; Hodges 1989:116-117). Soft materials (e.g. leather) can be pierced by pressing and turning a pointed tool against them. Harder materials, e.g. wood, bone, shell and stone, are pierced by drilling with a pointed tool. The enlarging, or reaming, of a hole is done with a relatively thick tool with strong edges (and not necessarily with a pointed end).

Notches (F, O)
Archaeological functional interpretations

In general notched blades and flakes are interpreted as concave scrapers (Semenov 1964:108), which were used to shave or whittle relatively hard materials of small diameter, especially wood (construction of shafts, axe and adze handles, stakes, etc). These implements may also have been used for smoothening and finishing of bone artefacts (e.g. Voigt 1983:244). Microwear studies have shown that notches were used for cutting wood (Unger-Hamilton 1988:150).

Microwear analysis has suggested that the notched blades and flakes of Arjoune were most likely used for scraping or shaving wood and bone (ibid.:153).

Copeland (1989:243) regards the Sabi Abyad notches as tools for scraping rods of bone or wood.

Denticulates (F, O)
Archaeological functional interpretations

Regarding the function of denticulates at Arjoune, Unger-Hamilton concludes (1988:156) from her microwear studies that: "... denticulates had been used as multi-purpose tools in a variety of ways: as notches to shave or scrape wood; as scrapers perhaps on wood, hide and stone; as "combs" of fibrous materials; as perforators, perhaps of wood; as gravers, perhaps of wood and stone".
Experiments

Denticulates have of several notches on the same edge. Unger-Hamilton (1988:150) points out that denticulates may have been notched pieces which were used consecutively. More likely a denticulate represents a tool on which all the notches were used at the same time. It has, for instance, experimentally been shown that for dehairing hides a denticulate is more suited than an end-scraper (Moss 1983b:72). Microwear analyses of denticulates from the Natufian sites of El Wad and Ain Mallaha (Israel) have suggested that notches were used to scrape and plane wood (Unger-Hamilton 1988:150). In experiments, however, Unger-Hamilton (1988:154) found that denticulates were not very efficient for scraping wood. The dehairing of hides with denticulates also proved difficult. Wool, however, was combed easily with these implements.

Composites (F)

These artefacts with different areas of retouch probably had double purposes. The following composites have been distinguished: scraper/notch; scraper/borer; scraper/burin; scraper/denticulate; denticulate/borer; borer/notch; truncation/notch; abrupt backing retouch/notch. For interpretations of the function of these tools the reader is referred to the various tool classes discussed above.

Splintered pieces (F)

These pieces with squamous bipolar retouch have been interpreted by Copeland as wedges in the process of pressure flaking (Copeland 1996:295).

Retouch: fine and abrupt (F, O)
Experimental evidence

Semenov (1964:152) established that sawing of bones was most effectively done with retouched bladelets. Sawing the bone through half or to a third and then breaking it, resulted in an uneven toothy end at the broken edge. In order to get a smooth end the bone had to be sawn through on all sides the way round.
Archaeological functional interpretations

In general, retouched flakes and blades are interpreted as cutting and scraping tools, used on meat, leather, reed and wood (e.g. Hole 1961:91; Moss 1983b:43-46; Voigt 1983:244).

Steeply retouched edges on blades are commonly interpreted as the backs of blades: the hafted or hand-held part. However, the retouched back of blades may also have been used as rasps. The sharp edge could have been inserted into a haft (Unger-Hamilton 1988:139).

Retouched (and unretouched) flakes and blades can be used as meat knives. Ethnographic observations of Aborigines of western Australia (Gould et al. 1971) made it clear that flakes were only used during butchering: to cut off the skin and ligaments of the killed animal. Subsequently the carcass was butchered with stone choppers, bone tools and wooden wedges. Most often the flake knives were used for day-to-day activities and were discarded after a few uses. Microwear analysis of Paleolithic blades at Pincevent (France) showed that they were used to butcher and cut meat, or to cut and bore hides (Moss 1983b).

Cores (F, O)
Archaeological functional interpretations

Moss (1983b) suggests that cores may have been used secondarily as chisels, scrapers and gouges for working wood and bone.

Copeland (1989:244) mentions that some of the Sabi Abyad cores showed battered ridges which indicate secondary use as a hammerstone.

Unretouched pieces: flakes and blades/bladelets (F, O)
Ethnographic accounts

Evidently, blades and flakes can be used as knives. There are, however, virtually no ethnographic accounts about the use of flint knives. A stone knife is said to have been used by the Copper Eskimo to scrape off the meat and sinews from caribou skin (Jenness 1970). The Cocopa Indians in California used an unhafted blade of stone to shape a pottery vessel (Gifford 1934).

In the Near East knives are often used in the context of shearing sheep (Van Der Kooij 1976:32) Knives are, of course, further used in the process of butchering animals: slitting throats, skinning (and scraping the hide), and cutting meat (ibid.:108-109, see also Nicolaisen 1963: figs. 171-172; Van Der Kooij 1976: photo 64).

Flint blades (and flakes), and fagments of blades, may be inserted into wooden boards to act as threshing equipment (e.g. Ataman 1992; Hole 1983:259; Seeden 1985:298).

Experiments

Unger-Hamilton (1988:141) found that unretouched blades were suited to cut hides, especially when it is dry (after five minutes, though, the blade was blunt). Meat was easily cut with long blades and several blades inserted into a haft. Blades were further used to cut fat off hide, tendon off bone and preparing fish. Unger-Hamilton concludes that for cutting acutely angled blades with edges that are straight in section and profile are preferable. Soft materials were conveniently cut with unretouched edges, and medium-hard materials with denticulated edges.

Semenov (1964:101-107) has argued that flint blades and flakes with sharp edges were generally used as meat knives, used for skinning and cutting meat. Flint blades were furthermore used as whittling knives (for wood and bone). According to Semenov these tools typically show the following characteristics: (1) they are made on blades; (2) they can be trimmed or formed by retouch and burin blows, but sometimes they are not shaped and have no retouch; (3) the main wear is confined to one side of the implement; (4) the worn side is the ventral side of the blade, the smooth side faced the material; (5) polishing is along the edge of the tool; (6) the microwear scratches appear as lines at right angles to the edge, or slightly inclined from this.

The activities which can be carried out with flakes that are mentioned in Grace's experimental study (1989) are: sawing wood, whittling wood, sawing bone and cutting grasses.

Archaeological functional interpretations

The importance of unretouched tools has been indicated by numerous ethnoarchaeological studies (see Juel Jensen 1988:74; Van Gijn 1990:144). Microwear research has furthermore revealed that in the archaeological record unmodified flakes and blades were also used as tools.

In the literature the following functions have been ascribed to blades which have been found at virtually every Neolithic site in the Near East: cutting grass and reeds, cutting and shaving bone, antler, and especially wood, meat, skin and leather (e.g. Beuker 1983:103; Hole 1977:157; Mortensen 1982; Moss 1983:43). Voigt assumes that the retouched and used blades of Hajji Firuz Tepe were used for cutting meat, leather, reed and wood (1983:244). These materials may also have been lightly scraped with blades according to Voigt.

Hole (1983:259), discussing the Jarmo chipped stone, mentions that flint knives, drills, scrapers, chisels and notched flakes were used for the manufacture of wooden and bone tools.

Drenkhahn (1976:7-18) reports that in Ancient Egypt knives with wide blades were used to cut the leather. For making holes in the skin (e.g. for the manufacture of sandals and belts) bone awls or horns were used.

It has already been indicated that the microwear study of Unger-Hamilton (1988:147) showed that identification of worked materials on blades and flakes was very difficult. Nevertheless something can be said. It seems that most of the blades on which it was possible to identify wear-traces were used as wood-whittling knives, as wood-saws, and as cutters of various soft (e.g. hide, meat) or hard (e.g. bone, shell, stone) materials (ibid.:209). Backing and truncations seem to indicate that many of these tools were once hafted.

According to Bergman et al. (1983) snapped blades, such as found at Sabi Abyad, were deliberately broken in order to increase the number of right-angled working edges and also to make strong edges. Copeland (1989:267), discussing the snapped obsidian blades from Sabi Abyad, has also suggested that obsidian blades may have been deliberately broken to overcome fragility. Microwear analysis of the snapped blades form Hengistbury Head (England) suggested that the breaks had been used to work bone. Microwear analysis showed that the end of two snapped blades from Arjoune were possibly used as scrapers, in one instance perhaps wood was scraped (Unger-Hamilton 1988:149).

Polished or ground pieces (O)
A few pieces of obsidian were ground or polished. It appears that they are fragments of larger objects such as beads, pendants or palettes (Copeland 1996:296).

Corner-thinned blades (O)
Archaeological functional interpretations
Nishiaki (1990) has suggested that the removal of corners from these small blades represents a method of hafting these pieces in series to knives or sickles. As yet it is, however, far from clear if corner-thinned blades are really tools. Copeland has, for example, suggested that the tiny inverse removal scars on the corner of snapped blades were due to a specific blade-breaking method (Copeland 1996:297).

"Core for side-blow blade-flakes" (O)
Archaeological functional interpretations
Evidently, these are the cores from which the side-blow blade-flake had been struck. Voigt (1983), however, regarded the 'cores' as tools, and the side-blow blade-flakes as debitage.

"Side-blow blade-flakes" (O)
Archaeological functional interpretations
As yet the function of these "sliver-like artefacts" (Copeland 1996:298) is still enigmatic (e.g. Copeland 1979:267-268, 1989:260; Hole 1983). Mortensen (1982:50) has noticed that side-blow blade-flakes often have secondary retouch, which according to him indicates that they were produced as artefacts for their own sake. However, convincing evidence affirming this view has not been presented, and Copeland (1979:267) concludes that: "For the moment we do not know whether the flake itself or the "core" was the desired result of the method - or whether both are waste-products of, for example, sickle-blade manufacture".

APPENDIX 2

EXCAVATION AND POST-EXCAVATION RESEARCH
METHODOLOGY

In the prehistoric excavations on the southeastern mound at Tell Sabi Abyad the emphasis has been on broad-scale exposure of the Halaf occupation levels and their immediate predecessors to obtain an insight into Late Neolithic settlement organization and developments in the use of space. The excavations were conducted primarily in 9 x 9 m squares, designated from west to east with capital letters and from north to south with cardinal numbers (fig. 1.2). The squares are separated from each other by 1 m wide baulks.

The excavations in each square are led by a 'square supervisor', who directs a team of ca. five workmen from the nearby village of Hammam et-Turkman. These workmen perform the 'heavy' work such as breaking the ground with a large pick, removing soil from the excavation square and driving the wheelbarrow. The supervisor is responsible for the administration; he/she makes drawings, takes height measurements, writes labels for find bags, etc.

When the work in a square is started the actual excavation consists of: (1) breaking the ground with a pick (large or small picks, depending on the conditions); (2) planing with a shovel (or brooming) to create an even surface which can be 'read'; (3) division of the square in so-called loci, which are real (i.e. an oven or a room), or arbitrary spatial units; (4) sampling of finds from the loci. When in a square a level, i.e. a coherent building phase, has been excavated (i.e. when the floorlevel has been reached), a final plan is drawn, every architectural feature is described in detail and all is photographed. When necessary, the architecture is broken down in order to excavate the level underneath.

The objects recovered are: sherds, flint and obsidian implements, animal bones, botanical remains, 'small finds' such as pestles, mortars, beads, figurines, etc., and human skeletal remains or burials. The finds are sampled in so-called lots. Per locus, the finds from each of the above mentioned assemblages are separately sampled in lots. In principle, lots contain the material of deposits of ca. 10 cm of soil. Point-provenience data (i.e. exact coordinates) are only noted for the small finds and complete objects such as intact vessels. As yet, no systematic programme of sieving or microstratigraphical analysis of the deposits (see Matthews et al. 1996, 1997) has been carried out.

Each day a 1:50 plan of the square is drawn, and in principle all four sections of the excavation square are drawn (scale 1:20). Often 1:10 detail drawings of features such as ovens are produced. Apart from being drawn, all features are also photographed. In the 'daynotes' the work in a day is described in detail, and first tentative interpretations may be given. A series of other forms is used for describing finds and features. On the 'deposition form' the characteristics of the deposits from which the finds were retrieved are noted: colour, hardness, consistency of layers, etc. The 'feature form' serves to indicate the dimensions and other attributes of architectural features such as walls, ovens and pits. On the 'object form' the context and properties of the small finds are noted. The 'burial form' is reserved for describing the human burials. 'Sample forms', finally, are filled in when samples (e.g. charcoal for radiocarbon age determination, botanical samples, etc.) are taken. Apart from these forms, which each square supervisor has to deal with, the 'specialists' working (often in the field) on pottery, animal bones, etc., of course use standardized code sheets, etc.

In the post-excavation analysis of the stratigraphy and architecture the sequence of soil depositions and the architectural features are classified. In general the stratigraphical sequence strongly varies from square to square, due to a different use of space for various activities. Per square of excavation a number of strata are recognized, each characterized by consistent soil characteristics or coherent architectural features. Most strata are subdivided on the basis of (minor) differences in the nature and sequence of the depositions and constructions. Then the strata themselves are regrouped into main levels of occupation. These levels emphasize the overall relationships between the various trenches and delineate coherent building phases. All strata and levels have been numbered in order of excavation.

Subsequently, all the finds (i.e. the lots) and spatial units (i.e. the loci) are assigned to these strata and levels. Furthermore, they are assigned to spatial units such as rooms, open areas, ovens, pits, etc. On the basis of these classifications the persons working on the various assemblages, e.g. pottery, flint, animal bones, can classify their materials, i.e. they can study the vertical and horizontal distribution of their material. In the present study functional and other supposedly meaningful categories of these classifications have been used for a spatial analysis (chapter 6).

Thus the stratigraphical analysis provides the chronological and spatial context of the objects recovered in the field. In other words: the relationships between the various forms of material culture which were, necessarily, destroyed during excavation, are reconstructed. This in fact is the first stage of the spatial analysis.

Summarizing, the following steps were taken: excavation - sampling - basic administration - stratigraphical analysis - analysis of the various assemblages - spatial analysis.

APPENDIX 3

NUMBERS AND PERCENTAGES RELATED TO THE VISUAL
INSPECTION IN CHAPTER 6 (SECTION 2)[89]

Small finds

room 6 of building II: n= 373, 26.2%, room 2 of building IV: n= 46, 3.2%, rooms 3, 6 and 7 of building V: n= 37, 2.6%; n= 25, 1.8%; n= 101, 7.1% respectively, tholos IX: n= 57, 4%;

floor contexts: n= 783, 55.1%, buildings I, II, IV, XI, tholos VI: n= 81, 10.3%; 473, 60.4 %; n= 87, 11.1%; n= 10, 1.5%; n= 50, 6.4% respectively;

fills: n= 639, 44.9%, building V: n= 272, 42.6%;

complete: n= 548, 38.5%, broken: n= 874, 61.5%;

objects of unbaked clay: n= 912, 64.1%, building IV: n= 50, 5.5%, room 6 of building II: n= 336, 36.8%, rooms 1 and 2 of tholos VI: n= 18, 2%, tholos IX: n= 28, 3.1%;

clay objects compared to stone objects: n= 428, 30.1%;

stone objects: room 6 of building II: n= 35, 8.2%, room 2 of building IV: n= 41, 9.6%, rooms 6 and 7 of building V: n= 24, 5.6%, tholos IX: n= 22, 5.1%, area 3 in building I: n= 11, 2.5%;

bone tools: n= 70, 5%, room 6 of building II: n= 11, 15.7%, building V: n= 9, 12.9%;

grinding slabs: n= 85, 6%s, grinders: n= 66, 4.6%s, pestles: n= 34, 2.4%s, room 6 of building II: n= 13, 7%s, rooms 17 and 18 of building II: n= 9, 5%s, room 2 of building IV: n= 15, 8.1%s, area 3 of building I: n= 10, 5.4%s, pestles: IV and V: n= 13, 7%s, mortar fragments: n= 12, 0.8%s, building II: n= 6;

hammerstones: n= 7, 0.5%;

stone bowl fragments: n= 35, 2.5%;

axes: n= 15, 1.1%, building II: n= 5, building IV: n= 2;

objects of unmodified stones: building II: n= 16, 31.4%, building IV: n= 13, 25.5%, building V: n= 6, 11.8%;

sling missiles: n= 104, 7.3%, room 6 of building II: n= 18, 17.3%; spindle whorls: n= 38, 2.7%, room 6 of building II: n= 19, 50%;

pierced discs: n= 73, 5.1%, building IV: n= 24, 32.9%;

personal adornment: beads: n= 19, 1.3%, labrets: n= 31, 2.2%, room 6 of building II: n= 10, 32.3%, room 2 of building IV: n= 5, 16.1%;

figurines (human and animal): n= 50, 3.4%, room 6 of building II: n= 17, 35.4%, tholos IX: n= 6, 14.5%, building V: n= 17, 35.4%;

tokens: n= 182, 12.8%, room 6 of building II: n= 69, 37.9%, rooms 6 and 7 of building V: n= 28, 5.4%, tholos IX: n= 6, 4.3%

Ceramics

Standard Ware: n= 20,401, 89.5%, Dark-Faced Burnished Ware: n= 1092, 4.8%, Grey-Black Ware: n= 269, 1.2%, Standard Fine Ware: n= 785, 3.4%;

Standard Ware: building II: n= 3602, 17.7%, building IV: n= 3293, 16.1%, building V: n= 2690, 13.2%, building I: n= 1510, 7.4%, building X: n= 1131, 5.5%, buildings XI and XII: n= 3636, 17.8%;

---

[89] Percentages: when an s is added to the percentage (e.g. 5%s) this means that the percentage refers not to the whole assemblage (in that case nothing is added), but to the relevant class within the assemblage (i.e. grinding implements to ground-stone tools, and awls to bone tools).

Dark-Faced Burnished Ware: building IV: n= 273, 25%, open area 2: n= 14, 5.2%;
Grey-Black Ware: building IV: n= 101, 37.5%, tholos IX: n= 15, 5.6%, open area 2: n= 14, 5.2%;
Standard Fine Ware: room 6 of building II: n= 54, 6.6%, building IV: n= 179, 22.8%, rooms 1 and 2 of tholos VI: n= 19, 2.4%;
Orange Fine Ware: n= 38: building II: n= 6, buildings IV and V: n= 25;
vessel shapes Orange Fine Ware: n= 38, building V: bowls: n= 5, jars: n= 8;
Standard Ware husking trays: building II (rooms 16 to 18): n= 10, 13.7%, area 3 of building I: n= 3, 4.1%, building II: n= 10, 13.7%;
vessel shapes of Standard Ware: n= 35: bowls: n= 22, 62.9%, jars: n= 12, 34.3%, bowls from building I: n= 5, 22.7%, jars from buildings II, IV and V: n= 4, n= 2 and n= 2 respectively;
Standard Fine Ware: majority of the shapes: n= 87, 76.3%

Flint/obsidian artefacts
distinction flint: n= 2196, 85.9% and obsidian artefacts: n= 359, 14.1%;
complete vs. broken tools: n= 489, 73.3%; n= 167, 25%, respectively;
tools vs. debitage: n= 452, 17.7%, n= 2103, 82.3% respectively;
cores: n= 71, 3.4%: buildings IV and V: n= 19, 26.7%; n= 9, 12.7% respectively;
by-products: n= 64, 3%;
racloirs: n= 14, 3.1%;
raclettes: n= 6, 1.3%;
pieces with fine retouch: n= 77, 17%, buildings IV and V: n= 21, 27.3%, n= 10, 13% respectively;
burins: n= 17, 3.8%;
borers: n= 37, 8.2%;
side-blow blade flakes: n= 17, 3.8%;
corner-thinned blades: n= 11, 2.4%;
sickles: n= 46, 10.2%, building II: n= 11, 2.4%

APPENDIX 4

EXAMPLE OF FUNCTIONAL ASSESSMENT OF SPATIAL UNITS
(ROOM 1, BUILDING I)

The numbers between brackets refer to the legend below.

1 Visual inspection

2 Architectural analysis
Dimensions: 2.75 x 1.55 m (relatively large area)
Dooropenings: in south, marked by two small buttresses
Architectural features: absent
Probable function: activity area (well-accessible, relatively large) (1)

3 Depositional analysis
Stratigraphy: R13: 3E, 3C/3E (2)
- loam deposition upon floor (3E), covered by burnt layer (3C)
Condition of objects/nature of assemblage:
- small finds: two broken, three complete (n= 5)
- ceramics: sherds (n= 227)
- flint/obsidian artefacts: one tool, nine pieces of debris (n= 10) (3)
Formation processes: provisional refuse, primary de facto refuse: in-situ context

4 Determination of object context
Object context:
- broken objects: discard
- complete objects: interactive/depositional

5 Determination of general function
General function: ACTIVITY AREA (A, B, D) (4)

6 Functional analysis
Activities:
- SUBSISTENCE: food preparation: **food deposition**? (Standard Ware, one Standard Ware jar fragment, one Standard Ware bowl fragment, one fragment of a stone bowl); **grinding** (fragment of a grinder and fragment of a grinding slab); **serving**? (sherds: Standard Fine Ware, Fine Painted Ware, Grey-Black Ware); (5)
- MANUFACTURE/MAINTENANCE: **spinning** (pierced disc).

Flint/obsidian artefacts: (6)

Artefact	Activity Category: mwa	Function: mwa	Function: eth./arch.
3 flakes	manufacture	flint knapping?	working the fields; butchering; plant-processing; man./main. of: wooden, bone and leather artefacts; skin-processing
1 blade	subsistence: food preparation; manufacture/ maintenance	plant-processing: working of plants; skin-processing: working of dry and hard skin	see above
1 end-scraper	subsistence: food procurement; manufacture/ maintenance	plant-processing: cutting of grain?; working of hard-silicious plants	see above, and: harvesting; man./main of shell artefacts

Scenario 2

Complete objects: (7)
pierced disc
flint/obsidian: see table above

Determination of general function
General function: ACTIVITY AREA (A, B, D)

Functional analysis
Activities:
- MANUFACTURE/MAINTENANCE: **spinning** (pierced disc); **plant-processing**: working of plant, working of hard silicious plant; **skin-processing**: working of dry and hard skin (see table above).

Remarks
Room 1 contained a relatively large percentage (20.3%) of all ceramics of building 1 (n= 1116). This may indicate that pots were regularly used in room 1. In addition to Standard Ware (20.9% of all Standard Ware in building I), Grey-Black Ware and Dark-Faced Burnished Ware are well represented (26.7 and 26.7% respectively of these wares in building I). Both wares may have had a special meaning: Dark-Faced Burnished Ware as a 'luxury good' and Grey-Black Ware as a display ware (cf. chapter 5, section 3). The vessel shapes that were reconstructed were one Standard Ware bowl and one Standard Ware Jar; food may have been prepared as well as kept in room 1 (serving is not likely, considering the small size of room 1).

Legend
(1) Function based on architectural information alone.
(2) First square, then stratum (see Verhoeven and Kranendonk 1996).
(3) The working edges of the majority of the flint and obsidian tools were complete, i.e. most often they were not broken, indicating that the tools were still in a useable condition. Most often, however, they were not fresh, but damaged and abraded due to usage and perhaps trampling (Copeland 1996:288). Since this holds for the large majority of tools, not artefact condition but the proportions of tools versus debitage will be given as an indication of the nature of the assemblage.
(4) See table 4.5 in chapter 4.
(5) With *food deposition* is meant the deposition or temporary storage of food during processing.

(6) Within the class of flint/obsidian *artefacts* of Sabi Abyad Copeland (1996:290) distinguishes *tools* and *debitage*. Tools are modified, retouched, pieces. Debitage represents unmodified, unretouched, pieces and cores. All tools (scrapers, borers, etc.) are listed. Of the debitage, when present, cores, flakes, blades and by-products have been represented, since these artefacts may (1) have been used (flakes, blades), or (2) indicate knapping activities (cores, by-products). Of the debitage those classes that cannot be used in a functional assessment (i.e. fragments, debris, chips, chunks) have been omitted (therefore the number given in point 3 may differ from the total as appears from the table).

In the tables 'Function' has been subdivided in 'Function: mwa': the function of flint artefacts according to the microwear analysis (chapter 5, section 4), and 'Function: eth./arch': the function of flint/obsidian artefacts according to the ethnographic and archaeological analysis (chapter 5). The function according to the microwear analysis indicates the most likely function: the function has been assessed on the basis of the analysis of a selection of flint artefacts. If an artefact itself has been analysed, this has been indicated in the table (mwa). In the discussion of the results of the present analysis I will mainly deal with the function according to the microwear analysis, but the function according to the ethnographic and archaeological analysis wil also be incorporated.

(7) In scenario 1 (points 1-6) all objects, complete as well as broken, are taken into account. In scenario 2, however, only the complete objects are taken into account.

BIBLIOGRAPHY

ADAMS M.J.

1977   "Style in Southeast Asian Material Processing - Some Implications for Ritual and Art", in: H. Lechtmann, ed., *Material Culture - Styles, Organization and Dynamics of Technology*, St. Paul: West Publishing, pp. 21-52.

AKKERMANS P.A. et al.

1983   "Bouqras Revisited - Preliminary Report on a Project in Eastern Syria", *Proceedings of the Prehistoric Society* 49:335-372.

AKKERMANS P.M.M.G.

1989a   "Tell Sabi Abyad - Stratigraphy and Architecture", in: P.M.M.G. Akkermans, ed., *Excavations at Tell Sabi Abyad*, Oxford: BAR-IS 468, pp. 17-75.

1989b   "The Other Small Finds of Tell Sabi Abyad", in: P.M.M.G. Akkermans, ed., *Excavations at Tell Sabi Abyad*, Oxford: BAR-IS 468, pp. 285-294.

1989c   "The Prehistoric Pottery of Tell Sabi Abyad", in: P.M.M.G. Akkermans, ed., *Excavations at Tell Sabi Abyad*, Oxford: BAR-IS 468, pp. 77-214.

1991   "New Radiocarbon Dates for the Later Neolithic of Northern Syria", *Paléorient* 17:121-126.

1993   *Villages in the Steppe - Later Neolithic Settlement and Subsistence in the Balikh Valley, Northern Syria*, Ann Arbor: International Monographs in Prehistory.

AKKERMANS P.M.M.G., ed.

1996   *Tell Sabi Abyad - The Late Neolithic Settlement. Report on the Excavations of the University of Amsterdam (1988) and the National Museum of Antiquities Leiden (1991-1993) in Syria*, Istanbul: Nederlands Historisch-Archaeologisch Instituut.

AKKERMANS P.M.M.G. and DUISTERMAAT K.

1997   "Of Storage and Nomads - The Sealings from Late Neolithic Sabi Abyad, Syria", *Paléorient* 22/2:17-44.

AKKERMANS P.M.M.G. and VERHOEVEN M.

1995   "An Image of Complexity - The Burnt Village at Late Neolithic Sabi Abyad, Syria", *American Journal of Archaeology* 99/1:5-32.

in prep.   "Tell Sabi Abyad II in Perspective", in: M. Verhoeven and P.M.M.G. Akkermans, eds., *Tell Sabi Abyad II - The Pre-Pottery Neolithic B Settlement* (provisional titles), Istanbul: Nederlands Historisch-Archaeologisch Instituut.

AKKERMANS P.M.M.G. and WITTMANN B.

1993   "Khirbet esh-Shenef 1991 - Eine späthalafzeitliche Siedlung im Balikhtal, Nordsyrien", *Mitteilungen der Deutschen Orient-Gesellschaft* 125:143-166.

AKKERMANS P.M.M.G., LIMPENS J. and SPOOR R.H.

1993   "On the Frontier of Assyria - Excavations at Tell Sabi Abyad, 1991", *Akkadica* 84/85:1-52.

ALDENFELDER M. and MASCHNER H.D.G., eds.

1996   *Anthropology, Space and Geographic Information Systems*, Oxford: Oxford University Press.

ALLEN G.A.

1891   "Manners and Customs of the Mohaves", *Smithsonian Institution Annual Report* (1890):615-616.

ALLEN N.J.

1985   "Hierarchical Opposition and some other Types of Relation", in: R.H. Barnes, D. de Coppet and R.J. Parkin, eds., *Contexts and Levels - Anthropological Essays on Hierarchy*, Oxford: JASO, pp. 21-32.

ALLEN K.M.S., GREEN S.W. and ZUBROW E.B.W., eds.

1990   *Interpreting Space - GIS and Archaeology*, London: Taylor & Francis.

ALFORD J. and DUGUID N.

1995   "On the Flatbread Trail", *Aramco World* September/October:16-25.

AL-KHALESI Y.M.

1977   "The *Bit Kispim* in Mesopotamian Architecture - Studies of Form and Function", *Mesopotamia* XII:53-81.

1978   *The Court of the Palms - A Functional Interpretation of the Mari Palace*, Bibliotheca Mesopotamica 8, Malibu (California): Undena.

ALLISON D.M.

1995   "House Contents in Pompeii - Data Collection and Interpretative Procedures for a Reappraisal of Roman Domestic Life and Site Formation Processes", *Journal of European Archaeology* 3.1:145-176.

ANDERSON P.C.

1980   "A Testimony of Prehistoric Tasks - Diagnostic Residues on Stone Tool Working Edges", *World Archaeology* 12/2:181-194.

1992   "Experimental Cultivation, Harvest and Threshing of Wild Cereals and their Relevance for Interpreting the Use of Epipaleolithic and Neolithic Artefacts", in: P.C. Anderson, ed., *Préhistoire de l'agriculture - Nouvelles approches expérimentales et ethnographiques*, Paris: Monographie du CRA 6, Éditions du CNRS, pp. 179-210.

1994   "Reflections on the Significance of Two PPN Typological Classes in Light of Experimentation and Microwear Analysis - Flint "Sickles" and Obsidian "Çayönü Tools", in: H.G. Gebel and S.K. Kozlowski, eds., *Neolithic Chipped Stone Industries of the Fertile Crescent - Studies in Early Near Eastern Production, Subsistence, and Environment* 1, Berlin, *ex oriente*, pp. 61-82.

ANDERSON P.C., ed.

1992   *Préhistoire de l'agriculture - Nouvelles approches expérimentales et ethnographiques*, Paris: Monographie du CRA 6, Éditions du CNRS.

ANTOUN R.

1972   *Arab Village - A Social Structural Study of a Transjordanian Peasant Community*, London: Indiana University Press.

ARCHI A.

1993   "Les archives royales D'Ebla", in: *Syrie, Mémoire et Civilisation*, Paris: Flammarion, pp. 108-111.

ARDENER S., ed.

1993   *Women and Space - Ground Rules and Social Maps*, Oxford: Berg.

ARNOLD P.J.

1991   *Domestic Ceramic Production and Spatial Organization - A Mexican Case Study in Ethnoarchaeology*, Cambridge: Cambridge University Press.

ASCHER R.

1961   "Analogy in Archaeological Interpretation", *Southwestern Journal of Anthropology* 17:317-325.

1968   "Time's Arrow and the Archaeology of a Contemporary Community", in: K.C. Chang, ed., *Settlement Archaeology*, Palo Alto: National Press Books, pp. 43-52.

ATAMAN K.

1992   "Treshing Sledges and Archaeology", in: P.C. Anderson, ed., *Préhistoire de l'agriculture - Nouvelles approches expérimentales et ethnographiques*, Paris: Monographie du CRA 6, Éditions du CNRS:305-320.

ATEN N.

1996   "Note on the Human Skeletal Remains", in: P.M.M.G. Akkermans, ed., *Tell Sabi Abyad - The
            Late Neolithic Settlement. Report on the Excavations of the University of Amsterdam (1988)
            and the National Museum of Antiquities Leiden (1991-1993) in Syria*, Istanbul: Nederlands
            Historisch-Archaeologisch Instituut, pp. 114-118.

ATEN A., FARADAY INNES R. and KNEW E.

1955   *Flaying and Curing of Hides and Skins as a Rural Industry*, Rome: FAO.

AURENCHE O.

1981   *La maison orientale - l'architecture du Proche-Orient Ancien des origines au milieu du
            quatrième millinaire*, Paris: Paul Geuthner.

1984   *Nomades et sédentaires - Perspectives ethnoarchéologiques*, Paris: Éditions Recherche sur les
            Civilisations.

AURENCHE O., BAZIN M. and SADLER S.

1997   *Villages engloutis - Enquête ethnoarchéologique à Cafer Höyük (vallée de l'Euphrate)*, Lyon:
            Travaux de la Maison de l'Orient 26.

AUSTIN S.J.

1995   *An Analysis of Architectural Variability in Levantine Settlements during the Late Pleistocene
            and Early Holocene*, Ann Arbor: U.M.I. Dissertation Services.

AZAR G., CHIMIENTI G., HADDAD H. and SEEDEN H.

1985   "Busra - Housing in Transition", *Berytus* 33:103-142.

BACHELARD G.

1964   *The Poetics of Space*, New York: Orion Press.

BADER N.O.

1993a   "Tell Maghzaliyah - An Early Neolithic Site in Nothern Iraq", in: N. Yoffee and J.J. Clark, eds.,
            *Early Stages in the Evolution of Mesopotamian Civilization - Soviet Excavations in Northern
            Iraq*, Tucson/London: The University of Arizona Press, pp. 7-40.

1993b   "Results of the Excavations at the Early Agricultural Site of Kültepe in Northern Iraq", in: N.
            Yoffee and J.J. Clark, eds., *Early Stages in the Evolution of Mesopotamian Civilization - Soviet
            Excavations in Northern Iraq*, Tucson/London: The University of Arizona Press, pp. 55-61.

BAHLOUL J.

1996   *The Architecture of Memory - A Jewish-Muslim Household in Colonial Algeria 1937-1962*,
            Cambridge: Cambridge University Press.

BANNING E.B.

1997   "Differing Trajectories in the Late Neolithic of Mesopotamia and Jordan", *Bulletin of the
            Canadian Society of Mesopotamian Studies* 32:43-52.

BARBER E.J.W.

1991   *Prehistoric Textiles - The Development of Cloth in the Neolithic and Bronze Ages - With Special
            Reference to the Aegean*, Princeton: Princeton University Press.

BARGATZKY T.

1994   "Introduction", in: T. Bargatzky and R. Kuschel, eds., *The Invention of Nature*, Frankfurt: Peter
            Lang, pp. 9-25.

BARRET J.C.

1981   "Aspects of the Iron Age in Atlantic Scotland - A Case Study in the Problems of Archaeological
            Interpretation", *Proceedings of the Society of Antiquaries of Scotland* 111:205-219.

BARTH F.

1961   *Nomads of South Persia - The Basseri Tribe of the Khamseh Confederacy*, London: George
            Allen & Unwin Ltd.

1975   *Ritual and Knowledge among the Baktaman of New Guinea*, Oslo: Universitets Forlaget.

BAR-YOSEF O.

1983   "The Natufian in the Southern Levant", in: T. Cuyler Young Jr. et al., eds., *The Hilly Flanks and Beyond - Essays on the Prehistory of Southwestern Asia presented to Robert J. Braidwood*, Chicago: the Oriental Institute of the University of Chicago (SAOC 36), pp. 11-42.

BAR-YOSEF O. and KHAZANOV A., eds.

1992   *Pastoralism in the Levant - Archaeological Materials in Anthropological Perspectives*, Madison: Prehistory Press.

BAR-YOSEF O. and MEADOW R.H.

1995   "The Origins of Agriculture in the Near East", in: T.D. Price and A.B. Gebauer, eds., *Last Hunters, First Farmers -New Perspectives on the Prehistoric Transition to Agriculture*, Santa Fe: School of American Research Press, pp. 39-94.

BAZELMANS J.

1996   *Tacitus' Germania, de Oudengelse Beowulf en het Ritueel-Kosmologische Karakter van de Relatie tussen Heer en Krijger-Volgeling in Germaanse Samenlevingen*, Amsterdam: University of Amsterdam (Ph.D. thesis).

BAZIN M. and BROMBERGER C.

1982   *Gilân et Azarbâyjân Oriental - Cartes et Documents ethnographiques*, Paris: Institut Français d'Iranologie de Téhéran - Bibliothèque Iranienne, Éditions Recherche sur les Civilisations.

BEALS R.L.

1934   *Material Culture of the Pima, Papago, and western Apache*, Berkeley: US Department of the Interior, National Park Service.

BEAZLY E. and HARVERSON M.

1982   *Living with the Desert - Working Buildings of the Iranian Plateau*, Warminster: Aris and Philips Ltd.

BECK L.

1986   *The Qashqa'i of Iran*, New Haven & London: Yale University Press.

1991   *Nomad - A Year in the Life of a Qashqa'i Tribesman in Iran*, London: I.B. Tauris & Co. Ltd.

BELL C.

1992   *Ritual Theory - Ritual Practice*, Oxford: Oxford University Press.

BERGMAN C.A., BARTON R.N.E., COLLCUTT S.N. and MORRIS G.

1983   "La Fracture Volontaire dans une Industrie du Paléolithique Supérieur Tardiff du Sud de l'Angleterre", *l'Antropologie* 87/3:323-337.

BERNBECK R.

1994   *Die Auflösung der häuslichen Produktionsweise - Das Beispiel Mesopotamiens*, Berlin: Dietrich Reimer.

1995   "Dörfliche Kulturen des keramischen Neolithicums in Nord-und Mittelmesopotamien - Vielfalt der Kooperationsformen", in: K. Bartl, R. Bernbeck and M. Heinz, eds., *Zwischen Euphrat und Indus - Aktuelle Forschungsprobleme in der Vorderasiatischen Archäologie*, Hildesheim/ Zürich/New York: Georg Olms, pp. 28-43.

BEUKER J.R.

1983   *Vakmanschap in Vuursteen - De Vervaardiging en het Gebruik van Vuurstenen Werktuigen in de Prehistorie*, Assen: Provinciaal Museum van Drenthe.

BEYRES S. and INIZAN M.L.

1982   "Typologie, ochre, fonction", Paper presented at 'Recent Progress in Microwear Studies', Tervuren: *Studia Praehistorica Belgica* 2:313-322.

BINFORD L.R.

1979   "Organization and Formation Processes - Looking at Curated Technologies", *Journal of Anthropological Research* 35/3:255-273.

1981   "Behavioral Archaeology and the "Pompeii Premise"", *Journal of Anthropological Research* 37/3:195-208.

1983   *In Pursuit of the Past - Decoding the Archaeological Record*, London: Thames & Hudson.

BLANKHOLM H.P.

1991   *Intrasite Spatial Analysis in Theory and Practice*, Aarhus: Aarhus University Press.

BLANTON R.E.

1994   *Houses and Households - A Comparative Study*, New York/London: Plenum Press.

BONNET H.

1926   *Die Waffen der Völker des Alten Orients*, Leipzig: J.C. Hinrichs'sche Buchhandlung.

BOTTEMA S.

1989   "Notes on the Prehistoric Environment of the Syrian Djezireh", in: O.M.C. Haex, H.H. Curvers and P.M.M.G. Akkermans, eds., *To the Euphrates and Beyond*, Rotterdam: Balkema, pp. 1-16.

BOTTEMA S., ENTJES G., VAN ZEIST W., eds.

1990   *Man's Role in the Shaping of the Eastern Mediterranean Landscape*, Rotterdam: Balkema.

BOURDIEU P.

1973   "The Berber House", in: M. Douglas, ed., *Rules and Meanings*, Suffolk: Penguin, pp. 98-110.

1977   *Outline of a Theory of Practice*, Cambridge: Cambridge University Press (translated from the French by R. Nice, originally published in 1972).

1984   *Distinction - A Social Critique of the Judgement of Taste*, Cambridge, MA: Harvard University Press.

1989   *Opstellen over Smaak, Habitus en het Veldbegrip*, Amsterdam: Van Gennep (translated from the French by various authors).

1990   *The Logic of Practice*, Cambridge: Polity Press.

1992   *Argumenten voor een Reflexieve Maatschappijwetenschap*, Amsterdam: Van Gennep (translated from the French by R. Hofstede).

BRAIDWOOD L.S. et al.

1983   *Prehistoric Archaeology along the Zagros Flanks*, Chicago: University of Chicago Press (OIP 105).

BRAIDWOOD R.J. and BRAIDWOOD L.S.

1960   *Excavations in the Plain of Antioch I*, Chicago: University of Chicago Press (OIP LXI).

BRAIDWOOD R.J. and HOWE B.

1960   *Prehistoric Investigations in Iraqi Kurdistan*, Chicago: University of Chicago Press (SAOC 31).

BRAIDWOOD R.J., BRAIDWOOD L., SMITH G. and LESLIE C.

1952   "Matarrah - A Southern Variant of the Hassunan Assemblage, excavated in 1948", *Journal of Near Eastern Studies* 11:2-75.

BRAITHWAITE M.

1982   "Decoration as Ritual Symbol - A Theoretical Proposal and an Ethnographic Study in Southern Sudan", in: I. Hodder, ed., *Symbolic and Structural Archaeology*, Cambridge: Cambridge University Press, pp. 80-88.

BREKELMANS T.

1979   *Korfvlechten - De Spiraalvlechttechniek van Schalen, Bijenkorven en Manden*, de Bilt: Cantecleer BV.

BRENIQUET C.

1996   *La disparition de la culture de Halaf - les origines de la culture d'Obeid dans le nord de la Mésopotamie*, Paris: Éditions Recherches sur les Civilisations.

BRESENHAM M.F.

1985   "Descriptive and Experimental Study of Contemporary and Ancient Pottery Techniques at Busra", *Berytus* 33:89-101.

BROMAN-MORALES V.

1983   "Jarmo Figurines and other Clay Objects", in: L.S. Braidwood et al., *Prehistoric Archaeology along the Zagros Flanks*, Chicago: University of Chicago Press (OIP 105), pp. 369-426.

1990   *Figurines and other Clay Objects from Sarab and Çayönü*, Chicago: University of Chicago (OIC 25).

BUCCELLATI G.

1990   ""River Bank", "High Country", and "Pasture Land" - the Growth of Nomadism on the Middle Euphrates and the Khabur", in: S. Eichler, M. Wäfler and D. Warburton, eds., *Tall Al-Hamidiya* 2, Göttingen: Vandenhoeck & Ruprecht, pp. 87-117.

BURROUGH P.A.

1986   *Principles of Geographical Information Systems for Land Resource Assessment*, Oxford: Clarendon Press.

BYRD B.F.

1994   "Public and Private, Domestic and Corporate - The Emergence of the Southwest Asian Village", *American Antiquity* 59/4:639-666.

CALDWELL J.R., ed.

1967   *Investigations at Tal-i-Iblis*, Illinois: Illinois State Museum Preliminary Reports no. 9.

CAMERON C.M. and TOMKA S., eds.

1993   *Abandonment of Settlements and Regions - Ethnoarchaeological and Archaeological Approaches*, Cambridge: Cambridge University Press.

CAMPANA D.V.

1989   *Natufian and Protoneolithic Bone Tools - The Manufacture and Use of Bone Implements in the Zagros and the Levant*, Oxford: BAR-IS 494.

CAMPBELL S.

1992   *Culture, Chronology and Change in the Later Neolithic of North Mesopotamia*, Edinburgh: University of Edinburgh (Ph.D. thesis).

1995   "Death for the Living in the Late Neolithic in North Mesopotamia", in: S. Campbell and A. Green, eds., *The Achaeology of Death in the Ancient Near East*, Oxford: Oxbow, pp. 29-34.

1998   "Problems of Definition - The Origins of the Halaf in North Iraq", *Subartu* IV/1: 39-52.

CAMPS-FABRER H. and COURTIN J.

1982   "Essaie d'approche technologique des faucilles préhistoriques dans le basin Méditerranéen", Université de Provence: *Travaux du laboratoire d'Antropologie des Pays de la Méditerranée Occidentale*, Étude 8, pp. 1-26.

CANAAN T.

1932   "The Palestinian Arab House - Its Architecture and Folklore", *Journal of the Palestine Oriental Society* 12:223-247.

CARMICHAEL D.L., HUBERT J., REEVES B. and SCHANCHE A., eds.

1997   *Sacred Sites - Sacred Spaces*, London/New York: Routledge.

CARNEIRO R.L.

1979   "Tree felling with the Stone Axe - An Experiment carried out among the Yanomamo Indians of Southern Venezuela", in: C. Kramer, ed., *Ethnoarchaeology - Implications of Ethnography for Archaeology*, New York: Columbia University Press, pp. 21-58.

CARR C.

1984   "The Nature of Organization of Intrasite Archaeological Records and Spatial Analytic Approaches to their Investigation", in: M.B. Schiffer, ed., *Advances in Archaeological Method and Theory* 7, pp. 103-222.

CARSTEN J. and HUGH-JONES S.
1995   "Introduction", in: J. Carsten and S. Hugh-Jones, eds., *About the House - Lévi Strauss and Beyond*, Cambridge: Cambridge University Press, pp. 1-46.
CARSTEN J. and HUGH-JONES S., eds.
1995   *About the House - Lévi Strauss and Beyond*, Cambridge: Cambridge University Press.
CAUVIN J.
1968   *Les outillages néolithiques de Byblos et du littoral libanais*, Fouilles de Byblos IV, Paris: Maisonneuve.
1972a   "Sondage à Tell Assouad (Djézireh, Syrie)", *Annales Archéologiques Arabes Syriennes* 22:85-88.
1972b   *Religions néolithique de Syro-Palestine*, Paris: Maisonneuve.
1994   *Naissance des Divinités, naissance de l'agriculture - la revolution des symboles au néolithique*, Paris: CNRS Editions.
CAUVIN M.-C.
1973   "Problèmes d'emmanchement des faucilles au Proche-Orient - les documents de Tell Assouad (Djézireh, Syrie)", *Paléorient* 1:103-108.
1983   "Les faucilles préhistoriques au Proche-Orient, données morphologiques et fonctionnelles", *Paléorient* 9/1:63-80.
CAUVIN M.-C. and BALKAN N.
1983   "Çafer Höyük - analyse de l'outillage lithique (campagnes 1982-1983): problèmes typologiques et chronologiques", *Cahiers de l'Euphrate* 4:53-86.
CAVALLO C.
1996   "The Animal Remains - A Preliminary Account", in: P.M.M.G. Akkermans, ed., *Tell Sabi Abyad - The Late Neolithic Settlement. Report on the Excavations of the University of Amsterdam (1988) and the National Museum of Antiquities Leiden (1991-1993) in Syria*, Istanbul: Nederlands Historisch-Archaeologisch Instituut, pp. 475-520.
1997   *Animals in the Steppe - A Zooarchaeological Analysis of Later Neolithic Tell Sabi Abyad, Syria*, Amsterdam: University of Amsterdam (Ph.D. thesis).
in press "Animal Remains enclosed in Oval Clay Objects from the "Burnt Village" of Tell Sabi Abyad, Northern Syria", *Anthropozoologica*.
CHAMPAULT D.
1969   *Une Oasis du Sahara Nord-Occidental, Tabelbala*, Paris: CNRS.
CHAPMAN J.
1990   "Social Inequality on Bulgarian Tells and the Varna Problem", in: R. Samson, ed., *The Social Archaeology of Houses*, Edinburgh: Edinburgh University Press, pp. 49-93.
CHAUVET J.M., BRUNEL DESCHAMPS E. and HILLAIRE C.
1996   *Chauvet Cave - The Discovery of the World's oldest Paintings*, London: Thames & Hudson.
CHAVALAS M.W.
1988   *The House of Puzurum - A Stratigraphic, Distributional, and Social Analysis of Domestic Units from Tell Ashara/Terqa, Syria, from the Middle of the Second Millennium B.C.*, Ann Arbor: University Microfilms International.
CHRISTENSEN N.
1967   "Haustypen und Gehöftbildung in Westpersien", *Anthropos* 62:89-138.
CIOLEK-TORRELLO R.
1984   "An Alternative Model of Room Function from Grasshopper Pueblo, Arizona", in: H. Hietala, ed., *Intrasite Spatial Analysis in Archaeology*, Cambridge: Cambridge University Press, pp. 127-154.
1985   "A Typology of Room Function at Grasshopper Pueblo, Arizona", *Journal of Field Archaeology* 12:41-63.

CLARK J.G.D.

1954   *Excavations at Starr Carr*, Cambridge: Cambridge University Press.

CLARKE D.L., ed.

1977   *Spatial Archaeology*, London: Academic Press.

CLIFFORD J.

1986   "Introduction - Partial Truths", in: J. Clifford and G. Marcus, eds., *Writing Culture - The Poetics and Politics of Ethnography*, Berkely/Los Angeles/London: University of California Press, pp. 1-26.

COLLET P.

1996   "The Figurines", in: P.M.M.G. Akkermans, ed., *Tell Sabi Abyad - The Late Neolithic Settlement. Report on the Excavations of the University of Amsterdam (1988) and the National Museum of Antiquities Leiden (1991-1993) in Syria*, Istanbul: Nederlands Historisch-Archaeologisch Instituut, pp. 403-415.

COLLET P. and SPOOR R.H.

1996   "The Ground-Stone Industry", in: P.M.M.G. Akkermans, ed., *Tell Sabi Abyad - The Late Neolithic Settlement. Report on the Excavations of the University of Amsterdam (1988) and the National Museum of Antiquities Leiden (1991-1993) in Syria*, Istanbul: Nederlands Historisch-Archaeologisch Instituut, pp. 415-439.

COPELAND L.

1979   "Observations on the Prehistory of the Balikh Valley, Syria, during the 7th to 4th Millennia B.C.", *Paléorient* 5:251-275.

1982   "Prehistoric Tells in the Lower Balikh Valley, Syria - Report on the Survey of 1978", *Annales Archéologiques Arabes Syriennes* 32:251-271.

1989   "The Flint and Obsidian Artefacts of Tell Sabi Abyad", in: P.M.M.G. Akkermans, ed., *Excavations at Tell Sabi Abyad - Prehistoric Investigations in the Balikh Valley, Northern Syria*, Oxford: BAR-IS 468, pp. 237-284.

1996   "The Flint and Obsidian Industries", in: P.M.M.G. Akkermans, ed., *Tell Sabi Abyad - The Late Neolithic Settlement. Report on the Excavations of the University of Amsterdam (1988) and the National Museum of Antiquities Leiden (1991-1993) in Syria*, Istanbul: Nederlands Historisch-Archaeologisch Instituut, pp. 285-339.

in press   "The Lithic Material of Tell Arjoune (Syria)", in: P. Parr, ed., *The Excavations at Tell Arjoune in Syria*.

COPELAND L. and VERHOEVEN M.

1996   "Bitumen-Coated Sickle-Blade Elements at Tell Sabi Abyad II, Northern Syria", in: S.K. Kozlowski and H.G. Gebel, eds., *Neolithic Chipped Stone Industries of the Fertile Crescent, and their Contemporaries in Adjacent Regions - Studies in Early Near Eastern Production, Subsistence, and Environment* 3, Berlin *ex oriente*, pp. 327-330.

COQUEUGNIOT E.

1983   "Analyse tracéologique d'une série de grattoirs et herminettes de Mureybit, Syrie (9ème-7ème millénaires)", in: M.-C. Cauvin, ed., *Traces d'utilisation sur les Outils néolithiques du Proche-Orient*, Lyon: Travaux de la Maison de l'Orient 5, pp. 163-172.

CORBEY R.

1994   "Gift en Transgressie - Kanttekeningen bij Bataille", *Tijdschrift voor Filosofie* 56/2:272-312.

COSTA P.M.

1991   *Musan dam - Architecture and Material Culture of a little known Region of Oman*, London: Immel Publishing.

COWGILL G.L., ALTSCHUL J.H. and SLOAD R.S.

1984   "Spatial Analysis of Teotihuacan - A Mesoamerican Metropolis", in: H. Hietala, ed., *Intrasite Spatial Analysis in Archaeology*, Cambridge: Cambridge University Press, pp. 127-154.

CRIBB R.

1991   *Nomads in Archaeology*, Cambridge: Cambridge University Press.

CUNNINGHAM C.E.

1973   "Order in the Atoni House", in: R. Needham, ed., *Right and Left - Essays on Dual Symbolic Classification*, Chicago: University of Chicago Press, pp 204-238 (originally published in 1964).

DALMAN G.

1964   *Arbeit und Sitte in Palestina*, Hildesheim: Georg Olms (7 vols., reprint from original 1928-1942 version).

DALTON G.

1981   "Anthropological Models in Archaeological Perspective", in: I. Hodder, G. Isaac and N. Hammond, eds., *Patterns of the Past*, Cambridge: Cambridge University Press, pp. 17-49.

DARVILL T. and THOMAS J., eds.

1996   *Neolithic Houses in Northwest Europe and Beyond*, Oxford: Oxbow.

DAVIAU P.M.M.

1993   *Houses and their Furnishings in Bronze Age Palestine - Domestic Activity Areas and Artefact Distribution in the Middle and Late Bronze Ages*, Sheffield: Academic Press.

DAVID N.

1971   "The Fulani Compound and the Archaeologist", *World Archaeology* 3:111-131.

DAVID N., STERNER J. and GAVUA K.

1988   "Why Pots are Decorated", *Current Anthropology* 29/3:365-389.

DAVIDSON T.E.

1977   *Regional Variation within the Halaf Ceramic Tradition*, Edinburgh: University of Edinburgh (Ph.D. thesis).

DAVIS D.P.

1983   "Investigating the Diffusion of Stylistic Innovations", in: M.B. Schiffer, ed., *Advances in Archaeological Method and Theory* 6, pp. 53-89.

DAVIS M.K.

1982   "The Çayönü Ground Stone", in: L.S. Braidwood and R.J. Braidwood, eds., *Prehistoric Village Archaeology in South-Eastern Turkey*, Oxford: BAR-IS 138, pp. 73-174.

DE BOER W.R. and LATHRAP D.

1979   "The making and breaking of Shipibo-Conibo Ceramics", in: C. Kramer, ed., *Ethnoarchaeology - Implications of Ethnography for Archaeology*, New York: Columbia University Press, pp. 102-138.

DIAMANT S. and RUTTER J.

1969   "Horned Objects in Anatolia and the Near East and possible Connexions with the Minoan "Horns of Consecration"", *Anatolian Studies* XIX:147-177.

DIAMOND G.

1979   "The Nature of so-called polished Surfaces on Stone Artefacts", in: B. Hayden, ed., *Lithic Use-Wear Analysis*, New York: Academic Press, pp. 159-166.

DOLLFUS G.

1983   "Remarques sur l'organisation de l'espace dans quelques agglomérations de la Susiane du V millénaire", *Studies in Ancient Oriental Civilization* 36:283-313.

DONLEY-REID L.W.

1990   "A Structuring Structure - The Swahili House", in: S. Kent, ed., *Domestic Architecture and the Use of Space - An Interdisciplinary Cross-Cultural Study*, Cambridge: Cambridge University Press, pp. 114-126.

DORAN J.E. and HODSON F.R.

1975   *Mathematics and Computers in Archaeology*, Edinburgh: Edinburgh University Press.

DORREL P.G.

1983   "Stone Vessels, Tools, and Objects", in: K.M. Kenyon and T.A. Holland, eds., *Excavations at Jericho* 5, London: British School of Archaeology in Jerusalem, pp. 487-509.

DOUGLAS M.

1966   *Purity and Danger - An Analysis of Concepts of Pollution and Taboo*, London: Routledge & Kegan Paul.

DRENKHAHN R.

1976   *Die Handwerker und Ihre Tätigkeiten im Alten Ägypten*, Wiesbaden: Harrasowitz.

DUISTERMAAT K.

1994   *The Clay Sealings from Late Neolithic Sabi Abyad*, Leiden: University of Leiden (M.A. thesis).

1996   "The Seals and Sealings", in: P.M.M.G. Akkermans, ed., *Tell Sabi Abyad - The Late Neolithic Settlement. Report on the Excavations of the University of Amsterdam (1988) and the National Museum of Antiquities Leiden (1991-1993) in Syria*, Istanbul: Nederlands Historisch-Archaeologisch Instituut, pp. 339-403.

DUMONT L.

1980   *Homo Hierarchicus - The Caste System and its Implications*, Chicago/London: University of Chicago Press (originally published in 1966).

DU PLAT TAYLOR J., SETON-WILLIAMS M.V. and WAECHTER J.

1949   "The Excavations at Sakçe Gözü", *Iraq* 12:53-138.

DURAND J.-M.

1987   "L'organisation de l'Espace dans le Palais de Mari", in: E. Lévy, ed., *Le système palatial en Orient, en Grèce et à Rome*, Leiden: Actes du colloque de Strasbourg.

EHLERS E.

1980   *Iran - Grundzüge einer geographischen Landeskunde*, Darmstadt: Wissenschaftliche Buchgesellschaft.

EICHMANN R.

1991   *Aspekte Prähistorischer Grundrissgestaltung in Vorderasien - Beiträge zum Verständnis bestimmter Grundrissmerkmale in ausgewählten neolitischen und chalkolitischen Siedlungen des 9. - 4. Jahrtausends v. Chr. (mit Beispielen aus der europäischen Prähistorie)*, Mainz am Rhein: Philipp von Zabern.

ELLISON R.

1984   "The Uses of Pottery", *Iraq* 46:63-68.

ESS M. von and PEDDE F.

1992   *Uruk Kleinfunde II*, Mainz am Rhein: Philipp von Zabern.

FALCONER S.E.

1995   "Rural Responses to Early Urbanism - Bronze Age Households and Village Economy at Tell el-Hayyat, Jordan", *Journal of Field Archaeology* 22:399-419.

FISKE J.

1990   *Introduction to Communication Studies*, London/New York: Routledge.

FLANNERY K.V. and WINTER M.

1976   "Analyzing Household Activities", in: K.V. Flannery, ed., *The Early Mesoamerican Village*, New York: Academic Press, pp. 34-45.

FLETCHER R.

1989   "The Messages of Material Behaviour - A Preliminary Discussion of Non-Verbal Meaning", in: I. Hodder, ed., *The Meanings of Things - Material Culture and Symbolic Expression*, London: Harper Collins Academic, pp. 33-39.

FLETCHER M. and LOCK G.R.

1994   *Digging Numbers - Elementary Statistics for Archaeologists*, Oxford: Oxbow.

FONTIJN D.

1995   *Nijmegen Kops Plateau - De Lange-Termijn Geschiedenis van een Prehistorisch Dodenlandschap*, Leiden: University of Leiden (M.A. thesis).

FORBES R.J.

1955   *Studies in Ancient Technology* 3, Leiden: Brill.

1956   *Studies in Ancient Technology* 4, Leiden: Brill.

1957   *Studies in Ancient Technology* 5, Leiden: Brill.

1958   *Studies in Ancient Technology* 6, Leiden: Brill.

FOX J.J.

1993   "Comparative Perspectives on Austronesian Houses - An Introductory Essay", in: J.J. Fox, ed., *Inside Austronesian Houses - Perspectives on Domestic Designs for Living*, Canberra: ANU printing service, pp. 1-30.

FOX J.J., ed.

1993   *Inside Austronesian Houses - Perspectives on Domestic Designs for Living*, Canberra: ANU printing service.

GALLERY J.A.

1976   "Town Planning and Community Structure", *Bibliotheca Mesopotamica* 6, Malibu (California): Undena.

GIDDENS A.

1984   *The Constitution of Society - Outline of a Theory of Structuration*, Cambridge: Polity Press.

GIFFORD E.W.

1934   *The Cocopa*, University of California Publications in American Archaeology and Ethnology 31 (1931-1933):257-334.

GLENNIE G.D. and LIPE W.D.

1984   *Replication of an Early Anasazi Pithouse*, Paper presented at the 49th Annual Meeting of the Society for American Archaeology, Portland (Oregon).

GNIVECKI P.L.

1983   *Spatial Organization in a Rural Akkadian Farmhouse - Perspectives from Tepe Al-Atiqeh*, Iraq, Ann Arbor: University Microfilms International.

GOLDBERG P., NASH D.T. and PETRAGLIA, M.D., eds.

1993   *Formation Processes in Archaeological Context*, Madison: Prehistory Press.

GOODALL B.

1987   *Dictionary of Human Geography*, Harmondsworth: Penguin.

GOULD R.A.

1980   *Living Archaeology*, New York/Cambridge: Cambridge University Press.

GOULD R.A., KOSTER D.A. and SONTZ A.H.L.

1971   "The Lithic Assemblage of the Western Desert Aborigines of Australia", *American Antiquity* 36/2:149-169.

GRACE R.

1989   *Interpreting the Function of Stone Tools*, Oxford: BAR-IS 474.

GRAMSCH A.

1996   "Landscape Archaeology - Of Making and Seeing", *Journal of European Archaeology* 4:19-38.

GRAMSCH A. and REINHOLD S.

1996   "Analogie und Archäologie", *Ethnografische und Archäologische Zeitschrift* 37:237-244.

GREGORY D. and URRY J., eds.

1985   *Social Relations and Spatial Structures*, Basingstroke: Macmillan.

GRINSELL L.V.

1961   "The breaking of Objects as a Funerary Rite", *Folk-lore* 72:475-491.

GRÖN O., ENGELSTAD E. and LINDHOLM I., eds.

1991   *Social Space - Human Spatial Behaviour in Dwellings and Settlements*, Odense: Odense University Press.

GULLINI G.

1970/71   "Struttura e spazio nell'architettura mesopotamica arcaica, da Eridu alle soglie del proto dinastico", *Mesopotamia* V-VI:181-279.

GUNN M.

1997   *Ritual Arts of Oceania - New Ireland*, Milan: Skira editore.

HALLAQ D.

1994   "Les sceaux des Grottes du Jebel el Akhdar", in: P. Ferioli et al., eds., *Archives Before Writing*, Torino: Scriptorium, pp. 394-402.

HANSEN H.H.

1976   *An Ethnographical Collection from the Region of the Alawites*, Copenhagen: Publications of the Carlsberg Expedition to Phoenicia 4.

HATTULA M.N.

1983   "Jarmo Artefacts of Pecked and Groundstone and of Shell, in: L.S. Braidwood et al., *Prehistoric Archaeology along the Zagros Flanks*, Chicago: University of Chicago Press (OIP 105), pp 289-346.

HAYDEN B. and CANNON A.

1983   "Where the Garbage goes - Refuse Disposal in the Maya Highlands", *Journal of Anthropological Archaeology* 2:117-163.

HEINZ M.

1997   *Der Stadtplan als Spiegel der Gesellschaft - Siedlungsstrukturen in Mesopotamien als Indikator für Formen wirtschaftlicher und gesellschaftlicher Organisation*, Berlin: Dietrich Reimer.

HENRICKSON E.F.

1981   "Non-religious Residential Settlement Patterning in the Late Early Dynastic of the Diyala Region", *Mesopotamia* XVI:43-140.

1982   "Functional Analysis of Elite Residences in the Late Early Dynastic of the Diyala Region", *Mesopotamia* XVII:5-34.

HENRICKSON E.F. and McDONALD M.M.A.

1983   "Ceramic Form and Function - An Ethnographic Search and an Archaeological Application", *American Anthropologist* 85:630-643.

HIETALA H.J., ed.

1984   *Intrasite Spatial Analysis in Archaeology*, Cambridge: Cambridge University Press.

HILL J.N.

1970   *Broken K Pueblo - Prehistoric Social Organization in the American Southwest*, Tucson (Arizona): University of Arizona Press, Anthropological Papers of the University of Arizona 18.

HILLIER B. and HANSON J.

1984   *The Social Logic of Space*, Cambridge: Cambridge University Press.

HINGLEY R.

1990   "Domestic Organisation and Gender Relations in Iron Age and Romano-British Households", in: R. Samson, ed., *The Social Archaeology of Houses*, Edinburgh: Edinburgh University Press, pp. 125-147.

HIRSCH E.

1995   "Landscape - Between Place and Space", in: E. Hirsch and M. O' Hanlon, eds., *The Anthropology of Landscape - Perspectives on Place and Space*, Oxford: Clarendon Press, pp. 1-30.

HODDER I.

1972   "Locational Models and the Study of Romano-British Settlement", in: D.L. Clarke, ed., *Models in Archaeology*, London: Methuen & Co. Ltd., pp. 887-909.

1982a   *Symbols in Action - Ethnoarchaeological Studies of Material Culture*, Cambridge: Cambridge University Press.

1982b   *Symbolic and Structural Archaeology*, Cambridge: Cambridge University Press.

1982c   *The Present Past - An Introduction to Anthropology for Archaeologists*, London: Batsford.

1986   *Reading the Past - Current Approaches to Interpretation in Archaeology*, Cambridge: Cambridge University Press.

1987   "The Meaning of Discard - Ash and Domestic Space in Baringo", in: S. Kent, ed., *Method and Theory for Activity Area Research - An Ethnoarchaeological Approach*, New York: Columbia University Press, pp. 424-448.

1990   *The Domestication of Europe - Structure and Contingency in Neolithic Societies*, Oxford: Blackwell.

1992   *Theory and Practice in Archaeology*, London: Routledge.

1994   "Architecture and Meaning - The Example of Neolithic Houses and Tombs", in: M. Parker-Pearson and C. Richards, eds., *Architecture and Order - Approaches to Social Space*, London/New York: Routledge, pp. 73-86.

1996a   "Re-opening Çatalhöyük", in: I. Hodder, ed., *On the Surface - Çatalhöyük 1993-95*, London/Cambridge: British Institute of Archaeology at Ankara/McDonald Institute for Archaeological Research, pp. 1-18.

1996b   "Conclusions", in: I. Hodder, ed., *On the Surface - Çatalhöyük 1993-95*, London/Cambridge: British Institute of Archaeology at Ankara/McDonald Institute for Archaeological Research, pp. 359-366.

HODDER I., ed.

1996   *On the Surface - Çatalhöyük 1993-95*, London/Cambridge: British Institute of Archaeology at Ankara/McDonald Institute for Archaeological Research.

HODDER I. and ORTON C.

1976   *Spatial Analysis in Archaeology*, Cambridge: Cambridge University Press.

HODDER I. et al., eds.

1995   *Interpreting Archaeology - Finding Meaning in the Past*, London/New York: Routledge.

HODGES H.

1970   *Technology in the Ancient World*, London: Penguin.

1989   *Artifacts - An Introduction to Early Materials and Technology*, London: Duckworth.

HOFFMAN M.A.

1974   "The Social Context of Trash Disposal in an Early Dynastic Egyptian Town", *American Antiquity* 39/1:35-50.

HOLE F.

1961   *Chipped Stone Analysis and the Early Farming Community*, Chicago: Unversity of Chicago (Ph.D. thesis).

1977   *Studies in the Archaeological History of the Deh Luran Plain*, Ann Arbor: University of Michigan (Museum of Anthropology Memoirs no. 9).

1983   "The Jarmo Chipped Stone", in: L.S. Braidwood et al., *Prehistoric Archaeology along the Zagros Flanks*, Chicago: University of Chicago Press (OIP 105), pp 233-288.

HOLE F., FLANNERY K.V. and NEELEY J.A.

1969   *The Prehistory and Human Ecology of the Deh Luran Plain*, Ann Arbor: University of Michigan (Museum of Anthropology Memoirs no. 1).

HOLLAND T.

1983   "Stone Maceheads", in: K.M. Kenyon and T.A. Holland, eds., *Excavations at Jericho* 5, London: British School of Archaeology in Jerusalem, pp. 804-813.

HORNE L.

1980   "Village Morphology - The Distribution of Structures and Activities in Turan Villages", *Expedition* 22/4:18-23.

1983   "Recycling in an Iranian Village - Ethnoarchaeology in Baghestan", *Archaeology* 36/4:16-21.

1994   *Village Spaces - Settlement and Society in Northeastern Iran*, Washington D.C./London: Smithsonian Institution Press.

HOWE B.

1983   "Karim Shahir", in: L.S. Braidwood et al., *Prehistoric Archaeology along the Zagros Flanks*, Chicago: University of Chicago Press (OIP 105), pp 23-154.

HUBERT H. and MAUSS M.

1964   *Sacrifice - Its Nature and Function*, London: Cohen & West (translated from the French by W.D. Halls, originally published in 1898).

INGOLD T.

1994   "The Temporality of Landscape", *World Archaeology* 25/2:152-174.

JACOBS L.

1979   "Tell-i Nun - Archaeological Implications of a Village in Transition", in: C. Kramer, ed., *Ethnoarchaeology - Implications of Ethnography for Archaeology*, New York: Columbia University Press, pp. 175-191.

JACQUES-MEUNIÉ D.

1949   "Greniers collectifs", *Hespéris* 36:97-137.

JENNESS D.

1970   *The Life of the Copper Eskimo, Part A of Volume 12 - A Report of the Canadian Arctic Expedition 1913-1918*, New York.

JOPLING C.F.

1977   "Yalalag Weaving - Its Aesthetic, Technological, and Economic Nexus", in: H. Lechtman and R.S. Merill, eds., *Material Culture - Styles, Organisation, and Dynamics of Technology*, St. Paul: West Publishing, pp. 211-236.

JOYCE A.A. and JOHANNESSEN S.

1993   "Abandonment and the Production of Archaeological Variability at Domestic Sites", in: C.M. Cameron and S. Tomka, eds., *Abandonment of Settlements and Regions - Ethnoarchaeological and Archaeological Approaches*, Cambridge: Cambridge University Press, pp. 138-153.

JUEL JENSEN H.J.

1988   "Functional Analysis of Prehistoric Flint Tools by High-Power Microscopy - A Review of West European Research", *Journal of World Prehistory* 2/1:53-88.

1994   *Flint Tools and Plant Working - Hidden Traces of Stone Age Technology*, Aarhus: Aarhus University Press.

KADOUR M. and SEEDEN H.

1983a   "Busra 1980 - Reports of an Archaeological and Ethnographic Campaign", *Damaszener Mitteilungen* 1 (1983):77-101.

1983b   "Space, Structures and Land in Shams ed-Din Tannira on the Euphrates - An Ethnoarchaeological Perspective", in: T. Khalidi, ed., *Land Tenure and Social Transformations in the Near East*, American University of Beirut, pp. 495-526.

KALTER J.

1992   *The Arts and Crafts of Syria*, London: Thames & Hudson.

KAMMINGA J.

1979   "The Nature of Use-Polish and Abrasive Smoothing on Stone Tools", in: B. Hayden, ed., *Lithic Use-Wear Analysis*, New York: Academic Press, pp. 143-157.

KANA'AN R. and McQUITTY A.

1994   "The Architecture of Al-Qasr on the Kerak Plateau - An Essay in the Chronology of Vernacular Architecture", *Palestine Exploration Quarterly* 126:127-151.

KEELEY L.H.

1980   *Experimental Determination of Stone Tool Uses - A Microwear Analysis*, Chicago/London: University of Chicago Press.

1983a   "Microscopic Examination of Adzes", in: K.M. Kenyon and T.A. Holland, eds., *Excavations at Jericho* 5, London: British School of Archaeology in Jerusalem, p. 759.

1983b   "Neolithic Novelties - The View from Ethnography and Microwear Analysis", in: M.-C. Cauvin, ed., *Traces d'utilisation sur les Outils néolithiques du Proche-Orient*, Lyon: Travaux de la Maison de l'Orient 5, pp. 251-256.

KENT S.

1984   *Analyzing Activity Areas - An Ethnoarchaeological Study of the Use of Space*, Albuquerque: University of New Mexico Press.

1987   "Parts as Wholes - A Critique of Theory in Archaeology", in: S. Kent, ed., *Method and Theory for Activity Area Research - An Ethnoarchaeological Approach*, New York: Columbia University Press, pp. 513-546.

1990   "A Cross-Cultural Study of Segmentation, Architecture, and the Use of Space", in: S. Kent, ed., *Domestic Architecture and the Use of Space - An Interdisciplinary Cross-Cultural Study*, Cambridge: Cambridge University Press, pp. 127-152.

KENT S., ed.

1987   *Method and Theory for Activity Area Research - An Ethnoarchaeological Approach*, New York: Columbia University Press.

1990   *Domestic Architecture and the Use of Space - An Interdisciplinary Cross-Cultural Study*, Cambridge: Cambridge University Press.

KHAZANOV A.M.

1994   *Nomads and the Outside World*, Cambridge: Cambridge University Press.

KIDDER A.V.

1932   *The Artifacts of Pecos*, New Haven: Yale University Press (Papers of the Southwestern Expedition 6).

KING A.D., ed.

1980   *Buildings and Society*, London: Routledge & Kegan Paul.

KLEINDIENST M.R.

1960   "Note on a Surface Survey at Baghouz (Syria)", *Anthropology Tomorrow* VI:65-72.

KÖHLER-ROLLEFSON I.

1987   "Ethnoarchaeological Research into the Origins of Pastoralism", *Annual of the Department of Antiquities of Jordan* 31:535-539.

1988   "The Aftermath of the Levantine Neolithic Revolution in the Light of Ecological and Ethnographic Evidence", *Paléorient* 14/1:87-93.

1992   "A Model for the Development of Nomadic Pastoralism on the Transjordanian Plateau", in: O. Bar-Yosef and A. Khazanov, eds., *Pastoralism in the Levant - Archaeological Materials in Anthropological Perspectives*, Madison: Prehistory Press, pp 11-19.

KOHLMEYER K.

1981   ""Wovon man nicht sprechen kan" - Grenzen der Interpretation von bei Oberflächen gewonnenen archäologischen Informationen", *Mitteilungen der Deutschen Orient-Gesellschaft* 113:53-79.

KORFMANN M.

1972   *Schleuder und Bogen in Südwestasien - Von den frühesten Belegen bis zum Beginn der historischen Stadtstaaten*, Bonn: Rudolf Habelt.

1973   "The Sling as a Weapon", *Scientific American* 229:35-42.

KOZAK V.

1972   "Stone Age revisited", *Natural History* 81/8:14-24.

KRAFELD-DAUGHERTY M.

1994   *Wohnen im Alten Orient - Eine Untersuchung zur Verwendung von Räumen in altorientalischen Wohnhäusern*, Münster: Ugarit Verlag (Altertumskunde des Vorderen Orients Band 3).

KRAMER C.

1979   "An Archaeological View of a Contemporary Kurdish Village - Domestic Architecture, Household Size, and Wealth", in: C. Kramer, ed., *Ethnoarchaeology - Implications of Ethnography for Archaeology*, New York: Columbia University Press, pp. 139-163.

1982   *Village Ethnoarchaeology - Rural Iran in Archaeological Perspective*, New York/London: Academic Press.

KUBBA S.A.A.

1987   *Mesopotamian Architecture and Town Planning from the Mesolithic to the end of the Proto-Historic Period, c. 10.000 - 3500 B.C.*, Oxford: BAR-IS 367 (i, ii).

KUECHLER S.

1987   "Malangan - Art and Memory in a Melanesian Society", *Man* 22/2:238-255.

LAMBERG-KARLOVSKY C.C. and BEALE T.

1986   *Excavations at Tepe Yahya, Iran, 1967-1975 - The Early Periods*, Cambridge/Massachusets: Harvard University Press.

LANCASTER W. and LANCASTER F.

1997   "Jordanian Village Houses in their Contexts - Growth, Decay and Rebuilding", *Palestine Exploration Quarterly* 129:38-53.

LANGE F.W. and RYDBERG C.R.

1972   "Abandonment and Post-Abandonment Behaviour at a Rural Central American House Site", *American Antiquity* 37/3:419-432.

LARSSON S. and SAUNDERS T.

1997   "Order and Architecture in the Age of Transition - A Social Analysis of the Archbishop's Palace in Trondheim, Norway", *Norwegian Archaeological Review* 30/2:79-102.

LAWRENCE R.J.

1987   *Housing, Dwelling, Homes - Design Theory, Research and Practice*, New York: Wiley.

LEBLANC S.A. and WATSON P.J.

1973   "A Comparative Statistical Analysis of Painted Pottery from seven Halafian Sites", *Paléorient* 1:117-133.

LE MIÈRE M. and PICON M.

1991   "Early Neolithic Pots and Cooking", in: R.B. Wartke, ed., *Handwerk und Technologie im Alten Orient - Ein Beitrag zur Geschichte der Technik des Altertums*, Mainz am Rhein: Philipp von Zabern, pp. 67-70.

LE MIÈRE M. and NIEUWENHUYSE O.

1996   "The Prehistoric Pottery", in: P.M.M.G. Akkermans, ed., *Tell Sabi Abyad - The Late Neolithic Settlement. Report on the Excavations of the University of Amsterdam (1988) and the National Museum of Antiquities Leiden (1991-1993) in Syria*, Istanbul: Nederlands Historisch-Archaeologisch Instituut, pp. 119-285.

LÉVI-STRAUSS C.

1963   *Structural Anthropology*, London: Penguin (translated from the French by C. Jacobsen and B.G. Schoepf, originally published in 1958).

LEWIS N.N.

1988   "The Balikh Valley and its People", in: M.N. van Loon, ed., *Hammam et-Turkman I*, Istanbul: Nederlands Historisch-Archaeologisch Instituut, pp. 683-695.

LIGHTFOOT R.R.

1993   "Abandonment Processes in Prehistoric Pueblos", in: C.M. Cameron and S. Tomka, eds., *Abandonment of Settlements and Regions - Ethnoarchaeological and Archaeological Approaches*, Cambridge: Cambridge University Press, pp. 165-177.

LITTAUER M.A. and CROUWEL J.H.

1979   *Wheeled Vehicles and Ridden Animals in the Ancient Near East*, Leiden: Brill.

LLOBERA M.

1996   "Exploring the Topography of Mind - GIS, Social Space and Archaeology", *Antiquity* 70:612-622.

LLOYD S. and SAFAR F.

1945   "Tell Hassuna - Excavations by the Iraq Government Directorate General of Antiquities in 1943 and 1944", *Journal of Near Eastern Studies* 4:255-289.

LLOYD S. and MELLAART J.

1957   "An Early Bronze Age Shrine at Beyçesultan", *Anatolian Studies* 7:27-36.

1958   Beyçesultan Excavations - Fourth Preliminary Report, 1957", *Anatolian Studies* 8:93-113.

LOCOCK M., ed.

1994   *Meaningful Architecture - Social Interpretations of Buildings*, Avebury: Aldershot.

LOWIE R.H.

1954   *Indians of the Plains*, New York.

LUTFIYYA A.M.

1966   *Baytin - A Jordanian Village. A Study of Social Institutions and Social Change in a Folk Community*, The Hague: Mouton & Co.

LYFORD C.A.

1943   *Ojibwa Crafts (Chipewa)*, Washington D.C.: U.S. Department of the Interior.

MAKAL M.

1987   *Unser Dorf in Anatolien*, Berlin.

MALEK-SHAHMIRZADI S.

1977   *Tepe Zagheh - A Sixth Millennium B.C. Village in the Qazvin Plain of the Central Iranian Plateau*, Ann Arbor: University Microfilms International, University of Pennsylvania (Ph.D. thesis).

MALLOWAN M.E.L.

1946   "Excavations in the Balikh Valley (1938)", *Iraq* 8:111-156.

MALLOWAN M.E.L. and ROSE J. C.

1935   "Excavations at Tell Arpachiyah 1933", *Iraq* 2:1-178.

MANN I.

1962   *Animal By-Products - Processing and Utilization*, Rome: FAO.

MARFOE L. et al.

1986   "The Chicago Euphrates Archaeological Project 1980-1984 - An Interim Report", *Anatolica* 13:37-148.

MARINELLI M.G.

1994   *The Bone Artifacts of Ilipinar - A Preliminary Report on the Osteological and Typological Aspects*, Leiden: University of Leiden (M.A. thesis).

1995   "The Bone Artifacts of Ilipinar", in: J. Roodenberg, ed., *Five Seasons of Fieldwork in NW Anatolia, 1987-91*, Istanbul: NHAI, pp. 121-142.

MARSHALL D.N.

1982   "Jericho Bone Tools and Objects", in: K.M. Kenyon and T.A. Holland, eds., *Excavations at Jericho* 5, London: British School of Archaeology in Jerusalem, pp. 570-576.

MASON O.T.

1891   "Aboriginal Skin Dressing - A Study based on Material in the U.S. National Museum", *Smithsonian Institution, Annual Report for 1888-1889*:553-589.

MASSON V.M. and SARIANIDI V.I.

1972   *Central Asia - Turkmenia before the Achaemenids*, New York/Washington D.C.: Praeger.

MATSUTANI T., ed.

1991   *Tell Kashkashok - The Excavations at Tell No. II*, Tokyo: The University of Tokyo (The Institute of Oriental Culture).

MATTHEWS R.J.

1989   *Clay Sealings in the Early Dynastic Mesopotamia - A Functional and Contextual Approach*, Cambridge: University of Cambridge (Ph.D. thesis).

MATTHEWS W., FRENCH C., LAWRENCE T. and CUTLER, D.

1996   "Multiple Surfaces - The Micromorphology", in: I. Hodder, ed., *On the Surface - Çatalhöyük 1993-95*, London/Cambridge: British Institute of Archaeology at Ankara/McDonald Institute for Archaeological Research, pp. 301-342.

MAUSS M.

1990   *The Gift - The Form and Reason for Exchange in Archaic Societies*, London: Routledge (translated from the French by W.D. Halls, originally published in 1923-24).

McGOVERN P., GLUSKER D., EXNER L.J. and VOIGT M.M.

1996   "Neolithic resinated Wine", *Nature* 381:480-481.

MEIJER D.J.W.

1989   "Ground Plans and Archaeologists - On Similarities and Comparisons", in: O.M.C. Haex, H.H. Curvers and P.M.M.G. Akkermans, eds., *To the Euphrates and Beyond*, Rotterdam: Balkema, pp. 221-236.

MELLAART J.

1967   *Çatal Hüyük - A Neolithic Town in Anatolia*, London: Thames & Hudson.

1975   *The Neolithic of the Near East*, London: Thames & Hudson.

MERPERT N. and MUNCHAEV R.M.

1969   "The Investigation of the Soviet Archaeological Expedition in Iraq in the Spring 1969", *Sumer* 25: 128 and Pl. III.

1971   "Early Agricultural Settlements in Northern Mesopotamia", *Soviet Anthropology and Archaeology* 10:203-252.

1973   "Early Agricultural Settlements in the Sinjar Plain, Northern Iraq", *Iraq* 35:93-113.

1993a   "Yarim Tepe I", in: N. Yoffee and J.J. Clark, eds., *Early Stages in the Evolution of Mesopotamian Civilization - Soviet Excavations in Northern Iraq*, Tucson/London: The University of Arizona Press, pp. 73-114.

1993b   "Yarim Tepe II - The Halaf Levels", in: N. Yoffee and J.J. Clark, eds., *Early Stages in the Evolution of Mesopotamian Civilization - Soviet Excavations in Northern Iraq*, Tucson/London: The University of Arizona Press, pp. 129-162.

1993c   "Yarim Tepe III - The Halaf Levels", in: N. Yoffee and J.J. Clark, eds., *Early Stages in the Evolution of Mesopotamian Civilization - Soviet Excavations in Northern Iraq*, Tucson/London: The University of Arizona Press, pp. 163-205.

MERPERT N. et al.

1977   "The Investigations of Soviet Expedition in Iraq, 1974", *Sumer* 33:65-104.

METCALF P. and HUNTINGTON R.
1991   *Celebrations of Death - The Anthropology of Mortuary Ritual*, Cambridge: Cambridge University Press.

MILLER D.
1987   *Material Culture and Mass Consumption*, Oxford: Blackwell.

MOLLESON T.
1994   "The Eloquent Bones of Abu Hureyra", *Scientific American* 271/2:60-65.

MONTAGNE R.
1930   *Un Magasin collectif de l'Anti-Atlas - l'Agadir des Ikounka*, Paris: Larosse.

MONTGOMERY B.K.
1993   "Ceramic Analysis as a Tool for Discovering Processes of Pueblo Abandonment", in: C.M. Cameron and S. Tomka, eds., *Abandonment of Settlements and Regions - Ethnoarchaeological and Archaeological Approaches*, Cambridge: Cambridge University Press, pp. 157-164.

MOORE A.M.T.
1983   "The First Farmers in the Levant", in: T. Cuyler Young Jr. et al., eds., *The Hilly Flanks and Beyond - Essays on the Prehistory of Southwestern Asia presented to Robert J. Braidwood*, Chicago: the Oriental Institute of the University of Chicago (SAOC 36), pp. 91-111.

MOORE H.L.
1982   "The Interpretation of Spatial Patterning in Settlement Residues", in: I. Hodder, ed., *Symbolic and Structural Archaeology*, Cambridge: Cambridge University Press, pp. 74-79.

1986   *Space, Text and Gender - An Anthropological Study of the Marakwet of Kenya*, Cambridge: Cambridge University Press.

MOOREY P.R.S.
1994   *Ancient Mesopotamian Materials and Industries - The Archaeological Evidence*, Oxford: Clarendon Press.

MORTENSEN I.D.
1993   *Nomads of Luristan - History, Material Culture, and Pastoralism in Western Iran*, London: Thames & Hudson.

MORTENSEN P.
1964   "Early Village-Farming Occupation", in: J. Meldgaard, P. Mortensen, H. Thrane, Excavations at Tepe Guran, Luristan, *Acta Archaeologica* 34:110-121.

1982   "Patterns of Interaction between Seasonal Settlements and Early Villages in Mesopotamia", in: T. Cuyler Young Jr. et al., eds., *The Hilly Flanks and Beyond - Essays on the Prehistory of Southwestern Asia presented to Robert J. Braidwood*, Chicago: the Oriental Institute of the University of Chicago (SAOC 36), pp. 207-230.

MOSS E.H.
1983a   "The Functions of Burins and Tanged Points from Tell Abu Hureyra, Syria", in: M.-C. Cauvin, ed., *Traces d'utilisation sur les Outils néolithiques du Proche-Orient*, Lyon: Travaux de la Maison de l'Orient 5, pp. 143-161.

1983b   *The Functional Analysis of Flint Implements*, Oxford: BAR-IS 177.

MUNCHAEV R.M. and MERPERT N.
1971   "The Archaeological Research in the Sinjar Valley (1971)", *Sumer* 27:23-32.

1973   "Excavations at Yarim Tepe 1972 - Fourth Preliminary Report", *Sumer* 29:3-16.

NARROL R.
1962   "Floor Area and Settlement Population", *American Antiquity* 27/4:587-589.

NAS P.J.M. and PRINS W.J.M., eds.
1991   *Huis, Cultuur en Ontwikkeling*, Leiden: DSWO Press.

NEGAHBAN E.O.

1979   "A Brief Report on the Painted Building of Zaghe (late 7th - early 6th Millennium B.C.)", *Paléorient* 5:239-250.

NETTING R.McC.

1982   "Some Thoughts on Household Size and Wealth", *American Behavioral Scientist* 25:641-662.

1990   "Population, Permanent Agriculture, and Polities - Unpacking the Evolutionary Portmanteau", in: S. Upham, ed., *The Evolution of Political Systems - Sociopolitics in Small-Scale Sedentary Societies*, Cambridge: Cambridge University Press, pp. 21-61.

NEWCOMER M.H.

1972   *An Analysis of a Series of Burins from Ksar Akil (Lebanon)*, London: University of London (Ph.D. thesis).

NICHOLAS I.M.

1980   *A Spatial/Functional Analysis of Late Fourth Millennium Occupation at the TUV Mound, Tal-E Malyan, Iran*, Ann Arbor: University Microfilms International, University of Pennsylvania (Ph.D. thesis).

NICOLAISEN J.

1963   *Ecology and Culture of the Pastoral Tuareg - With particular Reference to the Tuareg of Ahaggar and Ayr*, Copenhagen: The National Museum of Copenhagen.

NIEUWENHUYSE O.

1997   "Following the Earliest Halaf - Some Later Halaf Pottery from Tell Sabi Abyad, Syria", *Anatolica* XXIII:227-242.

NIPPA A.

1991   *Haus und Familie in arabischen Ländern - Vom Mittelalter zur Gegenwart*, Darmstadt: Wissenschaftliche Buchgesellschaft Darmstadt.

NISHIAKI Y.

1990   "Corner-Thinned Blades - A New Tool-Type from a Pottery Neolithic Mound in the Khabur Basin, Syria", *Bulletin of the American Society of Oriental Research* 280:1-14.

NOLL E.

1996   "Ethnographische Analogien - Forschungsstand, Theoriediskussion, Anwendungsmöglich-keiten", *Ethnografische und Archäologische Zeitschrift* 37:245-252.

OATES J.

1978   "Religion and Ritual in Sixth-Millennium B.C. Mesopotamia", *World Archaeology* 10/2:117-124.

OATES D. and OATES J.

1976   *The Rise of Civilization*, Oxford: Elsevier.

O'CONNEL J.F.

1987   "Alyawara Site Structure and its Archaeological Implications", *American Antiquity* 52/1:74-108.

OCHSENSCHLAGER E.L.

1974   "Mud Objects from Al-Hiba - A Study in Ancient and Modern Technology", *Archaeology* 27:162-174.

1993   "Village Weavers - Ethnoarchaeology at Al-Hiba", *Bulletin on Sumerian Agriculture* 7:43-62.

ODELL G.H.

1977   *The Application of Microwear Analysis to the Lithic Component of an Entire Prehistoric Settlement - Methods, Problems and Functional Reconstructions*, Harvard (Ph.D. thesis).

ORTON C.R.

1982   "Stochastic Process and Archaeological Mechanism in Spatial Analysis", *Journal of Archaeological Science* 9:1-23.

OSGOOD C.

1937   *The Ethnography of the Tanaina*, New Haven: Yale University Publications in Anthropology 16.

PARKER-PEARSON M. and RICHARDS C., eds.

1994   *Architecture and Order - Approaches to Social Space*, London/New York: Routledge.

PARKER-PEARSON M. and RICHARDS C.

1994   "Ordering the World - Perceptions of Architecture, Space and Time", in: M. Parker-Pearson and C. Richards, eds., *Architecture and Order - Approaches to Social Space*, London/New York: Routledge, pp. 1-38.

PARKIN R.

1997   *Kinship - An Introduction to the Basic Concepts*, Oxford: Blackwell.

PERKINS A.L.

1949   *The Contemporary Archaeology of Early Mesopotamia*, Chicago: University of Chicago Press (SAOC 25).

PFÄLZNER P.

1995   *Mittanische und Mittelassyrische Keramik - Eine Chronologische, Funktionale und Produktionsökonomische Analyse*, Berlin: Dietrich Reimer.

1996   "Activity Areas and the Social Organisation of Third Millennium B.C. Households", in: K.R. Veenhof, ed., *Houses and Households in Ancient Mesopotamia*, Istanbul: Nederlands Historisch-Archaeologisch Instituut, pp. 117-127.

POLLOCK S., POPE, M. and COURSEY C.

1996   "Household Production at the Uruk Mound, Abu Salabikh, Iraq", *American Journal of Archaeology* 100:683-698.

QUESADA F., BAENA J. and BLASCO C.

1995   "An Application of GIS to Intra-Site Spatial Analysis - The Iberian Iron Age Cemetery at El Cigarralejo (Murica, Spain)", in: J. Hugget and N. Ryan, eds., *Computer Applications and Quantitative Methods in Archaeology 1994*, Oxford: BAR-IS 600, pp. 137-146.

RAPOPORT A.

1969   *House Form and Culture*, Englewood Cliffs: Prentice-Hall.

1982   *The Meanings of the Built Environment - A Nonverbal Communication Approach*, Beverly Hills: Stage Publications.

RAPPAPORT R.A.

1971   "Ritual Sanctity and Cybernetics", *American Anthropologist* 73:59-76.

RATHJE W. and MURPHY C.

1992   *Rubbish! - The Archaeology of Garbage*, New York: Harper Collins.

REDMAN C.L.

1973   "Early Village Technology - A View through a Microscope", *Paléorient* 1:249-261.

1986   *Qasr es-Seghir - An Archaeological View of Medieval Life*, New York: Academic Press.

REED R.

1972   *Ancient Skins, Parchments and Leathers*, London/New York: Seminar.

REID J.J.

1973   *Growth and Response to Stress at Grasshopper Pueblo, Arizona*, Ann Arbor: University Microfilms, University of Arizona, Tucson (Ph.D. thesis).

REINHOLD S. and STEINHOF M.

1995   "Die Neolithisierung im Vorderen Orient - Neue Fragen an ein altes Thema", in: K. Bartl, R. Bernbeck and M. Heinz, eds., *Zwischen Euphrat und Indus - Aktuelle Forschungsprobleme in der Vorderasiatischen Archäologie*, Hildesheim/Zürich/New York: Georg Olms, pp. 7-27.

REYNOLDS B.

1968   *The Material Culture of the Peoples of the Gwembe Valley*, Manchester: Manchester University Press (Kariba Studies 3).

RICE P.M.

1987   *Pottery Analysis - A Source Book*, Chicago: University of Chicago Press.

RICHARDS C.

1996   "Monuments as Landscape - Creating the Centre of the World in Late Neolithic Orkney", *World Archaeology* 28/2:190-208.

RIPPENGAL R.

1993   "Villas as a Key to Social Structure - Some Comments on Recent Approaches to the Romano-British Villa and some Suggestions toward an Alternative", in: E. Scott, ed., *Theoretical Roman Archaeology - First Conference Proceedings*, Aldershot: Avebury, pp. 79-114.

ROAF M.

1989   "Social Organization and Social Activities at Tell Madhur", in: E.F. Henrickson and I. Thuesen, eds., *Upon this Foundation - The 'Ubaid Reconsidered*, Copenhagen: Museum Tusculanum Press, pp. 91-146.

ROBBEN A.C.G.M.

1989   "Habits of the Home - Spatial Hegemony and the Structuration of House and Society in Brazil", *American Anthropologist* 91:570-588.

ROLLEFSON G.O.

1986   "Neolithic 'Ain Ghazal (Jordan) - Ritual and Ceremony II", *Paléorient* 12:45-52.

ROODENBERG J.J.

1983   "Traces d'utilisation sur les Haches polies de Bouqras (Syrie)", in: M.-C. Cauvin, ed., *Traces d'utilisation sur les Outils néolithiques du Proche-Orient*, Lyon: Travaux de la Maison de l'Orient 5, pp. 177-186.

ROSEN A.M.

1986   *Cities of Clay - The Geoarchaeology of Tells*, Chicago/London: University of Chicago Press.

ROSMAN A. and RUBEL P.G.

1971   *Feasting with Mine Enemy - Rank and Exchange among Northwest Coast Societies*, New York: Columbia University Press.

1976   "Nomad-Sedentary Relations in Iran and Afghanistan", *Journal of Middle East Studies* 7:545-570.

ROWLANDS M.

1993   "The Role of Memory in the Transmission of Culture", *World Archaeology* 25/2:141-151.

SAHLINS M.

1972   *Stone Age Economics*, New York: Aldine.

SAMSON R., ed.

1990   *The Social Archaeology of Houses*, Edinburgh: Edinburgh University Press.

SCHIFFER M.B.

1972   "Archaeological Context and Systemic Context", *American Antiquity* 37/2:156-165.

1976   *Behavioral Archaeology*, New York: Academic Press.

1983   "Towards the Identification of Formation Processes", *American Antiquity* 48/4:675-706.

1985a   "Is there a "Pompeii Premise" in Archaeology?", *Journal of Anthropological Research* 41/1:18-41.

1985b   Review of Binford's 'Working at Archaeology', *American Antiquity* 50/1:191-193.

1987   *Formation Processes of the Archaeological Record*, Albuquerque: University of New Mexico Press.

SCHMANDT-BESSERAT D.
1977   "An Archaic Recording System and the Origin of Writing", *Syro-Mesopotamian Studies* 1/
        2:31-70.
1992   *Before Writing - From Counting to Cuneiform*, Austin: University of Texas Press.
SCHECHNER R.
1994   "Ritual and Performance", in: T. Ingold, ed., *Companion Encyclopedia of Anthropology -
        Humanity, Culture and Social Life*, London/New York: Routledge, pp. 613-647.
SCHIEFFELIN E.L.
1985   "Performance and the Cultural Construction of Reality", *American Anthropologist* 30/1:707-
        724.
SCHIRMER W.
1990   "Some Aspects of Building at the 'Aceramic-Neolithic' Settlement of Çayönü Tepesi", *World
        Archaeology* 21/3:363-387.
SCHOLZ F. and JANZEN J., eds.
1982   *Nomadismus - Ein Entwicklungsproblem?*, Berlin: Dietrich Reimer.
SEEDEN H.
1982   "Ethnoarchaeological Reconstruction of Halafian Occupational Units at Shams ed-Din
        Tannira", *Berytus* 30:55-97.
1985   "Aspects of Prehistory in the Present World - Observations gathered in Syrian Villages from
        1980 to 1985", *World Archaeology* 17/2:289-303.
SEEDEN H. and WILSON J.
1988   "Processes of Site Formation in Villages of the Syrian Gazira", *Berytus* 36:169-188.
SEMENOV S.A.
1964   *Prehistoric Technology - An Experimental Study of the Oldest Tools and Artefacts from Traces
        of Manufacture and Wear*, London: Cory, Adams & Mackay.
SERJEANT R.B. and LEWCOCK R., eds.
1983   *Sana - An Arabian Islamic City*, London: World of Islam Trust.
SHANKS M. and TILLEY C.
1987   *Social Theory and Archaeology*, Oxford: Polity Press.
SHARER R.J. and ASHMORE A.
1987   *Archaeology - Discovering Our Past*, Paolo Alto: Mayfield Publishing Company.
SHEFFER A.
1981   "The Use of Perforated Clay Balls on the Warp-Weighted Loom", *Tel Aviv* 8:81-83.
SHRYOCK H.S. et al.
1976   *The Methods and Materials of Demography*, New York: Academic Press.
SMITH P.E.L.
1976   "Reflections on Four Seasons of Excavations at Tappeh Ganj Dareh", in: F. Bagherzadeh, ed.,
        Proceedings of the IVth Annual Symposium on Archaeological Research in Iran, pp. 11-22.
1990   "Architectural Innovation and Experimentation at Ganj Dareh, Iran", *World Archaeology* 21/
        3:323-335.
SKIBO J.M.
1992   *Pottery Function - A Use-Alteration Perspective*, New York/London: Plenum Press.
SOJA E.W.
1985   "The Spatiality of Social Life - Towards a Transformative Retheorisation", in: D. Gregory and
        J. Urry, eds., *Social Relations and Spatial Structures*, Basingstroke: Macmillan, pp. 90-127.
SOLECKI R.L. and SOLECKI R.S.
1970   "Grooved Stones from Zawi Chemi Shanidar, a Protoneolithic Site in Northern Iraq", *American
        Anthropologist* 72:831-841.

SOUTH S.

1979   "Historic Site Content, Structure, and Function", *American Antiquity* 44/2:213-236.

SPAULDING A.C.

1968   "Explanations in Archaeology", in: S. Binford and L. Binford, eds., *New Perspectives in Archaeology*, Chicago: Aldine, pp. 33-40.

SPOOR R.H. and COLLET P.

1996   "The other Small Finds", in: P.M.M.G. Akkermans, ed., *Tell Sabi Abyad - The Late Neolithic Settlement. Report on the Excavations of the University of Amsterdam (1988) and the National Museum of Antiquities Leiden (1991-1993) in Syria*, Istanbul: Nederlands Historisch-Archaeologisch Instituut, pp. 439-475.

STAGER L.E.

1985   "The Archaeology of the Family in Ancient Israel", *Bulletin of the American School of Oriental Research* 260:1-35.

STARKEY J.L.

1936   "Excavations at Tell ed-Duweir 1935-6", *Palestine Exploration Quarterly* 68:178-189.

STEENSBERG A.

1943   *Ancient Harvesting Equipment*, Copenhagen: Nordisk Verlag.

STEVANOVIĆ M.

1997   "The Age of Clay - The Social Dynamics of House Destruction", *Journal of Anthropological Archaeology* 16:334-395.

STEVENSON M.G.

1982   "Toward an Understanding of Site Abandonment Behavior - Evidence from Historic Mining Camps in the Southwest Yukon", *Journal of Anthropological Archaeology* 1:237-265.

STÖBER G.

1978   *Die Afshar - Nomadismus im Raum Kerman (Zentraliran)*, Marburg/Lahn: C. Schott.

STRATHERN M.

1988   *The Gender of Gift - Problems with Women and Problems with Society in Melanesia*, Berkely: University of California Press.

STUIVER et al.

1993   "Calibration 1993", *Radiocarbon* 35:1-244.

SULLIVAN A.P.

1978   "Inferences and Evidence in Archaeology - A Discussion of the Conceptual Problems", in: M.B. Schiffer, ed., *Advances in Archaeological Method and Theory* 1, pp. 183-222.

SUTER K.

1964   "Die Wohnhöhlen und Speicherburgen des tripolitanisch-tunesischen Berglandes", *Zeitschrift für Ethnologie* 89:216-275.

SWEET L.

1960   *Tell Toqa'an - A Syrian Village*, Ann Arbor: University of Michigan Anthropological Papers no. 14.

TALALAY L.E.

1987   "Rethinking the Function of Clay Figurine Legs from Neolithic Greece: An Argument by Analogy", *American Journal of Archaeology* 91:161-169.

TANI M.

1995   "Beyond the Identification of Formation Processes - Behavioral Inference based on Traces left by Cultural Formation Processes", *Journal of Archaeological Method and Theory* 2/3:231-252.

THOUMIN R.

1932   *La Maison Syrienne dans la Plaine Hauranaise, le Bassin du Barada et sur les Plateaux du Qalamun*, Paris: Librairie E. Leroux.

TILLEY C.

1989   "Interpreting Material Culture", in: I. Hodder, ed., *The Meanings of Things - Material Culture
         and Symbolic Expression*, London: Harper Collins Academic, pp. 185-194.

1991   *Material Culture and Text - The Art of Ambiguity*, London: Routledge.

1994   *A Phenomenology of Landscape - Places, Paths and Monuments*, Oxford/Providence: Berg.

TILLEY C., ed.

1993   *Interpretative Archaeology*, Providence/Oxford: Berg.

TOBLER A.J.

1950   *Excavations at Tepe Gawra, II*, Philadelphia: University of Pennsylvania Press.

TODD I.A.

1976   *Çatal Hüyük in Perspective*, Menlo Park: Cummings Publishing Company.

TOSI M. and PIPERNO M.

1973   "Lithic Technology behind the Ancient Lapis Lazuli Trade", *Expedition* 16/1:15-23.

TOWNSEND W.

1969   "Stone and Shell Tool Use in a New Guinea Society", *Ethnology* 8:199-205.

TUAN Y.F.

1977   *Space and Place - The Perspective of Experience*, London: Edward Arnold.

TURNER V.W.

1967   *The Forest of Symbols - Aspects of Ndembu Rituals*, Ithaca: Cornell University Press.

1969   *The Ritual Process - Structure and Anti-Structure*, Chicago: Aldine.

1986   *On the Edge of the Bush - Anthropology as Experience*, Tucson: University of Arizona Press.

1990   "Are there Universals of Performance in Myth, Ritual and Drama?", in: R. Schechner and W.
         Appel, eds., *By Means of Performance*, Cambridge: Cambridge University Press.

UCKO P.J.

1968   *Anthropomorphic Figurines of Predynastic Egypt and Neolithic Crete - With comparative
         Material from the Prehistoric Near East and Mainland Greece*, London: Andrew Szmilda.

1969   "Ethnography and Archaeological Interpretation of Funerary Remains", *World Archaeology*
         1:262-277.

UMEOV A.I.

1970   "The Basic Forms and Rules of Inference by Analogy", in: P.V. Tavence, ed., *Problems in the
         Logic of Scientific Knowledge*, Dordrecht.

UNGER-HAMILTON R.

1983   "An Investigation into the Variables affecting the Development and Appearance of Plant Polish
         on Flint Blades", in: M.-C. Cauvin, ed., *Traces d'utilisation sur les Outils néolithiques du
         Proche-Orient*, Lyon: Travaux de la Maison de l'Orient 5, pp. 243-250.

1988   *Method in Microwear Analysis - Prehistoric Sickles and other Stone Tools from Arjoune, Syria*,
         Oxford: BAR-IS 435.

UNWIN S.

1997   *Analysing Architecture*, London/New York: Routledge.

VAN DER KOOIJ G.

1976   *Notities over enkele Dorpen nabij de Jebel Aruda*, Leiden: University of Leiden.

VAN DER PLICHT J.

1993   "The Groningen Radiocarbon Calibration Program", *Radiocarbon* 35/1:231-237.

VAN DER STEEN E.J.

1995   "Aspects of Nomadism and Settlement in the Central Jordan Valley", *Palestine Exploration
         Quarterly* 127:141-158.

VAN GENNEP A.

1960   *The Rites of Passage*, London: Routledge & Kegan Paul (translated from the French by M.B. Vizedom and G.L. Caffee, originally published in 1909).

VAN GIJN A.

1990   *The Wear and Tear of Flint - Principles of Functional Analysis applied to Dutch Neolithic Assemblages*, Leiden: Analecta Praehistoria Leidensia 22.

1992   "The Interpretation of Sickles - A Cautionary Tale", in: P.C. Anderson, ed., *Préhistoire de l'agriculture - Nouvelles approches expérimentales et ethnographiques*, Paris: Monographie du CRA 6, Éditions du CNRS, pp. 363-372.

VAN ZEIST W. and WATERBOLK-VAN ROOIJEN W.

1996   "The Cultivated and Wild Plants", in: P.M.M.G. Akkermans, ed., *Tell Sabi Abyad - The Late Neolithic Settlement. Report on the Excavations of the University of Amsterdam (1988) and the National Museum of Antiquities Leiden (1991-1993) in Syria*, Istanbul: Nederlands Historisch-Archaeologisch Instituut, pp. 521-551.

VAUGHAN P.

1985   "Funktionsbestimmung von Steingeräten anhand mikrokospischer Gebrauchsspuren", *Germania* 63:309-329.

VEENHOF K.R., ed.

1996   *Houses and Households in Ancient Mesopotamia - Papers read at the 40th Rencontre Assyriologique Internationale Leiden. July 5-8, 1993*, Istanbul: Nederlands Historisch-Archaeologisch Instituut.

VERTOVEC S.

1993   "Potlatching and the Mythic Past - A Re-Evaluation of the Traditional Northwest Coast American Indian Complex", *Religion* 13:323-344.

VERHOEVEN M.

1990   *Activity Areas - Vorming, Vervorming en Vernietiging*, Amsterdam: A.E van Giffen Instituut voor Prae- en Protohistorie (M.A. thesis).

1994   "Excavations at Tell Sabi Abyad II, a Later Pre-Pottery Neolithic Village in the Balikh Valley, Northern Syria", *Orient Express* 1994/1:9-12.

1997a   "The 1996 excavations at Tell Sabi Abyad II, a Later PPNB Village in the Balikh Valley, Syria", *Neo-Lithics* 1/97:1-3.

1997b   "Excavated Spaces, Prehistoric Places - Spatial Analysis at Tell Sabi Abyad I, a Neolithic Settlement in Syria", *Neo-Lithics* 3/97.

in press "A Preliminary Report on Tell Sabi Abyad II, a Later PPNB Settlement in the Balikh Valley, Syria", *Archív orientální - Supplementa* VIII.

        "Excavated Spaces, Prehistoric Places - Spatial Analysis at Tell Sabi Abyad, a Neolithic Settlement in Syria", in: S. De Martino, F.M. Fales, G.B. Lanfranchi and L. Milano, eds., *Landscapes - Territories, Frontiers and Horizons in the Ancient Near East*. Papers presented at the XLIV Rencontre Assyriologique Internationale (Venezia 7-11 July 1997), Padova.

in prep. "Betwixt and Between - An Interpretative Essay about Ritual Practice and Social Relations at Later Neolithic Tell Sabi Abyad, Syria" (provisional title), *Archaeological Dialogues*.

        "Microwear Analysis of Later Neolithic Flint Tools of Tell Sabi Abyad, Syria" (provisional title).

        "Spatial Analysis of the Burnt Village at Later Neolithic Tell Sabi Abyad, Syria" (provisional title).

        "The Excavations - Stratigraphy and Architecture", in: M. Verhoeven and P.M.M.G. Akkermans, eds., *Tell Sabi Abyad II - The Pre-Pottery Neolithic B Settlement* (provisional titles), Istanbul: Nederlands Historisch-Archaeologisch Instituut.

"The Small Finds", in: M. Verhoeven and P.M.M.G. Akkermans, eds., *Tell Sabi Abyad II - The Pre-Pottery Neolithic B Settlement* (provisional titles), Istanbul: Nederlands Historisch-Archaeologisch Instituut.

VERHOEVEN M. and KRANENDONK P.

1989   *Twee Federmesser Sites nabij Bakel-Milheeze - Een Analyse*, Amsterdam: A.E van Giffen Instituut voor Prae- en Protohistorie.

1996   "The Excavations - Stratigraphy and Architecture", in: P.M.M.G. Akkermans, ed., *Tell Sabi Abyad - The Late Neolithic Settlement. Report on the Excavations of the University of Amsterdam (1988) and the National Museum of Antiquities Leiden (1991-1993) in Syria*, Istanbul: Nederlands Historisch-Archaeologisch Instituut, pp. 25-118.

VERHOEVEN and AKKERMANS P.M.M.G.

in prep.   "Foreword", in: M. Verhoeven and P.M.M.G. Akkermans, eds., *Tell Sabi Abyad II - The Pre-Pottery Neolithic B Settlement* (provisional titles), Istanbul: Nederlands Historisch-Archaeologisch Instituut.

VERHOEVEN M. and AKKERMANS P.M.M.G., eds.,

in prep.   *Tell Sabi Abyad II - The Pre-Pottery Neolithic B Settlement* (provisional title), Istanbul: Nederlands Historisch-Archaeologisch Instituut.

VILLA P.

1982   "Conjoining Pieces and Site Formation Processes", *American Antiquity* 47/2:276-290.

VILLA P. and COURTIN J.

1983   "The Interpretation of Stratified Sites - A View from Underground", *Journal of Archaeological Science* 10:267-281.

VOORRIPS A.

1987   "Spatial Analysis in Archaeology", *Pact* 16/3.6.1.:423-445.

VOIGT M.M.

1983   *Hajji Firuz Tepe, Iran - The Neolithic Settlement*, Pennsylvania: University of Pennsylvania Monograph 50.

VOSSEN R. and EBERT W.

1986   *Marokkanische Töpferei - Töpferorte und Zentren - Eine Landesaufnahme (1980)*, Bonn: Dr. Rudolf Habelt GmbH.

WANDSNIDER L.

1996   "Describing and comparing Archaeological Spatial Structures", *Journal of Archaeological Method and Theory* 3/4:319-384.

WANSLEEBEN M.

1988   "Applications of Geographical Information Systems in Archaeological Research", in: S.P.Q. Rahtz, ed., *Computer and Quantitative Methods in Archaeology 1988*, Oxford: BAR-IS 446 (ii), pp. 435-451.

WATKINS T.

1990   "The Origins of House and Home?", *World Archaeology* 21:336-347.

WATSON P.J.

1979   *Archaeological Ethnography in Western Iran*, Tucson (Arizona): The University of Arizona Press.

1980   "The Theory and Practice of Ethnoarchaeology with special Reference to the Near East", *Paléorient* 6:55-64.

1983a   "The Soundings at Banahilk", in: L.S. Braidwood et al., *Prehistoric Archaeology along the Zagros Flanks*, Chicago: University of Chicago Press (OIP 105), pp. 545-613.

1983b   "Jarmo Worked Bone", in: L.S. Braidwood et al., *Prehistoric Archaeology along the Zagros Flanks*, Chicago: University of Chicago Press (OIP 105), pp. 347-368.

WATSON P.J. and LeBLANC S.A.

1990   *Girikihaçyan - A Halafian Site in Southeastern Turkey*, Los Angeles: UCLA Institute of Archaeology.

WENDRICH W.

1991   *Who is afraid of Basketry? - A Guide to recording Basketry and Cordage for Archaeologists and Ethnographers*, Leiden: CNWS.

WHALLON R.

1984   "Unconstrained Clustering for the Analysis of Spatial Distributions in Archaeology", in: H. Hietala, ed., *Intrasite Spatial Analysis in Archaeology*, Cambridge: Cambridge University Press, pp. 242-277.

WILK R.R. and NETTING R. McC.

1984   "Households - Changing Forms and Functions", in: R. McC. Netting, R.R. Wilk and E. Arnould, eds., *Households - Comparative and Historical Studies of the Domestic Group*, Berkely: University of California Press, pp. 1-28.

WILSHUSEN R.A.

1986   "The Relationships between Abandonment Mode and Ritual Use in Pueblo I Anasazi Protokivas", *Journal of Field Archaeology* 13:245-254.

WILSON P.

1988   *The Domestication of the Human Species*, New Haven: Yale University Press.

WOOD W.R. and JOHNSON D.L.

1978   "A Survey of Disturbance Processes in Archaeological Site Formation", in: M.B. Schiffer, ed., *Advances in Archaeological Method and Theory* 1, pp. 315-381.

WRIGHT G.R.H.

1985   *Ancient Building in South Syria and Palestine*, Leiden: Brill.

WRIGHT H.T., ed.

1981   *An Early Town on the Deh Luran Plain - Excavations at Tepe Farukhabad*, Ann Arbor: University of Michigan.

WRIGHT K.

1992   "A Classification System for Ground Stone Tools from the Prehistoric Levant", *Paléorient* 18/2:53-81.

WULFF H.E.

1966   *The Traditional Crafts of Persia - Their Development, Technology, and Influence on Eastern and Western Civilizations*, London/Cambridge (Mass.): MIT Press.

WYLIE A.

1980   "Analogical Inference in Archaeology", *Paper presented at the Society for American Archaeology Conference*, Philadelphia.

1985   "The Reaction against Analogy", in: M.B. Schiffer, ed., *Advances in Archaeological Method and Theory* 8, pp. 63-111.

YAKAR J.

1974   "The Twin Shrines of Beyçesultan", *Anatolian Studies* XXIV:151-161.

YASIN W.

1970   "Excavations at Tell es-Sawwan 1969 - Report on the Sixth Season's Excavations", *Sumer* 26:3-35.

YELLEN J.E.

1977   *Archaeological Approaches to the Present - Models for reconstructing the Past*, New York/London: Academic Press.

YOFFEE N. and CLARK J.J., eds.

1993   *Early Stages in the Evolution of Mesopotamian Civilization - Soviet Excavations in Northern Iraq*, Tucson/London: The University of Arizona Press.

PRINTED ON PERMANENT PAPER • IMPRIME SUR PAPIER PERMANENT • GEDRUKT OP DUURZAAM PAPIER - ISO 9706

ORIENTALISTE, KLEIN DALENSTRAAT 42, B-3020 HERENT